NUREG-1021
Rev. 9, Supp. 1

Operator Licensing Examination Standards For Power Reactors

Final Report

Manuscript Completed: September 2007
Date Published: October 2007

Prepared by
D. Muller

Division of Inspection and Regional Support
Office of Nuclear Reactor Regulation
U.S. Nuclear Regulatory Commission
Washington, DC 20555-0001

SUPPLEMENT 1 TO REVISION 9 OF NUREG-1021,
"OPERATOR LICENSING EXAMINATION STANDARDS FOR POWER REACTORS,"
PAGE REPLACEMENT INSTRUCTIONS

Remove existing Revision 9 (with errata) pages and insert Supplement 1 pages as noted below. Individual changes within the body of the NUREG are indicated by a vertical bars in the margins. Note that since NUREG-1021 is printed double-sided, a single change results in a minimum of two replacement pages. Also note that changes can effect page breaks and the total number of pages in a section, resulting in a large number of pages to replace.

Section / Standard	Remove Page(s)	Replacement Page(s)	Correction Summary
Cover Pages	Cover pages	Cover pages	Added Supplement 1.
Abstract	iii - iv	iii - iv	Describes Supplement 1 changes.
ES-201	1 - blank (all)	1 - 28	Total number of pages changed, replace entire section.
Detailed Changes in ES-201:		Pages 3, 6 8 9, 24 9 12 15, 18 22, 23 26 6, 7, 9, 15, 19 - 25	References new Attachment 2 for procedure freezes. New guidance for delaying parts of an NRC examination. Revised examination approval letter; NRC and licensee agree that it meets the guidelines of NUREG-1021. Added Supplement 1 to the letter. Removed requirement to send list of applicants Form ES-201-4 to the NRR program office. New policy on using surrogate operators. New Attachment 2 for procedure freezes. Added Supplement 1, updated OMB date on Attachment 4. Restored * footnote to Form ES-201-2. Revised ES-201 Attachment numbers throughout ES-201, due to new Attachment 2.
ES-202	3 - 13 & blank	3 - 13 & blank	New text changes page breaks.
Detailed Changes in ES-202:		Pages 3 4 6 7 12 13	Updated availability of Forms 396, 398 and their instructions. Clarified medical exam timeliness for ROs applying for a SRO license. Added waiver for exam delays. Clarified electronic submittal of license applications. Updated Division of Inspection Program Management (DIPM) to Division of Inspection and Regional Support (DIRS). Updated block numbers for Form 396. Changed name of "Restricted Individuals List." Deleted research reactor start-ups for cold license eligibility. Updated DIPM to DIRS. Added address for express mail.
ES-204	3 - 4	3 - 4	Corrected name of Form ES-201-4. Clarified that the NRC regional office may grant certain waivers. Deleted cold license requirement for startups on a research reactor.

SUPPLEMENT 1 TO REVISION 9 OF NUREG-1021,
"OPERATOR LICENSING EXAMINATION STANDARDS FOR POWER REACTORS,"
PAGE REPLACEMENT INSTRUCTIONS

Section / Standard	Remove Page(s)	Replacement Page(s)	Correction Summary
ES-205	5 - 10	5 - 10	Update title and address for Chief, Operator Licensing and Human Performance Branch. Updated DIPM to DIRS.
ES-301	1 - 4	1 - 4	Changed Administrative Topics to conform to the new K/A catalogs. Added text "/plan" to agree with topic title in K/A catalogs ("Emergency Procedures / Plan"). Revised ES-201 attachment number reference.
	9 - 12	9 - 12	Added text to randomly select JPMs from past NRC exams, to conform with existing Forms ES-301-1 and 301-2. Changed Administrative Topics description to conform to the new K/A catalogs. Added text "/plan" to agree with topic title in K/A catalogs.
	21 - 22	21 - 22	Added text "/plan" to agree with topic title in K/A catalogs.
	23 - 24	23 - 24	Added a type code for engineered safety feature, to allow check of existing requirement.
	25 - 26	25 - 26	Added allowance for SROIs to credit a malfunctions if performing in an additional scenario as a BOP operator.
ES-302	3 - 12	3 - 12	New text changes page breaks.
Detailed Changes in ES-302:		Page(s) 4 11	New policy on using surrogate operators. Revised ES-201 attachment number.
ES-401	1 - 18 21 - 22 27 - 28 31 - 33 & blank	1 - 18 21 - 22 27 - 28 31 - 33 & blank	New text changes page breaks.
Detailed Changes in ES-401:		Page(s) 2 4 - 5 5 8 14 16 17 21 22 28 32 - 33	Revised ES-201 attachment number. Revised generic K/A sampling. Moved K/A elimination guidance out of Attachment 2. Deleted Attachment 2. Clarified what constitutes SRO-level questions. Added text to include revision/version number for references. Deleted Attachment 2 from list of Attachments. Changed reference from Attachment 2 to Section D.1.b. Added word "reactor" to K/A 295037; added new K/A 700000, "Generator Voltage and Electric Grid Disturbances." Changed reference from Attachment 2 to Section D.1.b. Added new K/A 700000, "Generator Voltage and Electric Grid Disturbances." Added text to include revision/version number for references. Added column for bank/modified/new questions.

Section / Standard	Remove Page(s)	Replacement Page(s)	Correction Summary
ES-402	5 - 6	5 - 6	Added collecting applicant comments and identifying docket numbers. Added reference to ES-403 prior to submitting comments.
ES-403	1 - blank (all)	1 - 6	Total number of pages changed, replace entire section.
Detailed Changes in ES-403:		Page(s) 3 3, 4	Qualified NRC's review of post-examination comments; provided examples of most likely and less likely changes. Clarified when two correct answers will not be accepted. Changed remaining lettering of Section D.1.
ES-501	3 - 6 7 - 8 11 - 12 13 - 14 17 - 18 19 - 20 21 - 22	3 - 6 7 - 8 11 - 12 13 - 14 17 - 18 19 - 20 21 - 22	Added reference to ES-403 to qualify NRC's review of post-examination comments. Corrected lettering typo - changed E.1.f to E.1.d. Added redacting applicant docket number in NRC exam report for post-examination comments. Revised ES-201 Attachment numbers. Corrected typo in Attachment 3 for Public Law 93-438. Corrected typo for Public Law 93-438. Added address for express mail to Attachment 4. Updated DIPM to DIRS. Added address for express mail to Attachment 4.
ES-502	1 - 4 (all)	1 - 5 & blank	Total number of pages changed, replace entire section. Changes in this section: Updated DIPM to DIRS; added address for express mail.
ES-601	5 - 6 19 - 20	5 - 6 19 - 20	Revised ES-201 Attachment number. Added Supplement 1; updated OMB date.
ES-603	3 - 6	3 - 6	Added text that JPMs shall not test solely for simple recall or memorization. New text changes page breaks.
ES-605	1 - 10 (all)	1 - 14	Total number of pages changed, replace entire section.
Detailed Changes in ES-605:		Page(s) 1 3 - 6 8 - 10 11 12 13	Clarified 24-month allotted time for completing a requalification program and the written examination. Added guidance for watch-standing proficiency and license reactivation. Clarified medical standards, including use of prescription medications. Added wording for the LSRO no-solo condition. Clarified license renewal for ROs in SRO upgrade training. Clarified electronic submittal of license renewals. Updated DIPM to DIRS and added address for express mail. Added address for express mail. Clarified license renewal for ROs in SRO upgrade training.

SUPPLEMENT 1 TO REVISION 9 OF NUREG-1021,
"OPERATOR LICENSING EXAMINATION STANDARDS FOR POWER REACTORS,"
PAGE REPLACEMENT INSTRUCTIONS

Section / Standard	Remove Page(s)	Replacement Page(s)	Correction Summary
ES-701	1 - 2	1 - 2	Reference ES-401 Section D.1 instead of deleted ES-401 Attachment 2.
	3 - 4	3 - 4	Clarified use of bank/modified/new questions. Added definition for significantly modified.
	7 - 10	7 - 10	Reference ES-401 Section D.1 instead of deleted ES-401 Attachment 2.
	13 - 16	13 - 16	Changed allowed repeat JPMs from past two NRC exams to last NRC exam. Restored * footnote.
App. C	1 - 6	1 - 6	Added text that JPMs shall not test solely for simple recall or memorization.
	9 - 10	9 - 10	
Inserted by: _____ Date: _____			

UNITED STATES
NUCLEAR REGULATORY COMMISSION
Washington, D.C. 20555-0001

ERRATA

May 13, 2005

Report Number:	NUREG-1021, Revision 9
Report Title:	Operator Licensing Examination Standards for Power Reactors
Prepared by:	Division of Inspection Program Management Office of Nuclear Reactor Regulation U.S. Nuclear Regulatory Commission Washington, DC 20555-0001
Date Published:	May 2005
Instructions:	Please remove and replace pages as instructed on the following table:

Publishing Services Branch
Office of Information Services

Instructions

Remove existing Revision 9 pages and insert replacement pages as noted below (corrections and clarifications have been made as indicated; forms identified with an "*" have been edited to support electronic completion in portable document format):

Section / Standard	Remove Page(s)	Replacement Page(s) / Correction Summary
Abstract	iii - iv	iii - iv / Correct OMB control number and regulatory citation.
ES-201	17 - 18 21 - 22 23 - 24	17 - 18, 21 - 22 / Attachments 2 & 3: Sensitive examination information should be marked and segregated. 23 - 24 / Form ES-201-1*: Items that do not apply to NRC-prepared exams are denoted with { }; Form ES-201-3 added to Item 8; audit requirements in Item 14 are clarified.
ES-202	13 - blank	13 - blank / Correct over-strikes.
ES-301	1 - 4 11 - 12 21 - 22 23 - 24 25 - 26 27 - blank	1 - 4 / B.1: Swap fuel handling and drawings; emphasize that listed administrative topics are examples. 11 - 12 / D.3.a: Remove reference to B.1 examples; select administrative subjects from Section 2 of the K/A catalog. D.3.d: move fuel handling from operations to equipment control. 21 - 22 / Form ES-301-1*: Added "Class(R)oom" as possible venue. 23 - 24 / Form ES-301-2*: Control room systems for SRO-U must include ESF; Type Code (L) includes shutdown. 25 - 26 / Form ES-301-5*: Minimum event requirements vary based on license level; signatures removed. 27 - blank / Form ES-301-6*: Signatures removed.
ES-303	9 - 10 11 - 12 13 - 14	9 - 10 / Form ES-303-1, Page 1: SRO Exam Types corrected; "Deny License" recommendation added. 11 - 12 / Form ES-303-1, Page 3.a: RO Competency 1 rating factors corrected (there is no "d"). 13 - 14 / Correct over-strikes.
ES-401	15 - 16 21 - 22 29 - 32	15 - 16, 21 - 22 / Forms ES-401-1&2: SRO Column "A2" split for fuel handling in Tier 2, Group 2; Note 8 clarified. 29 - 32 / Form ES-401-6*: Intent of Items 4 & 5 clarified; Form ES-401-7*.
ES-403	5 - blank	5 - blank / Form ES-403-1*.
ES-501	13 - 14 21 - 24	13 -14 / E.4.c & F.1.f: Reference SECY-04-0191 to ensure that sensitive information is withheld from public disclosure and require inclusion of handouts in the examination file. 21 - 24 / Correct over-strikes; add facility and date to Form ES-501-1; Form ES-501-2*.
ES-604	11 - 12	11 - 12 / Correct over-strikes.
ES-701	1 - 18 (all)	1 - 18 / Correct over-strikes; clarify intent of Form ES-701-6, Items 4 & 5; Form ES-701-8*.
Appendix C	1 - 10 (all)	1 - 10 / Correct over-strikes.
Appendix F	1 - 6 (all)	1 - 6 / Correct over-strikes.

Inserted by: _____ Date: _____

ABSTRACT

The U.S. Nuclear Regulatory Commission (NRC) publishes NUREG-1021, "Operator Licensing Examination Standards for Power Reactors," to establish the policies, procedures, and practices for examining licensees and applicants for reactor operator and senior reactor operator licenses at power reactor facilities pursuant to Title 10, Part 55, of the *Code of Federal Regulations* (10 CFR Part 55). The related guidance that was previously published in the "Examiners' Handbook for Developing Operator Licensing Written Examinations" (NUREG/BR-0122, Rev. 5, dated March 1990) has been incorporated herein. NUREG/BR-0122 is no longer in effect.

These examination standards are intended to help NRC examiners and facility licensees better understand the processes associated with initial and requalification examinations. The standards also ensure the equitable and consistent administration of examinations for all applicants. These standards are *for guidance purposes* and are not a substitute for the operator licensing regulations (i.e., 10 CFR Part 55), and they are subject to revision or other changes in internal operator licensing policy. Minor policy clarifications that become necessary before the next formal revision of these standards will be promulgated on the NRC's operator licensing Web page at http://www.nrc.gov/reactors/operator-licensing.html.

The NRC issued Revision 9 in July 2004 primarily to (1) improve efficiency by reducing the length of the reactor operator written examination, without sacrificing validity or reliability; (2) clarify and simplify the design of the senior reactor operator written examination; (3) better risk-inform both written examinations; (4) better balance the administrative and systems portions of the walk-through operating test; (5) clarify the grading criteria for the simulator operating test to improve objectivity and ensure proper emphasis on competence; and (6) incorporate guidance that was previously promulgated on the NRC's operator licensing Web page regarding the suppression of inappropriate knowledge and ability (K/A) statements and the conduct of peer checks. The changes are identified with bars in the margins and described in the Executive Summary.

Supplement 1 to Revision 9 is being issued to (1) clarify licensed operator medical requirements, including the use of prescription medications; (2) clarify the use of surrogate operators during dynamic simulator scenarios; (3) clarify the selection process for generic knowledge and ability (K/A) statements; (4) qualify the NRC review of post-examination comments; (5) provide additional guidance for maintaining an active license (watchstander proficiency) and license reactivation; and (6) conform with updates to NUREGs-1122 [and -1123], "Knowledge and Abilities Catalog[s] for Nuclear Power Plant Operators: Pressurized [and Boiling] Water Reactors." The changes are identified with bars in the margins and summarized in "Revision 9 Supplement 1 Page Replacement Instructions."

Revision 9, Supplement 1, will become effective for corporate notification letters issued 60 days after its publication is noticed in the *Federal Register*. This will provide facility licensees with at least 180 days notice that the examinations will be administered in accordance with the revised policies, procedures, and practices. Facility licensees may make arrangements for earlier implementation by contacting their NRC regional office.

PAPERWORK REDUCTION ACT STATEMENT

This NUREG contains information collection requirements that are subject to the Paperwork Reduction Act of 1995 (44 U.S.C. 3501 et seq.). These information collections were approved by the Office of Management and Budget, approval number 3150-0018.

Public Protection Notification

The NRC may neither conduct nor sponsor, and a person is not required to respond to, a request for information or an information collection requirement unless the requesting document displays a currently valid OMB approval number.

CONTENTS

ES-3xx   INITIAL OPERATING TESTS

ES-4xx   INITIAL WRITTEN EXAMINATIONS

ES-5xx   INITIAL POST-EXAMINATION ACTIVITIES

ES-6xx   REQUALIFICATION EXAMINATIONS

Appendices

EXECUTIVE SUMMARY

Title 10, Part 55, of the *Code of Federal Regulations* (10 CFR Part 55) requires that applicants for reactor operator (RO) and senior reactor operator (SRO) licenses must pass both a written examination and an operating test that are developed and administered in accordance with 10 CFR 55.41 and 55.45, or 10 CFR 55.43 and 55.45, respectively. The regulations (specifically 10 CFR 55.40) allow facility licensees to develop and submit, upon approval by an authorized representative of the facility licensee, proposed examinations for review and approval by the staff of the U.S. Nuclear Regulatory Commission (NRC). The NRC will prepare the examinations if requested in writing by a facility licensee, and may elect to prepare the examinations, in lieu of allowing a specific facility licensee to do so, as necessary to maintain the proficiency of its examiners or the quality of the examinations.

Facility licensees who elect to prepare their own examinations shall develop and submit their proposed examinations based on the guidelines and instructions contained herein. Section 107 of the *Atomic Energy Act of 1954*, as amended, requires the Commission to prescribe uniform licensing conditions for operators. Therefore, the NRC discourages facility licensees from using testing methodologies that do not conform to the policies, procedures, and practices defined in this NUREG-series report. Nonetheless, facility licensees may propose alternatives to specific guidance in NUREG-1021, and the NRC will review and rule on the acceptability of the alternatives.

The NRC will make a reasonable attempt to administer all license examinations on the dates requested by facility licensees. At times, however, resource limitations may compel the staff to prioritize its examination review and development activities based on need and safety considerations. Facility licensees are strongly encouraged to schedule their initial license examinations and to resolve any applicant eligibility questions with their NRC regional office *before* commencing a license training class.

The NRC staff developed the changes in Revision 9 during a series of public meetings with the nuclear power industry's Initial Licensed Operator Focus Group. Summaries of those meetings, which have taken place since the NRC published Revision 8, Supplement 1, in April 2001, are available through the NRC's operator licensing Web page at http://www.nrc.gov/reactors/operator-licensing/meetings.html.

Draft Revision 9, which is available in the NRC's public electronic reading room (http://www.nrc.gov/reading-rm/adams.html) under Accession Number ML030230303, was published for comment and voluntary trial use in February 2003 (68 FR 5312), and the comment period closed in December 2003. The public and internal comments and resolutions are summarized in ML041240004, which is also available in the NRC's public electronic reading room.

The following table summarizes the significant (but not all) changes from Revision 8, Supplement 1. New or modified text is also identified with vertical bars in the margins throughout this revision of NUREG-1021. Refer to pages xvii through xx for a definition of abbreviations used within this executive summary and throughout NUREG-1021.

Changes from NUREG-1021, Revision 8, Supplement 1	
Location	Change
ES-102	D.5 and F.4 have been revised to reflect the issuance of Revision 3 of Regulatory Guide (RG) 1.149, "Nuclear Power Plant Simulation Facilities for Use in Operator Training and License Examinations," dated October 2001.
	E.6 has been added to reference NUREG-1262, "Answers to Questions at Public Meetings Regarding Implementation of Title 10, *Code of Federal Regulations*, Part 55 on Operators' Licenses," as a historical document.
	E.7 has been edited to indicate that NUREG-1021 takes precedence over NUREG-1291, "BWR and PWR Off-Normal Event Descriptions," dated November 1987.
ES-201	C.1.i and C.2.c have been revised to address the simulator fidelity requirements in 10 CFR 55.46.
	C.2.i and Attachment 4 have been revised to require a formal examination approval letter instead of an assignment sheet.
	C.3.g has been revised to allow the agency to forward NRC-prepared exams to the facility before the formal review.
	C.3.j has been revised to recognize that the chief examiner can make or change simulator crew assignments (up to 2 weeks before the examination date).
	Attachment 3 has been edited to reference 10 CFR 2.390, "Public Inspections, Exemptions, Requests for Withholding."
	Form ES-201-1 has been edited to improve the task descriptions and cross-references.
	Form ES-201-2 has been edited to capture more of the walk-through criteria and to delete the note exempting NRC-prepared operating tests from duplicating scenarios and tasks from the applicants' audit test(s).
	Form ES-201-3 has been edited to more accurately reflect Section D.2.b.
	Form ES-201-4 replaces the assignment sheet (formerly Attachment 4).
ES-202	C.1.a has been revised to note that SRO-upgrade applicants in good standing can request a waiver of the RO written examination.
	C.1.a has been revised to clarify the guidelines on medical examinations, including the use of nurse practitioners and physician's assistants and the policy on waivers; it also adds a reference to the Form 396 instructions on the Web site.
	C.1.b has been revised to reflect that the generic fundamentals examination (GFE) must be taken within 24 months before the date of application or waived in accordance with ES-204.
	C.1.c and D.1.b(3) have been revised to reflect that the control manipulations required by 10 CFR 55.31(a)(5) can be done on the simulator; it also clarifies the NRC's expectations regarding magnitude.
	C.1.d has been added to note that U.S. citizenship is not required for licensure.
	C.1.f has been revised to clarify the definition of senior management representative on site.

Changes from NUREG-1021, Revision 8, Supplement 1	
Location	Change
	C.1.f has been revised to address the electronic submittal of forms.
	C.2.e has been revised to require the NRC's regional offices to audit 10 percent of applications.
	D.2.a(2) has been revised to delete the reference to the chief reactor watch as being equivalent to a licensed RO.
	D.2.a(4) has been clarified to reflect that the NRC's Office of Nuclear Reactor Regulation (NRR) will evaluate the eligibility of applicants who might otherwise be disqualified because they do not meet the strict definition of "responsible nuclear power plant experience."
	D.3.a has been clarified to reflect that limited senior reactor operators (LSROs) are required to perform five significant control manipulations in accordance with 10 CFR 55.31(a)(5).
ES-204	D.1.a has been revised to allow the NRC's regional offices to grant the SRO-only written and administrative walk-through waivers, to clarify the policy regarding SRO waivers, and to note that waivers are limited to 1 year after the date of the exam (rather than the final denial date).
	D.1.c has been clarified to indicate that medical waivers/exceptions will be coordinated with the NRC's contract physician.
	D.1.g has been edited to note that the written exam waiver would include the GFE.
	D.1.h has been revised to allow the NRC's regional offices to issue conditional licenses, regardless of the reason for not completing the control manipulations required by 10 CFR 55.31(a)(5).
	D.1.j has been added to allow the NRC's regional offices to waive the RO written exam for SRO-upgrade applicants in good standing.
	D.1.k has been added to establish criteria that will allow the NRC's regional offices to waive the need to retake the GFE for applicants who passed their original GFE more than 24 months before the date of license application.
ES-205	The Background discussion has been revised to limit the longevity of the GFE to 24 months without proficiency training and reflect four exam administrations per year.
	C.2.b and Attachment 1 have been revised to reflect the shift to an annual notification letter.
	C.3.b has been added to require the NRC's regional offices to informally remind facilities to submit their registration letters for the June, September, and December exams.
	C.3.e has been revised to indicate that the results letters will only be sent to participating licensees.
	C.4 has been added to address the criteria for the industry to use in reviewing and commenting on the GFEs before they are administered.

Changes from NUREG-1021, Revision 8, Supplement 1	
Location	Change
	Section D and Attachment 4 have been revised to reflect the shorter, 50-question examination and the question distribution guidelines.
	Attachment 1 has been revised to indicate that the GFEs will be available in the NRC's public electronic reading room and on the NRC's GFE Web page, and its enclosure has been clarified to indicate that the exams cannot be sent to home addresses.
ES-301	Throughout: Old Categories A and B have been combined into a section called "Walk-Through," and Category C is simply called "Simulator Test."
	Throughout: The "Walk-Through" now consists entirely of job performance measures (JPMs); questions will be used only for followup, as necessary.
	D.1.a has been revised such that the prohibition on duplicating test items from the applicants' audit test(s) applies to NRC-prepared operating tests (as well as those prepared by facility licensees).
	D.2 has been created from old Section D.1.k and parts of old Section D.3.b.
	D.3.a and Form ES-301-1 have been revised to reduce the number of RO administrative topics on the "Walk-Through" from five to four.
	D.4.a and Form ES-301-2 have been revised to increase the RO control room systems coverage from seven to eight.
	D.4.b and Form ES-301-2 have been revised to allow a 40- to 60-percent range of alternative path tasks, and to define "low-power" as 5 percent.
	D.5.a and Form ES-301-5 have been revised to allow the reactivity and normal evolutions to be replaced with additional instrument or component malfunctions, and to clarify crew rotation policies.
	D.5.d and Form ES-301-5 have been revised to require SRO applicants to perform two or more technical specification (TS) evaluations during the "Simulator Test"; applicants should be given multiple opportunities to demonstrate competence in each area.
	D.5.f has been edited to indicate that all "required" actions shall be documented on Form ES-D-2.
	Forms ES-301-1 and ES-301-2 have been revised to incorporate the acceptance criteria at the bottom of the form.
	Form ES-301-5 has been completely revised to make it more user friendly and to incorporate other changes (e.g., crew rotation, TS for SROs, and optional reactivity manipulations).
	Some of the competencies on Form ES-301-6 have been consolidated to conform with ES-303 and Section E of Appendix D.
ES-302	Throughout: Old Categories A and B have been combined into the "Walk-Through," and Category C is simply called "Simulator Test."
	Throughout: All references to questions have been edited to reflect that they will only be asked "for cause."

Changes from NUREG-1021, Revision 8, Supplement 1	
Location	Change
	D.1.j, D.2.g, and D.3.l have been revised to incorporate the "peer check" guidance previously issued on the NRC's operator licensing Web page.
	D.1.l has been added to ensure that uncorrected simulator performance deficiencies do not interfere with the planned tests.
	D.1.k has been revised to add guidance on protecting predecisional performance information.
	D.3.k has been added for consistency with Appendix E.
	D.3.o has been clarified to require an additional scenario, if necessary, to evaluate the required evolutions and competencies.
	D.3.q has been revised to prohibit "backtracking" when restarting an inoperable simulator.
ES-303	The Background definitions of "satisfactory" and "unsatisfactory" have been deleted.
	Throughout: Old Categories A and B have been combined into the "Walk-Through," and Category C is simply called "Simulator Test."
	Throughout: The "Walk-Through" now consists entirely of JPMs; questions will be used only for followup, as necessary.
	C.2 has been revised to allow examiners to recommend a passing grade even if the applicant made errors that would normally result in a failure.
	D.2.a has been revised to collectively grade the administrative and systems JPMs, with an overall 80-percent cut score and separate administrative cut scores of 60- and 50-percent for SRO and RO applicants.
	D.2.b and Forms ES-303-1, 3, and 4 have been revised to consolidate some of the rating factors and competencies, to allow examiners to assign "not observed" grades for some rating factors, to specify the number of errors allowed for each integral rating factor grade, and to eliminate the behavioral anchors.
	D.3.b has been edited to ensure that examiners document the potential and actual consequences of an applicant's action if the error contributes to a failure of the operating test.
	D.3.d has been added to provide guidance on documenting deviations from the nominal grading criteria.
ES-401	Throughout: The RO written exam has been shortened to 75 questions; SRO applicants will take that exam plus a 25-question exam focused on SRO-only knowledge and abilities (K/As); SRO-upgrade applicants may apply for a waiver of the RO examination pursuant to 10 CFR 55.47.
	D.1.b now references a new Attachment 2, which incorporates previous Web-based guidance regarding the elimination of inapplicable or inappropriate K/A statements; when selecting K/As, every item in the group should be sampled once before selecting a second K/A for any item in the group.

| Changes from NUREG-1021, Revision 8, Supplement 1 ||
Location	Change
	D.1.c has been revised to focus the SRO-only sample on those K/A categories that are linked to 10 CFR 55.43 and the fuel handling system.
	Old D.1.d, which allowed exam authors to propose 10 site-specific priority K/As, has been deleted.
	D.2.a has been revised to provide guidance on testing multi-part K/A statements.
	D.2.c and Form ES-401-6 have been edited to indicate that more than 60 percent of the questions in the overall SRO exam could assess higher cognitive level.
	D.2.g and Form ES-401-6 have been clarified to ensure that reference materials do not assist the applicants in eliminating incorrect distractors.
	Section E, E.2.d, and Form ES-401-9 have been clarified to indicate that distractors should always be plausible.
	E.2.d has been clarified to require the replacement of otherwise good questions that do not match the approved K/A statement.
	Attachment 1, Step 1, has been revised to require the addition of important systems and evolutions that are not included on the generic lists. Instructions have also been added for sampling SRO-only K/As.
	Attachment 2 incorporates K/A elimination guidance that was previously issued on the NRC's operator licensing Web page, with minor clarifications.
	Forms ES-401-1 and 2 have been revised to conform with the new RO/SRO-only formats. The notes have been revised to conform with the body of ES-401. The evolutions and systems have been divided into high- and low-risk-significance groups, and the sampling rates have been adjusted accordingly.
	Form ES-401-6, Item 4, has been revised to require NRC examiners to review the sampling process if too many questions are repeated.
	Form ES-401-8 has become the new cover sheet for SRO applicants.
ES-402	C.1.e has been revised to state that a dictionary should be available in the examination room.
	D.1.f has added a caution regarding the use of machine-gradable answer sheets.
	D.2.d has been edited to ensure that the applicants properly page-check their examinations.
	D.4.d has been revised to include a 3-hour time limit for the SRO-only exam taken alone and to allow licensees to extend the time upon notifying the NRC.
ES-403	D.2.d and Form ES-403-1 have been edited to accommodate the new SRO cover sheet and grading criteria.
ES-501	C.2.c has been revised to adjust the thresholds for conducting a validity review.
	D.2.c has been revised to trigger borderline reviews based on SRO-only grades.

Changes from NUREG-1021, Revision 8, Supplement 1	
Location	Change
	D.2.e has been revised to clarify guidance regarding licensing recommendations based on applicant grades. SRO-instant applicants who pass the written exam overall and the operating test do not automatically qualify for an RO license.
	D.3.b has been revised to ensure that NRR program office concurrence is obtained when operating test grading deviates from nominal guidance.
	D.3.c and Attachment 5 have been edited to hold licenses and trigger a performance analysis at 82 percent overall and 74 percent on the SRO-only items.
	E.3.a has been revised to include question deletions and answer key changes when counting unacceptable questions for documentation in the exam report; RO and SRO-only questions will be counted separately; negative comments will not be made only if it was the facility's first submittal; the regions may adjust the nominal 20-percent comment threshold with NRR program office concurrence; and the criteria for documenting security issues have been clarified.
	E.3.c has been revised to ensure that simulator fidelity issues are addressed during the next requalification program inspection.
	E.4.a has been revised to clarify the policy regarding SRO-upgrade applicants who received a waiver of the RO portion of the license exam and did not participate in RO requalification training and testing while in the upgrade training program; SROs who passed overall but scored below 80 percent on either the RO or SRO-only may require additional review and training.
	Attachment 4 and Form ES-501-2 have been edited to conform with examination format changes.
	Attachment 5 has been edited to accommodate medical waivers and exam format changes.
ES-502	C.1.b(1) and (2) have been edited to better conform with 10 CFR 55.35.
	C.2.a has been revised to indicate that the NRC may request a facility licensee who prepared its licensing examination to confirm the validity of any test items that are challenged by an applicant during an appeal.
	D.2.a has been clarified to address multiple appeals and generic corrections.
ES-601	Throughout: References to systematic assessment of licensee performance (SALP) have been removed, and the requalification program inspection procedure number has been updated to IP 71111.11.
	Section C has been clarified to indicate that the NRC will consider preferentially using the facility's exam process if it complies with 10 CFR 55.59 and is free of significant flaws.
	J.1 has been updated to reflect new record retention guidelines.
	Form ES-601-1 has been revised to parallel Form ES-201-3.

Changes from NUREG-1021, Revision 8, Supplement 1	
Location	Change
ES-604	E.2 has been revised to indicate that the individual evaluations will be done using the appropriate sections of Form ES-604-2 instead of the competency evaluation forms in ES-303.
ES-605	C.1.a has been added to elaborate on requalification testing requirements that were previously clarified on the NRC's operator licensing Web page.
	C.1.b has been added to clarify requalification testing requirements for newly licensed operators.
	C.2.a has been clarified to address the proficiency requirements for SROs who normally stand watch only as ROs.
	C.2.b has been clarified by incorporating guidance related to LSRO watch-standing proficiency, which was previously issued on the NRC's operator licensing Web page.
	C.3 has been extensively updated to clarify the regulatory requirements and guidelines, to note that temporary medical conditions may preclude operators from completing the requalification program, and to address conditional licenses.
	C.4 has been added to incorporate staff practice as it pertains to voluntarily down-grading an SRO license to the RO level.
	D.1.a has been edited to include the results of the most recent requalification written exam and operating test on license renewal applications.
	D.1.c has been clarified to address renewal applications that are received more than 60 days before the license expiration date.
ES-701	C.1.c and C.2.d have been revised, pursuant to 10 CFR 55.46(b), to require Commission approval to use the plant or something other than a plant-referenced simulator for the operating tests.
	The written examination has been revised to more closely follow the 3-Tier format in ES-401, with a total of 40 questions; separate outlines and quality checklists are included.
	The operating test has been revised to consist of 10 JPMs, with 3 in the administrative area, 4 in systems, and 3 related to emergency and abnormal plant evolutions; the discussion scenarios have been deleted; and separate outlines and quality checklists are included.
Appendix A	C.3.c has been added to elaborate on "level of knowledge" versus "level of difficulty."
Appendix D	B.3 has been edited to state that Form ES-D-2 should include every required, rather than expected, operator action.
	Section E has been revised to consolidate some of the competency descriptions.

Changes from NUREG-1021, Revision 8, Supplement 1	
Location	Change
Appendix E	Part B has been revised to incorporate separate SRO cut scores and time limits, to note that programable calculator memories must be erased, to add a caution regarding machine-gradable answer sheets, to note that applicant questions are taken into consideration when reviewing appeals, and to caution applicants not to make assumptions regarding operator actions.
	Part C, Item 4, has been clarified to prohibit discussions with other applicants who have not completed the applicable portion of the operating test.
	Part D has been edited to eliminate guidelines related to prescripted questions, adjust the length of the walk-through, and incorporate "peer check" guidance.
	Part E, Item 4, has been expanded to include "peer check" guidance.
Appendix F	The definition of "Category" has been eliminated.
	A definition of "Low-power" has been added.
	The definitions of "Plant-referenced simulator" and "Simulation facility" have been edited.

ABBREVIATIONS

AC	alternating current
ADAMS	Agencywide Documents Access and Management System (NRC)
ADS	automatic depressurization system
AFW	auxiliary feedwater
ANS	American Nuclear Society
ANSI	American National Standards
AO	auxiliary operator
AOP	abnormal operating procedure
APRM	average power range monitor
ARP	alarm (or annunciator) response procedure
ATC	at the controls (operator)
ATWS(T)	anticipated transient without scram (trip)
B&W	Babcock and Wilcox
BOP	balance of plant (operator)
BWR	boiling-water reactor
C	(degrees) Celsius
CAL	confirmatory action letter
CCP	centrifugal charging pump
CCW	component cooling water
CD-ROM	compact disk, read-only memory
CE	Combustion Engineering
CFPT	condensate feedwater pump turbine
CFR	*Code of Federal Regulations*
CRD	control rod drive
CRT	criterion-referenced test
CS	core spray
CT	critical task
CTMT	containment
CVCS	chemical and volume control system
DAS	dominant accident sequence
DC	direct current
DG	diesel generator
DHR	decay heat removal
DIPM	Division of Inspection Program Management (NRR)
EAL	emergency action level
E/APE	emergency/abnormal plant evolution
ECA	emergency contingency action (procedure)
ECCS	emergency core cooling system
ECP	estimated critical position
EDG	emergency diesel generator
EHC	electrohydraulic control
EIE	electronic information exchange

	EOL	end-of-life
	EOP	emergency operating procedure
	EPIP	emergency plan implementing procedure
	EQB	examination question bank
	ES	examination standard
	ESF	engineered safety feature
	F	(degrees) Fahrenheit
	FHE	fuel handling equipment
	FR	*Federal Register*
	FRP	functional recovery procedure
	FSAR	final safety analysis report
	GE	General Electric
	GFE	generic fundamentals examination
	GL	generic letter
	GUI	graphic user interface
	HCL	higher cognitive level
	HCU	hydraulic control unit
	HHSI	high head safety injection
	HP	health physics
	HPCI	high-pressure coolant injection
	HPCS	high-pressure core spray
	HPSI	high-pressure safety injection
	HVAC	heating, ventilation, and air conditioning
	IC	initial condition *or* instrumentation and control
	INPO	Institute of Nuclear Power Operations
	IP	inspection procedure
	IPE	individual plant examination
	IR	importance rating
	IRM	intermediate range monitor
	JPM	job performance measure
	JTA	job task analysis
	K/A	knowledge and ability
	KSA	knowledge, skill, and ability
	LAN	local area network
	LCO	limiting condition for operation
	LER	licensee event report
	LOCA	loss-of-coolant accident
	LOD	level of difficulty
	LOK	level of knowledge
	LOOP	loss of offsite power
	LPCI	low-pressure coolant injection

LPCS	low-pressure core spray
LPRM	local power range monitor
LSRO	limited senior reactor operator
LWR	light-water reactor
MC	Manual Chapter (NRC Inspection)
MCC	motor control center
MDAFW(P)	motor-driven AFW (pump)
MFP	main feedwater pump
MIP	master inspection plan
MSIV	main steam isolation valve
NANT	National Academy for Nuclear Training
NEI	Nuclear Energy Institute
NNAB	National Nuclear Accrediting Board
NRC	U.S. Nuclear Regulator Commission
NOP	normal operating procedure
NRR	Office of Nuclear Reactor Regulation (NRC)
NRT	norm-referenced test
NWPA	*Nuclear Waste Policy Act* (of 1982)
OJT	on-the-job training
OLA	operator licensing assistant
OLTS	operator licensing tracking system
OTSG	once-through steam generator
OMB	Office of Management and Budget (U.S.)
PARS	Publicly Available Records System
PCIS	primary containment isolation system
PDR	Public Document Room
PORV	power-operated relief valve
PPR	plant performance review
PRA	probabilistic risk assessment
PSI(A)(G)	pounds per square inch (absolute) (gauge)
PZR	pressurizer
PWR	pressurized-water reactor
QA	quality assurance
QPTR	quadrant power tilt ratio
RBCCW	reactor building closed cooling water
RBM	rod block monitor
RCA	radiologically controlled area
RCS	reactor coolant system
RCP	reactor coolant pump
RCIC	reactor core isolation cooling
RF	rating factor
RFP	reactor feed pump
RG	Regulatory Guide

RHR	residual heat removal	
RMCS	reactor manual control system	
RO	reactor operator	
ROI	report on interaction	
RM	radiation monitor	
RNPPE	responsible nuclear power plant experience	
RPIS	rod position indication system	
RPM	revolutions per minute	
RPS	reactor protection system	
RPV	reactor pressure vessel	
RWST	refueling water storage tank	
S(AT)	satisfactory	
SALP	systematic assessment of licensee performance	
SAT	systems approach to training	
S(B)GTS	standby gas treatment system	
SD	standard deviation	
SG(TR)	steam generator (tube rupture)	
SI	safety injection	
SLC	standby liquid control	
SO	senior operator	
SME	subject matter expert	
SPND	self-powered neutron detector	
SRO(I)(U)	senior reactor operator (instant) (upgrade)	
SRP	Standard Review Plan (NUREG-0800)	
SRV	safety relief valve	
SSW	standby service water	
STA	shift technical advisor	
TDAFW(P)	turbine-driven AFW (pump)	
T/F	true-false (statement/question)	
TPA	temporary plant alteration	
TS	technical specification (or other technical requirements document)	
U(NSAT)	unsatisfactory	
UPS	uninterruptible power supply	
U.S.C.	*United States Code*	
V(AC)(DC)	volts AC or DC	
VCT	volume control tank	
W	Westinghouse	
W/T	walk-through	

ES-101
PURPOSE AND FORMAT
OF OPERATOR LICENSING EXAMINATION STANDARDS

A.   Purpose

Title 10, Part 55, of the *Code of Federal Regulations* (10 CFR Part 55) requires that applicants
for reactor operator (RO) and senior reactor operator (SRO) licenses must pass both a written
examination and an operating test (both initially and for requalification). Moreover,
the regulations mandate that the license examinations must be developed and administered
in accordance with 10 CFR 55.41 and 55.45 for ROs, or 10 CFR 55.43 and 55.45 for SROs.

The "Operator Licensing Examination Standards for Power Reactors" (NUREG-1021) establish
the policies, procedures, and practices for administering the required initial and requalification
written examinations and operating tests. These standards describe the provisions of the
Atomic Energy Act of 1954 and the regulations on which the operator licensing program
is based. They also ensure the equitable and consistent administration of examinations
to all applicants and licensed operators at all facilities that are subject to the regulations.

B.   Format

Each examination standard (ES) explains the policies, procedures, and practices for a particular
aspect of the program. For ease of reference, each standard is assigned a three-digit number,
and related standards are grouped together in the sense that standards beginning with
the same digit apply to related aspects of the program, as follows:

ES-1xx: General
ES-2xx: Initial pre-examination activities
ES-3xx: Initial operating tests
ES-4xx: Initial written examinations
ES-5xx: Initial post-examination activities
ES-6xx: Requalification examinations
ES-7xx: Fuel handling examinations

ES-102
REGULATIONS AND PUBLICATIONS
APPLICABLE TO OPERATOR LICENSING

A.   Purpose

This standard lists the United States statutes and the regulations of the U.S. Nuclear Regulatory Commission (NRC) that establish the requirements for conducting operator licensing examinations. It also identifies the regulatory guides and NUREG-series reports that establish the procedures for implementing the regulations and administering the examinations, as well as industry standards promulgated by the American National Standards Institute/American Nuclear Society (ANSI/ANS), which may provide additional guidance.

Regulatory guides, NUREG-series reports, and industry standards do not constitute requirements, except as specified in Commission orders or as committed to by the facility licensee. NRC examiners and licensees should consult the appropriate revisions, as referenced in each facility's final safety analysis report (FSAR) or approved training program. The following paragraphs summarize the latest revisions of these documents.

B.   Statutes

1.   *Atomic Energy Act of 1954*

Section 107 of the *Atomic Energy Act of 1954* (42 U.S.C. 2137), as amended, requires that the NRC must prescribe uniform conditions for licensing individuals as operators of production and utilization facilities, determining the qualifications of these individuals, and issuing licenses to such individuals.

2.   *Nuclear Waste Policy Act of 1982*

Section 306 of the *Nuclear Waste Policy Act of 1982* (42 U.S.C. 10226, 96 Stat. 2201, at 2262–2263) directs the NRC to establish requirements governing (1) simulator training for applicants for operator licenses and for operator requalification training programs, (2) NRC administration of requalification examinations, and (3) operating tests at civilian nuclear power plant simulators.

C.   Regulations

1.   10 CFR Part 2, "Rules of Practice"

The regulations in 10 CFR Part 2 govern the conduct of all proceedings under the *Atomic Energy Act of 1954*, as amended, and the *Energy Reorganization Act of 1974* with regard to (a) granting, suspending, revoking, amending, or taking other action with respect to any license; (b) imposing civil penalties; and (c) public rulemaking.

10 CFR 2.103 establishes the applicant's right to demand a review of a proposed license denial, and defines the applicant's appeal and hearing rights.

Subpart L, "Informal Hearing Procedures for NRC Adjudications," governs proceedings for the issuance, renewal, or licensee-initiated amendment of an operator or senior operator license.

2.   10 CFR Part 9, "Public Records"

The regulations in 10 CFR Part 9 prescribe the rules governing the NRC's public records that relate to any proceeding subject to 10 CFR Part 2.

Subparts A and B describe and implement the requirements for balancing the public's rights to information under the *Freedom of Information Act* and the NRC's responsibility to protect personal information under the *Privacy Act*.

Subparts C and D implement the provisions of the *Sunshine Act*, concerning the opening of Commission meetings to public observation. They also describe the procedures governing the production of agency records, information, or testimony in response to subpoenas or demands of courts or other judicial authorities in State and Federal proceedings.

3.   10 CFR Part 20, "Standards for Protection Against Radiation"

The regulations in 10 CFR Part 20 establish standards for protection against radiation hazards arising from licensed activities. Some of the material is appropriate for inclusion in the examinations administered to candidates for RO or SRO licenses.

4.   10 CFR Part 50, "Licensing of Production and Utilization Facilities"

10 CFR 50.34(b)(8) requires that the FSAR must include a description of the operator requalification program. That description forms the basis for the inspection, audit, and approval of requalification programs.

10 CFR 50.54(l-1) requires facility licensees to implement an operator requalification program that meets the requirements of 10 CFR 55.59(c) within 3 months after receiving a facility operating license. Notwithstanding the provisions of 10 CFR 50.59, the licensee may not decrease the scope of its approved requalification program without authorization from the Commission.

10 CFR 50.54(k) – (m) contain regulations that restrict control manipulations to licensed operators. These regulations are conditions of all facility licenses issued under 10 CFR Part 50.

10 CFR 50.74 requires facility licensees to notify the Commission within 30 days if there is a change in the status of a licensed RO or SRO.

5.   **10 CFR Part 55, "Operators' Licenses"**

10 CFR Part 55 is the implementing regulation that establishes the requirements and the regulatory basis for licensing and requalifying ROs and SROs.

D.   **Regulatory Guides**

1.   **Regulatory Guide 1.8, "Qualification and Training of Personnel for Nuclear Power Plants," Revision 3, May 2000**

Section C of this regulatory guide (RG) currently endorses ANSI/ANS 3.1-1993, "American National Standard for Selection, Qualification, and Training of Personnel for Nuclear Power Plants," with additions, exceptions, and clarifications thereto. No backfitting is intended or required in connection with the issuance of the revised RG.

2.   **Regulatory Guide 1.33, "Quality Assurance Program Requirements: Operations," Revision 2, February 1978**

Appendix A to this RG contains a list of typical procedures for pressurized-water reactors and boiling-water reactors.

3.   **Regulatory Guide 1.114, "Guidance on Being an Operator at the Controls of a Nuclear Power Plant," Revision 2, May 1989**

This RG describes a method that the NRC staff finds acceptable for complying with the Commission's regulations in 10 CFR 50.54(k) – (m), which require the presence of an RO at the controls of a nuclear power unit and an SRO in the control room from which the nuclear power unit is being operated.

4.   **Regulatory Guide 1.134, "Medical Evaluation of Licensed Personnel for Nuclear Power Plants," Revision 3, March 1998**

This RG currently endorses ANSI/ANS 3.4-1996, "Medical Certification and Monitoring of Personnel Requiring Operator Licenses for Nuclear Power Plants," with exceptions. However, facility licensees may continue to use the 1983 version of ANSI/ANS 3.4, which was previously endorsed in its entirety by Revision 2 of RG 1.134, dated April 1987.

5.   Regulatory Guide 1.149, "Nuclear Power Plant Simulation Facilities
      for Use in Operator License Examinations," Revision 3, October 2001

      This RG currently endorses ANSI/ANS 3.5-1998, "Nuclear Power Plant Simulators
      for Use in Operator Training and Examination," with clarifications. However, facility
      licensees may continue to use the 1985 and/or 1993 versions of ANSI/ANS 3.5,
      which were previously endorsed, with exceptions, by Revisions 1 and 2 of RG 1.149,
      dated April 1987 and April 1996, respectively.

E.   NUREG-Series Reports

1.   NUREG-0660, Vol. 1, "NRC Action Plan Developed as a Result of the TMI-2
      Accident," May 1980

      Item I.A.4.2 of this document describes the guidelines for long-term simulator upgrades.

2.   NUREG-0737, "Clarification of TMI Action Plan Requirements," November 1980

      This document clarifies the following action plan items, which are intended to upgrade
      the training, licensing, education, and experience of operators on the basis of
      experience gained from the accident at Three Mile Island, Unit 2:
      •   Item I.A.2.1, "Immediate Upgrading of RO and SRO Training and Qualifications"
      •   Item 1.A.2.3, "Administration of Training Programs"
      •   Item 1.A.3.1, "Revised Scope and Criteria for Licensing Exams"
      •   Item 11.B.4, "Training for Mitigating Core Damage"

3.   NUREG-0800, "Standard Review Plan for the Review of Safety Analysis Reports
      for Nuclear Power Plants, LWR Edition," July 1981

      Section 13.2, "Reactor Operator Training," describes the training and licensing
      of operators and identifies information to be submitted by applicants for construction
      permits and operating licenses.

4.   NUREG-1122, "Knowledge and Abilities Catalog for Nuclear Power Plant
      Operators: Pressurized-Water Reactors," Revision 2, June 1998

      This document provides the basis for developing content-valid licensing examinations
      for operators at pressurized-water reactors. It contains knowledge and ability (K/A)
      statements that have been rated for their importance to ensuring that the plant
      is operated in a manner that is consistent with the health and safety of plant personnel
      and the public.

5.   NUREG-1123, "Knowledge and Abilities Catalog for Nuclear Power Plant
      Operators: Boiling-Water Reactors," Revision 2, June 1998

      This document provides the basis for developing content-valid licensing examinations
      for operators at boiling-water reactors. It contains K/A statements that have been rated

for their importance to ensuring that the plant is operated in a manner that is consistent with the health and safety of plant personnel and the public.

6. **NUREG-1262, "Answers to Questions at Public Meetings Regarding Implementation of Title 10, _Code of Federal Regulations_, Part 55 on Operators' Licenses," November 1987**

This report presents questions and answers based on the transcripts of four public meetings (and written questions submitted after the meetings) conducted by the NRC staff shortly after publication of the Part 55 rule change in 1987. Although many of the answers have been overtaken by events since 1987, this report remains useful in that it provides a historical perspective on many issues. If the report conflicts with any other guidance issued since 1987 (e.g., NUREG-1021 or the frequently asked questions on the NRC's operator licensing Web page) the more recent guidance would take precedence.

7. **NUREG-1291, "BWR and PWR Off-Normal Event Descriptions," November 1987**

The reactor event descriptions in this document previously served as a generic reference tool to help examiners develop simulator scenarios. Refer to ES-301 or ES-604 (as applicable for initial or requalification examinations) and Appendix D to NUREG-1021 for current guidance regarding the preparation of site-specific simulator scenarios.

8. **NUREG-1560, "Individual Plant Examination Program: Perspectives on Reactor Safety and Plant Performance," December 1997**

This report provides perspectives gained by reviewing 75 individual plant examination (IPE) submittals pertaining to 108 nuclear power plant units. Chapter 13, "Operational Perspectives," is of particular interest because it identifies a number of important human actions that should be considered for evaluation on licensing and requalification examinations for pressurized- and boiling-water reactors.

9. **NUREG-1600, "General Statement of Policy and Procedure for NRC Enforcement Actions," May 2000**

This report addresses the NRC's expectations regarding compliance with 10 CFR 55.49, "Integrity of Examinations and Tests," and possible enforcement actions against parties who are subject to that regulation (i.e., Part 55 license holders and applicants and Part 50 licensees).

10. **NUREG/BR-0122, "Examiners' Handbook for Developing Operator Licensing Written Examinations," Revision 5, March 1990**

This document, which presented a procedure for systematically constructing content-valid licensing examinations for nuclear power plant operators, has been incorporated into the examination standards in NUREG-1021, Revisions 8 and 9. It may be used for historical perspective, but is no longer used for developing examinations.

F. Industry Standards

1. ANSI/ANS 3.1, "American National Standard for Selection, Qualification, and Training of Personnel for Nuclear Power Plants"

This standard provides criteria for selecting and training nuclear power plant employees who perform a variety of functions at various levels of responsibility (e.g., managers, supervisors, operators, and technicians). RG 1.8, Revision 3 (May 2000) endorses the 1993 version of this standard, with additions, exceptions, and clarifications thereto.

2. ANS 3.2 (ANSI N18.7-1976), "Administrative Controls and QA for the Operational Phase of Nuclear Power Plants"

This standard provides guidance and recommendations for administrative rules of practice and related subjects and for preparing procedures and audit programs. See RG 1.33.

3. ANSI/ANS 3.4-1996, "Medical Certification and Monitoring of Personnel Requiring Operator Licenses for Nuclear Power Plants"

This standard is the basic document covering the general health and disqualifying conditions applicable to license applicants and licensed personnel. Revision 3 of RG 1.134 currently endorses this standard, with exceptions, but facility licensees may continue to use the 1983 version, which was previously endorsed in its entirety by Revision 2 of RG 1.134.

4. ANSI/ANS 3.5-1998, "Nuclear Power Plant Simulators for Use in Operator Training"

This standard establishes the minimum functional requirements and capabilities for nuclear power plant simulators for use in operator training. Revision 3 of RG 1.149 endorses this standard, with clarifications. However, facility licensees may continue to use the 1985 and 1993 versions, which were previously endorsed, with exceptions, by Revisions 1 and 2 of RG 1.149.

ES-201
INITIAL OPERATOR LICENSING EXAMINATION PROCESS

A.   Purpose

This standard describes the activities that must be completed to prepare for initial operator licensing examinations (including written examinations and operating tests) at power reactor facilities. As such, this standard includes instructions for scheduling and coordinating examination development, assigning NRC examiners and facility personnel, maintaining examination security, and obtaining reference and examination materials from the facility licensee.

B.   Background

Title 10, Part 55, of the *Code of Federal Regulations* (10 CFR Part 55) requires that applicants for reactor operator (RO) and senior reactor operator (SRO) licenses must pass both a written examination and an operating test. The regulation allows power reactor facility licensees to prepare the site-specific written examinations and operating tests, provided that (1) the facility licensee shall prepare the examinations and tests in accordance with the criteria contained herein; (2) the facility licensee shall establish, implement, and maintain procedures to control examination security and integrity; (3) an authorized representative of the facility licensee shall approve the examinations and tests before they are submitted to the NRC for review and approval; and (4) the facility licensee shall obtain NRC approval of its proposed written examinations and operating tests. Moreover, the regulation requires that the license examinations must be developed and administered in accordance with 10 CFR 55.41 and 55.45 for ROs, or 10 CFR 55.43 and 55.45 for SROs.

Facility licensees may propose alternatives to the examination criteria contained herein and evaluate how the proposed alternatives provide an acceptable method of complying with the Commission's regulations. The NRC staff will review any proposed alternatives and make a decision regarding their acceptability. The NRC will not approve any alternative that would compromise the agency's statutory responsibility to prescribe uniform conditions for the operator licensing examinations.

The NRC staff will continue to prepare the examinations (or discrete portions thereof, including the outline, written, or operating tests) upon written request by facility licensees (consistent with NRC staff availability) and retains the authority to develop the examinations on a case-by-case basis to certify new examiners or if the staff loses confidence that a facility licensee will develop examinations upon which the NRC can base its licensing decisions. If the staff determines that a facility is unable to develop acceptable examinations, the examinations could be delayed until the NRC can schedule sufficient resources to develop and conduct the examinations, or until the facility licensee can develop an acceptable examination. Each NRC regional office will also prepare at least one examination per calendar year to certify new examiners, as required, and to maintain examiner proficiency.

The NRC will make a reasonable attempt to administer all license examinations on the dates requested by facility licensees. At times, however, resource limitations may compel the staff to prioritize its examination review and development activities based on need and safety considerations. Examinations for fewer than three applicants should be scheduled only under extenuating circumstances, such as a shortage of licensed ROs or SROs at the facility. If a facility licensee has fewer than three license applicants, the examinations may be delayed until more applicants are trained. Moreover, facility licensees who elect to have the NRC prepare their licensing examinations should keep in mind that the NRC staff requires more time to prepare than to review an examination and that the NRC will require greater flexibility to schedule those services.

In accordance with 10 CFR 55.40(a), the NRC shall use the criteria in NUREG-1021 to prepare the written examinations required by 10 CFR 55.41 and 55.43 and the operating tests required by 10 CFR 55.45. The NRC shall also use the criteria in NUREG-1021 to evaluate the written examinations and operating tests prepared by power reactor facility licensees pursuant to 10 CFR 55.40(b). The NRC's regional offices shall obtain approval from the NRR operator licensing program office before knowingly deviating from the intent of NUREG-1021. Moreover, the regional offices shall obtain program office approval before undertaking any initiative that could undermine examination consistency among the regions.

Other pre-examination activities, such as submitting and reviewing license applications and eligibility waivers and administering the generic fundamentals examination program, are addressed in ES-202, ES-204, and ES-205. Specific instructions for developing, administering, and grading the written examinations and operating tests are found in ES-401 through ES-403 and ES-301 through ES-303, respectively. Post-examination administrative activities, including management review of the examination results and preparation of examination reports, are discussed in ES-501. Cross-references to each of these standards have been provided where appropriate.

C.   Responsibilities

Facility licensees and NRC staff should use Form ES-201-1, "Examination Preparation Checklist," to track the examination preparations. As noted on the form, the target due dates can be adjusted as necessary to accommodate a given situation. The NRC chief examiner will initial the items as they are completed and ensure that the original form is retained for the master examination file (refer to ES-501).

1.   <u>Facility Licensee</u>

If a facility licensee asks the NRC to prepare the licensing examinations, only those items identified with an asterisk (*) apply.

a*.   The facility licensee is expected to apprise its NRC regional office of changes in its examination requirements.

The facility licensee should respond in writing to the NRC's annual letter soliciting estimated operator licensing needs (including estimated numbers of applicants, examination dates, and the licensee's intended level of participation in developing all parts of the examination). The facility licensee should also notify its NRC regional office if its examination requirements change significantly from those stated in its response. The NRC strongly encourages facility licensees to schedule their examinations and to resolve any applicant eligibility questions with their NRC regional office *before* commencing an initial license training class.

In accordance with 10 CFR 55.40(c), facility licensees who elect to have the NRC prepare, proctor, and grade any portion of their operator licensing examinations shall submit written requests (to the responsible NRC regional office) for those examinations pursuant to 10 CFR 55.31(a)(3). A response to the NRC's annual letter will satisfy this requirement.

b*.   In accordance with 10 CFR 55.49, facility licensees and applicants shall not engage in any activity that compromises the integrity of any application, test, or examination that is required by 10 CFR Part 55. Attachment 1 to this examination standard summarizes several examination security and integrity considerations. NUREG-1600, "General Statement of Policy and Procedures for NRC Enforcement Actions," dated May 1, 2000, addresses possible enforcement actions against parties who are subject to the requirements in the regulation (i.e., Part 55 license applicants and licensees and Part 50 licensees).

c.   Pursuant to 10 CFR 55.40(b)(2), facility licensees who elect to prepare their own examinations shall establish, implement, and maintain procedures to control examination security and integrity. Attachment 1 discusses a number of examination security and integrity guidelines that may be appropriate for incorporation in those procedures.

d*.   All facility and contractor personnel involved with an examination are subject to the restrictions stated in Section D of this examination standard. Any questions regarding those restrictions should be resolved with the NRC chief examiner before granting an individual access to the licensing examination.

The facility licensee shall designate a point of contact to work with the NRC chief examiner and assign additional personnel as required to ensure that the examinations are developed, reviewed, administered, and graded in accordance with the applicable examination standards. The facility licensee may use contractors or other outside assistance to develop the examinations, but the licensee bears full responsibility for the product, including conformance with the examination criteria and maintenance of examination security and integrity.

e*.   The facility contact shall submit the required reference materials, examination outlines, and examinations, as applicable, based on the level of facility participation. Form ES-201-1 specifies target due dates for the various materials; the actual dates may be adjusted with prior agreement from the NRC regional office. For the purposes of operator training and examination, the facility licensee may "freeze" the plant procedures at a particular revision in order to facilitate examination development. The facility licensee shall discuss this option with the the NRC chief examiner in advance and refer to Attachment 2 for additional guidance on procedure freezes.

f. The examination outlines and examinations shall be prepared in accordance with the guidelines in ES-301, ES-401, and ES-701, as applicable. The proposed outlines and examinations shall cover all portions of the license examination (written, dynamic simulator, and walk-through) at all license levels relevant to the applicants (RO, SRO, and limited SRO) to be tested.

A facility supervisor or manager shall independently review the examination outline(s) and the proposed examination(s) before they are submitted to the NRC regional office in accordance with Item (g), below.

In conducting this review, the facility supervisor or manager shall use Forms ES-201-2, "Examination Outline Quality Checklist"; ES-301-3, "Operating Test Quality Checklist"; ES-301-4, "Simulator Scenario Quality Checklist"; and ES-401-6, "Written Examination Quality Checklist."

g. Pursuant to 10 CFR 55.40(b)(3), an authorized representative of the facility licensee shall approve the examination outline(s) and the proposed examination(s) before they are submitted to the NRC regional office for review and approval. The outline(s) and examination(s) should be forwarded to the NRC regional office with a cover letter signed by the facility representative. The materials must be complete and ready-to-use.

h. In its examination submittal to the NRC, the facility licensee (or its contractor) shall provide the following information for each test item proposed for use as part of the written examination and/or the operating tests:

  • State the source of each item (e.g., is the item taken directly, without changes, from the facility licensee's or *any* other bank; is the item a modified version of a bank item; or is the item new?). Facility licensees are encouraged to identify those bank items that were used on an NRC license examination at the facility since October 1995 because they will generally undergo less-rigorous review by the NRC.

  • For those items that were derived by modifying existing bank items, note the changes that were made or submit a copy of the item from which it originated.

i. The facility licensee shall make its simulation facility available, as necessary, for NRC examiners to prepare for and administer the operating tests. The NRC will make reasonable efforts to minimize the impact on other training activities.

Before developing or administering an initial licensing examination, facility licensees are encouraged to review the simulator examination security considerations in Appendix D to NUREG-1021 for applicability to their facility. Because facility licensees are more familiar than NRC examiners with the unique capabilities, limitations, and vulnerabilities of their simulators, the NRC staff expects licensees to take responsibility for determining and implementing whatever measures might be necessary to ensure the integrity of the operating tests.

Pursuant to 10 CFR 55.46(c)(1)(i) and 55.46(d), facility licensees must ensure sufficient simulator fidelity to allow conduct of the evolutions listed in 10 CFR 55.45(a)(1) – (13), as applicable to the design of the reference plant. In addition, facility licensees must make available for NRC review the results of any uncorrected performance deficiencies that may exist at the time of the operating test.

j.   The facility licensee shall meet with the NRC in the regional office or at the facility, as necessary and appropriate, to review the examinations and discuss potential changes.

If the NRC prepared the examination, the facility reviewers should make their comments and recommendations on a copy of the written examination(s) and operating test(s) provided to them by the NRC examiner. Simple editorial changes that do not change the intent of the question require no justification; however, *every* substantive change (e.g., deleting a question, replacing a distractor, or revising an answer) must be supported by approved facility reference material.

If the facility licensee has significant concerns with the content or difficulty of the NRC-prepared examination, the changes that the NRC has directed the facility licensee to make in its proposed examination, or the general implementation of the requirements and guidelines in this NUREG, the facility licensee is encouraged to communicate those concerns to the NRC and, if appropriate, to request a meeting with the NRC to address the concerns. The NRC chief examiner is normally the first point of contact for resolving any concerns regarding the examination. If the concerns are not resolved at that level, the facility licensee should contact NRC regional management and, if necessary, the chief of the NRR operator licensing program office for resolution.

k.   If the facility licensee developed the examination, the licensee will generally make any necessary changes as agreed upon with the NRC; however, the NRC retains final authority to approve the examinations.

l*. In accordance with ES-202, the facility licensee shall submit the license applications along with a letter requesting that licensing examinations be administered.

2. **NRC Regional Management, Supervision, and Designees**

a.   The regional office shall schedule the NRC's initial operator licensing examinations and shall arrange for the development, administration, and grading of those examinations as discussed below. The regional office shall periodically review each facility licensee's examination requirements and shall negotiate with the facility licensee's training representatives, as necessary, to schedule specific examination dates consistent with operational requirements and NRC resource availability. Each regional office shall plan to prepare at least one complete examination per calendar year.

b.   Approximately 6 months before each anticipated examination date, the regional office should contact the facility licensee and confirm the examination date(s) and the expected number of applicants to be examined. The regional office should use that information to estimate the required number of NRC examiners and to make preliminary work assignments.

c.   The regional office should contact the facility licensee by telephone at least 4 months before the scheduled examinations to reconfirm the expected number of applicants and the examination dates, and to make other preliminary arrangements for developing the examinations. The person who contacts the facility licensee shall discuss the following examination arrangements, as applicable, depending on the facility licensee's level of participation in the examination development process:

- the examination integrity and security requirements and considerations (refer to Attachment 1)

- the guidance related to freezing plant procedure changes (refer to Attachment 2)

- the requirement that an authorized representative of the facility licensee must approve the examination outlines and examinations before they are submitted to the NRC for review

- the need to have the examination outlines delivered to the NRC approximately 75 days before the scheduled examination date

- the need to have the reference materials necessary for the NRC to develop the examination (if applicable; refer to Attachment 3) delivered to the regional office at least 75, but preferably 90, days before the scheduled examination date

- the guidelines for developing, administering, and grading the written examinations, as applicable (i.e., the effective version of ES-401, ES-402, and ES-403, respectively)

- the need to ensure simulator fidelity in accordance with 10 CFR 55.46(c)(1)(i), and to have the simulator and a list of uncorrected performance deficiencies and deviations from the reference plant available at the time of the operating tests

- the guidelines for developing and administering the operating tests (i.e., the effective version of ES-301 and ES-302, respectively)

- the need to have the examinations and the supporting reference materials (refer to Attachment 3) delivered to the NRC regional office approximately 45 days before the scheduled examination date

- the option to submit some sample test items (e.g., 5 to 10 written questions, 1 scenario, and 1 to 2 job performance measures) for preliminary NRC review and comment (this could increase the efficiency of the examination review process by promoting early identification and correction of generic examination development concerns)

- the requirements (refer to 10 CFR 55.31) and guidelines (refer to ES-202) for submitting the license applications

The NRC regional office may negotiate earlier due dates with the facility contact, but should refrain from advancing the dates if it is unlikely that the review will begin promptly after the material arrives in the regional office. The regional office should also keep the facility contact informed of the dates by which the region expects to provide its comments regarding the licensee's submittals.

d. The NRC regional office shall normally issue a letter confirming the arrangements no later than 120 days before the examination begins. The letter should be addressed to the person at the highest level of corporate management who is responsible for plant operations (e.g., Vice President of Nuclear Operations). Attachment 4 is an example of such a letter; the exact wording may be modified, as necessary to reflect the situation.

e. Approximately 4 months before the scheduled examination, the NRC regional office will assign the required number of examiners to develop, prepare for, and administer the examination as arranged with the facility licensee. The regional office will also designate a chief examiner to coordinate the examination project with the facility licensee and other examiners assigned to the examination. When making assignments, the regional office should consider each examiner's certification status, other examination commitments, possible conflicts of interest (as discussed in Section D of this examination standard), and general availability.

Once the facility licensee has begun preparing the examination, the regional office shall avoid changing the chief examiner assignment unless absolutely necessary. If a change is unavoidable, the responsible supervisor shall attempt to minimize the impact on the facility licensee.

Regional management should try to assign a sufficient number of examiners so that no examiner will have to administer more than four operating tests per week.

f. The regional office will evaluate each examination assignment to determine if some or all of the assigned examiners should make a separate preparatory site visit. The purposes of such a visit may include providing examiner orientation, retrieving additional reference material, auditing the accuracy of the license applications per ES-202, or reviewing and validating the examinations. When making a decision, the regional office should carefully weigh the costs and benefits associated with each additional trip to the facility. The regional office should also consider such factors as the experience of the assigned examiners, the quality of the facility licensee's examinations (if applicable), the number of written examinations and operating tests to be validated, and the status of the simulation facility (e.g., is it new or recently upgraded?). In addition, the regional office should consider the alternative of reviewing the written examination(s) and operating test(s) with the facility licensee via telephone (if the examination quality is high) or in the regional office, as well as the alternative of validating the operating test(s) on-site at the beginning of the examination week.

g. Upon receiving the preliminary license applications, approximately 30 days before the examination date, the regional office shall review the applications in accordance with ES-202. In addition, the regional office shall evaluate any waiver requests in accordance with ES-204 to determine if the applicants meet the eligibility criteria specified in 10 CFR 55.31.

h. The responsible regional supervisor will review the examination outlines and the draft examinations and evaluate any recommended changes and corrections noted during the chief (and other) examiner's review. (Refer to ES-301 and ES-401 for additional guidance regarding examination reviews). The supervisory review is not intended to be another detailed review, but rather a check to ensure that all applicable administrative requirements have been implemented. If the outlines, examinations, and recommended changes are acceptable, the supervisor will authorize the chief examiner to resolve any noted deficiencies with the author or facility contact.

If any of the facility-developed examination materials (written, walk-through, or simulator) require substantive changes and cannot be made to conform with the examination standards by the end of the designated examination review week, regional management shall consult the NRR operator licensing program office and make a decision whether to proceed with the facility-developed examinations or develop the examinations in-house. If the regional office does not have the resources to ensure that acceptable examinations are prepared by the scheduled administration date, regional management shall negotiate with the facility licensee to reschedule the examinations as necessary. Although it is generally easier to postpone the written examination and focus on the operating tests so that they can be administered on schedule and without affecting examinations at other facilities, regional management may delay either part (written examination or operating test) of an examination for up to 30 days to allow additional time for examination development or to address other scheduling concerns. It is *not* appropriate to delay one part of an NRC examination based on license applicant performance on another part of an NRC examination that has already been administered, or based on applicant performance on facility-administered audit examinations. However, the entire NRC examination may be delayed for other reasons (e.g., applicant readiness) as agreed upon by the regional office. The regional office shall consult the NRR operator licensing program office regarding any delay and notify the facility licensee in writing of the reasons for delaying the examination(s).

The responsible supervisor will also ensure that any significant deficiencies and problems are addressed in the examination report in accordance with ES-501.

i. After the chief examiner has verified that the necessary changes and corrections have been made, the responsible supervisor will review and approve the examinations for administration. Before signing the applicable quality checklist (i.e., Form ES-301-3 and/or Form ES-401-6), the supervisor must be satisfied that the examination is acceptable for administration.

After approving the examination and license applications, including resolving all waiver requests, the region will prepare an "Examination Approval Letter" (in the format of Attachment 5) and a "List of Applicants" (Form ES-201-4). The letter will notify the facility licensee that the NRC has completed its review of the license applications, confirm that both the NRC and the facility licensee agree that the examination meets the guidelines of NUREG-1021, and provide authorization to the facility licensee to administer the written examinations, if applicable. Form ES-201-4 will identify the approved applicants by name, docket number, and type of examination to be administered (e.g., SRO upgrade, SRO-only written, RO written only). All applicants listed on the form will be administered complete examinations (written and operating) as indicated unless waivers have been granted in accordance with ES-204. A copy of Form ES-201-4 will be distributed to all assigned examiners; however, because it contains information that is protected by the *Privacy Act*, the form will not be attached to the approval letter, but will be provided separately to the facility licensee.

j. The responsible supervisor shall query the facility licensee management counterpart regarding the licensee's views on the examination sometime before the examination is administered. The following subjects should be considered for discussion, and corrective measures shall be implemented when necessary:

- whether the NRC test item comments were justified and clearly explained

- the licensee's assessment of the significant test item changes

- whether any of the examination changes are believed to render the test items or the examination/test as a whole unfair, and whether this concern was shared with the chief examiner

- whether the NRC asked the licensee to rework any "NRC-validated" questions

- whether the facility licensee requested and was permitted to defer the correction of test item flaws that were identified as minor in nature

k. If there is an indication that an examination may have been compromised, the responsible supervisor will take action as necessary to ensure and restore the integrity and security of the examination process. Actions may include not giving the examination, making additional changes to the examination, voiding the results if the examination has already been given, reevaluating the licensing decisions pursuant to 10 CFR 55.61(b), and possibly imposing enforcement action in accordance with NUREG-1600. The supervisor shall keep regional management and the NRR operator licensing program office informed of any concerns regarding examination integrity or security.

3. **Assigned NRC Examiners**

a. When assigned to administer operating tests for the first time at a particular facility, the examiner should inform the chief examiner and the responsible supervisor so that arrangements can be made to conduct an orientation trip to the facility as described in Item C.2.f, if deemed appropriate.

b. NRC examiners monitor and ensure the integrity of the examination process. If they perceive that a compromise has occurred, they must immediately report it to the responsible regional supervisor so that the necessary actions can be taken to restore the integrity of the examination. Attachment 1 summarizes several examination security and integrity considerations that examiners should note when reviewing the procedures that the facility licensee has established pursuant to 10 CFR 55.40(b)(2), as applicable.

c. The assigned examiners shall review and inventory the reference materials received from the facility licensee in response to the 120-day corporate notification letter. The purpose of this review is to determine if the materials are complete and adequate to enable the regional office to review or develop the examinations, as applicable. If not, the reviewer(s) shall inform the chief examiner and the responsible supervisor and request that the facility licensee send any additional materials that might be required. If necessary, an examiner may review and select additional reference materials during a site orientation trip (refer to Item C.2.f).

d. The chief examiner will work with the assigned examiners and the designated facility contact, as applicable, to ensure that the examination outlines and examinations are developed in accordance with the applicable examination standards. The chief examiner should adapt the level of oversight and coordination based upon the experience of the individuals who are preparing the examinations. Facility employees are generally less familiar with the examination standards and will require more oversight to ensure that a quality examination is ready on time.

e. The chief examiner will ensure that the examination outlines are independently reviewed using Form ES-201-2, "Examination Outline Quality Checklist," as a guide; if the chief examiner prepared any portion of the outline, another NRC examiner shall perform that part of the independent review. The NRC reviewer(s) will initial Column "c" of Form ES-201-2 for the specific items they reviewed. A thorough and timely review (i.e., within 5 working days) will minimize the potential for significant problems with the examinations.

The chief examiner will note/review any necessary changes and forward the outlines to the responsible supervisor for review and comment before resolving any deficiencies with the author or facility contact. The chief examiner will document his/her review/concurrence, as applicable, by signing the bottom

of the form. If the outlines are significantly deficient, refer to Item C.2.h
for additional guidance.

f.   The chief examiner will ensure that the written examinations and operating tests
     are independently reviewed for quality in accordance with the applicable
     checklists (refer to ES-301 and ES-401) forwarded with the examination.
     If the chief examiner wrote any portion of the examination, another NRC
     examiner shall perform the independent review of that portion. The NRC
     reviewer(s) will initial Column "c" of the applicable checklist for the specific item(s)
     they reviewed. The regional office may conduct additional reviews
     at its discretion if resources permit.

     It is especially important that facility-developed written examinations and
     operating tests be reviewed promptly because of the extra time that may be
     required if extensive changes are necessary. The written examination sampling
     review (as described in Section E of ES-401) should be completed within 1 week
     after receiving the examination, and the balance of quality reviews should be
     completed within 2 weeks after the written examinations and operating tests
     are received from the author or facility contact.

     The chief examiner will note any necessary changes and forward the written
     examinations and operating tests to the responsible supervisor for review and
     comment before reviewing the examinations with the author or facility contact.
     The chief examiner will document his/her review/concurrence, as applicable,
     by signing the bottom of each quality checklist. There are no minimum or
     maximum limits on the number or scope of changes the NRC may direct
     the facility licensee to make to its proposed examinations, provided that
     they are necessary to make the examinations conform with established
     acceptance criteria or to attain an appropriate level of examination difficulty.
     Chief examiners shall exercise their experience and judgement to ensure that
     the level of difficulty remains consistent with that expected on NRC-prepared
     examinations. If the examinations are significantly deficient, refer to Item C.2.h
     for additional guidance. The chief examiner shall document the responsible
     supervisor's authorization to proceed with for the facility review by initialing
     Item 11 on Form ES-201-1.

g.   Upon supervisory approval, generally about 2 weeks before the examinations
     are scheduled to be given, the chief examiner will review the written examinations
     and operating tests with the facility licensee.

     The chief examiner may conduct the examination review via telephone,
     in the regional office, or at the facility, as appropriate to the circumstances,
     depending on the extent of the changes, and as approved by the responsible
     regional supervisor (refer to Item C.2.f).

     If the NRC staff prepared the examination, the regional office will provide a copy
     of the written examination(s) and operating test(s) to the facility reviewers
     after they sign the security agreement (Form ES-201-3); if necessary to promote
     efficiency, this may be done before the actual review. The facility reviewers

should make their comments directly on the examination(s), return the marked-up copies to the NRC chief examiner, and ensure that he or she understands their comments and recommendations. The facility reviewers may retain a copy of the marked-up examination(s), subject to the physical security considerations in Attachment 1.

If the facility reviewers have significant disagreements with the chief examiner, the chief examiner will inform the responsible regional supervisor so that the disagreements can be resolved before the examinations are administered.

h.   After the examination corrections have been made, the chief examiner shall verify that the changes are appropriate and route the examinations and the marked-up drafts to the responsible supervisor for final approval.

i.   As soon as possible after the responsible supervisor has approved the operating tests for administration, the chief examiner shall distribute copies of the scenarios, job performance measures (JPMs), and questions to the other assigned examiners so that they can familiarize themselves with those materials and be better prepared to probe the applicants' deficiencies if required.

j.   The chief examiner should work with the designated facility contact to schedule the operating tests to optimize efficiency and the mix of RO and SRO applicants in the crews assembled for the simulator examinations. The chief examiner may elect to make or change the facility licensee's crew assignments; however, crew changes will generally not be made less than 2 weeks before the date on which the examinations are scheduled to begin, so that the affected applicants have some time to adapt to working as a crew. When assembling crews for the simulator examinations, surrogate operators should be used only when they are required to complete an operating crew. A facility licensee may not replace license applicants with surrogates solely because the applicants have performed the minimum required number of events or scenarios. If an applicant would be exposed to only *one* additional scenario above the minimum required, a surrogate operator should not be used in place of a license applicant. However, no applicant will be required to participate in *more* than one scenario above the minimum required, in which case, a surrogate operator should be used. If, at the discretion of the chief examiner, it is desired to use surrogate operators contrary to the above guidance, the operator licensing program office should be consulted prior to implementation.

The number of applicants on a crew shall not exceed the number of assigned examiners (i.e., one-on-one evaluations are mandatory), except as noted below. However, if the facility licensee's technical specifications routinely require more than two ROs to be stationed in the control room, the chief examiner may authorize the use of additional surrogates. Only one individual (applicant or surrogate) is allowed to fill a shift supervisor or manager position during the simulator operating test.

If a three-person operating crew consists entirely of SRO-upgrade applicants (who do not have to be evaluated on the control boards), the region may assign only two examiners to observe the crew. Although the applicants in the RO and balance-of-plant positions may not be individually evaluated, they will be graded and held accountable for any errors that occur as a result of their action(s) or inaction(s). SRO-instant applicants will always be individually evaluated, regardless what operating position they are filling during a given scenario.

Normally, for purposes of test integration and continuity, the same examiner should administer all three operating test categories to an applicant. However, under certain circumstances, the walk-through portion of the operating test may be divided among different examiners. Such division is appropriate if a facility licensee's simulator is not located near the plant, because of limitations in examiner resources or scheduling, or if a facility licensee requests examinations for an unusually large group of applicants. Refer to ES-302 for specific instructions regarding administration of the operating tests.

Operating tests will normally be administered on regular work days. If weekend or shift work is required to administer the operating tests, the chief examiner will coordinate the arrangements with the assigned examiners and the facility licensee.

The written examinations may be administered as soon as they and the license applications (including any applicable waivers) have been approved. The region shall not allow the written examination and operating test dates to diverge by more than 30 days without obtaining concurrence from the NRR operator licensing program office.

If, as an efficiency measure, the facility licensee prepared the written examinations or operating tests in conjunction with another facility, the two examinations/tests must be administered at the same time.

If the examination schedule has to be changed on short notice, the chief examiner will work with his or her supervisor and the designated facility contact to reschedule the examinations to a time when examiners are available and other examinations are not affected.

k.   If the facility licensee will administer the written examinations, the chief examiner shall review the ES-402 requirements (e.g., proctoring and responding to applicant questions) and confirm the applicant's status on Form ES-201-4 (i.e., examination type and waivers) with the facility contact before the examinations are given.

D.   Personnel Restrictions

It is impossible to define criteria that anticipate every possible conflict-of-interest issue. Supervisors must apply sound judgment to the facts of each case. If any doubt exists regarding a particular case, the supervisor should consult with regional management and/or the NRR operator licensing program office to resolve the issue.

1. **NRC Examiners**

   a. The regional office shall not assign an examiner who failed an applicant on an operating test to administer any part of that applicant's retake operating test.

   b. If an examiner was previously employed by a facility licensee (or one of its contractors) and was significantly involved in training the current license applicants, the regional office will not assign that examiner any direct responsibilities for developing or administering written examinations or operating tests at that facility. Regional management will control other in-office examination activities concerning the facility, such as technical consultation and quality reviews of examinations.

   c. If an examiner is assigned to an examination that might appear to present a conflict of interest, the examiner shall inform his or her immediate supervisor of the potential conflict. Such notifications should include the following information:

      • the nature and extent of previous personal and professional relationships with the applicants

      • anything that could affect the administration, performance, evaluation, or results of the examination

      • anything that could create the *appearance* of a conflict of interest

2. **Facility Personnel**

   a. Although there is no specific upper limit to the number of facility personnel who have access to the NRC licensing examination, the facility licensee shall ensure that access is limited on a need-to-know basis. Moreover, the facility licensee should limit each person's access to only those portions of the examination for which the individual bears responsibility (e.g., the individuals who prepare the simulator scenarios may not require access to the written examinations).

   b. All personnel who will receive detailed knowledge of any portion of the NRC licensing examination, including the examination outline, must acknowledge their responsibilities by reading and signing Form ES-201-3, "Examination Security Agreement," before they obtain detailed knowledge and again after the examinations are complete. Prohibited activities for personnel who have signed Form ES-201-3 include the following examples:

      • the design and administration of any classroom and simulator instruction (including scheduled sessions, individual coaching, and remedial training) specifically for the license applicants (Simulator booth operation is acceptable if the individual does not select the training content or provide direct or indirect feedback. Continued participation in requalification training for groups including SRO upgrade applicants is also acceptable, as long as it is documented on Form ES-201-3 and is limited to areas in which the instructor has no examination knowledge.)

- all on-the-job training, practice, coaching, and sign-offs

- the preparation, review, grading, and evaluation of periodic quizzes, examinations, and simulator exercises (Individuals on the security agreement may prepare and grade the audit examination, subject to an NRC review for test item duplication.)

Supervisors and managers having knowledge of the examination content may continue their general oversight of the training program for the license applicants, including the review of examinations, quizzes, and remedial training programs, as well as the counseling of applicants concerning non-technical issues. However, those supervisors and managers may not provide any technical guidance, training, or other direct feedback regarding the content of those examinations, quizzes, or programs in a manner that might compromise the integrity of the licensing examination as defined in 10 CFR 55.49.

The original security agreement forms must be submitted to the NRC's regional office for retention after the examinations are complete.

E.   Attachments/Forms

Attachment 1,	"Examination Security and Integrity Considerations"
Attachment 2,	"Guidelines for Freezing Plant Procedures"
Attachment 3,	"Reference Material Guidelines for Initial Licensing Examinations"
Attachment 4,	"Sample Corporate Notification Letter"
Attachment 5,	"Sample Examination Approval Letter"
Form ES-201-1,	"Examination Preparation Checklist"
Form ES-201-2,	"Examination Outline Quality Checklist"
Form ES-201-3,	"Examination Security Agreement"
Form ES-201-4,	"List of Applicants"

NRC and facility licensee personnel must be attentive to examination security measures to ensure compliance with 10 CFR 55.49. Moreover, pursuant to 10 CFR 55.40(b)(2), facility licensees who elect to prepare their own examinations must establish, implement, and maintain procedures to control examination security and integrity. At the time the examination arrangements are confirmed, an NRC examiner shall review the facility licensee's security procedures and brief the facility contact on the following examination security guidelines. Although these guidelines are not regulatory requirements, the NRC staff encourages facility licensees to consider them when establishing their own procedures.

Physical Security Guidelines

1. The NRC expects that personnel will be aware of the facility licensee's physical security measures and requirements (as documented in the facility licensee's approved procedures), sign the NRC's examination security agreement, and understand their security responsibilities, including the limits on their interaction with the license applicants (as discussed in Section D.2 of ES-201), before they are given knowledge or custody of any examination materials.

2. The examination outlines and final examinations shall be positively and continuously controlled and protected as sensitive information (i.e., under lock-and-key or in the custody of someone who has signed the security agreement). The number of copies of outlines and examinations should be limited, and each should be uniquely identified and controlled (e.g., with sign-out custody) at all times. Drafts, copies, and waste materials must also be controlled and disposed of properly.

    The NRC staff recommends that facility licensees should consider implementing additional security measures when they are developing, storing, or printing examinations using a computer network to which license applicants or other persons who have not signed the security agreement could gain access. Although the use of passwords should provide adequate security if normal computer security practices (e.g., selecting and changing passwords) are observed, special cases may need additional consideration. For example, if a trainee has extended access to the local area network (LAN) in his normal position, additional security measures might be appropriate.

3. The examination outlines, written examinations, and operating tests that are sent to the NRC's regional office shall be placed in a double envelope. The inner envelope shall be conspicuously marked "FOR OFFICIAL USE ONLY" and "TO BE OPENED BY ADDRESSEE ONLY." Furthermore, the cover letter forwarding the examination materials shall state that the materials must be withheld from public disclosure until after the examinations are complete.

    The facility licensee should follow up on its examination mailing by communicating with the NRC chief examiner to ensure that the package was received.

The examination outlines and examinations shall not be transmitted via non-secure electronic means. However, they may be transmitted via the NRC's "AUTOS" LAN in the resident inspector's office or as password-protected electronic files over the Internet if the licensee's word processing software provides adequate security and is compatible with the NRC's, and the password is separately provided to the NRC chief examiner by mail (*not* email), fax, or phone. The files do not need to be encrypted.

4.   The facility licensee is expected to immediately report to the NRC chief examiner any indications or suggestions that examination security may have been compromised, even if the situation is identified and corrected before the examination is submitted to the NRC for review and approval. The NRC will evaluate such situations on a case-by-case basis and determine the appropriate course of action.

5.   The facility licensee and the NRC should determine if examination security problems were noted in the past and ensure that corrective actions have been taken to preclude recurrence.

6.   The facility licensee and the chief examiner will review the simulator security considerations in Appendix D to ensure that the instructor station features, programmers' tools, and external interconnections do not compromise examination integrity. The primary objective is to ensure that the exam material cannot be read or recorded at other unsecured consoles, and that examination materials are either physically secured or electronically protected when not in use by individuals listed on the security agreement.

Examination Bank Limitations

1.   The facility licensee and chief examiner shall ensure that written examinations and operating tests conform with the guidelines in ES-301 and ES-401 regarding the use of items taken directly from the bank, modified items, and new items.

2.   If the facility licensee has an open bank, it will not place any new or modified test items (written questions, JPMs, or simulator scenarios) that will be used on the examination in its examination bank until after the last examination has been administered.

Other Considerations

1.   The NRC will consider an examination to be potentially compromised if any activity occurs that could affect the equitable and consistent administration of the examination, regardless of whether the activity takes place before, during, or after the examination is administered.

2.   The license applicants should not be able to predict or narrow the possible scope or content of the licensing examination based on the facility licensee's examination practices (other than those authorized by NUREG-1021 or in writing by the NRC).

3.   Facility licensees are responsible for the integrity, security, and quality of examinations prepared for them by contractor personnel.

The NRC understands that facility licensees may wish to train and examine their license applicants to the same version of plant procedures. At their discretion, facility licensees may "freeze" plant procedures to a particular revision for purposes of applicant training and examination development (either for facility-prepared examinations or as reflected in the reference materials submitted for NRC-prepared examinations). The NRC does not have any specific requirements related to the timing of procedure freezes, but offers the following general guidance and cautions:

- Clearly, the later the procedures are frozen the better, thereby limiting the disparity between training/testing and current plant operations. Alternatively, facility licensees could choose to not freeze procedures at all, but rather track any procedure changes and make adjustments to the training an examinations as required. However, depending on the nature and volume of changes, this alternative could impose a significant additional burden on the facility and NRC examiners to ensure that procedure revisions affecting test items are reconciled prior to exam administration.

- Note that applicants will be exposed to the current version of the procedures when they spend time in the control room. Therefore, freezing procedures for the exam has the potential to confuse applicants, by testing them on a different version of procedures than they have seen. There have been cases in which such confusion contributed to applicants' failure on the written examination, because they based their answer on the wrong version of procedures. If the procedures are frozen, the applicants must be informed of the date of the procedure freeze, such that they have a complete understanding of which versions of the procedures the NRC examination is based upon. Note that freezing different procedures at different times would probably just add to their confusion.

- Examination authors and NRC reviewers need to consider the implications of the freeze when they develop the examination; for example, the plausibility and correctness of a distractor should not hinge on a procedure change that has not yet been incorporated into the frozen version of the procedure. Another consideration is whether the simulator will support the implementation of both procedure versions - the new one for license holders and the old one for the applicants.

- If changes in the procedures occur after the freeze and before the licensing date, the NRC would expect the facility licensee to provide training to fill the gap; if the changes are significant, the NRC would likely request more information about the nature of such training and testing. In at least one instance, applicants were trained and tested on a new version of the emergency operating procedures (EOPs) that had not yet been implemented in the plant; this eliminated the need to retrain the applicants but prompted the NRC to delay their licensing until the new EOPs went into effect.

The facility contact should discuss the details of and basis for their freeze proposal with their NRC contact when confirming the examination arrangements as discussed in Section C.2.c of ES-201 of NUREG-1021. The chief examiner, in consultation with the regional operator licensing supervisor (and the operator licensing program office, if deemed necessary), will review the facility's proposal and negotiate a mutually acceptable plan and cut-off date.

This attachment discusses the reference materials that facility licensees are expected to provide for each NRC initial licensing examination. The regional office will customize the list of reference materials, as required, to support the specific examination assignment; the regional office may request additional materials at a later time, if necessary, to ensure the accuracy and validity of the examinations.

In determining the need for reference materials, the regional office will consider the facility licensee's level of participation in the examination development process. If the facility licensee will prepare the examinations, it may be sufficient to obtain only those references necessary to review and validate the items that appear on the examination, plus a set of key procedures and other documents required to prepare for the operating tests. The regional office will duly consider the administrative burden it places on the facility licensee and will request only those materials that are actually necessary for the NRC examiners to prepare for the examinations.

All reference materials provided for the license examinations should be approved, final issues and should be so marked; any personal, proprietary, sensitive, or safeguards information should be marked and submitted in a separate enclosure. If any of the material is expected to change before the scheduled examination date, the facility licensee should reach agreement with the NRC chief examiner regarding changes before the examinations are administered.

The facility licensee may submit reference materials on computer diskettes (in a format that is compatible with the NRC's word processing software), as hard copy, or a combination, as arranged with the NRC chief examiner. If the facility licensee prepares the examinations, the hard-copy references should normally be limited to those materials required to validate the selected test items. All procedures and reference materials should be bound with appropriate indices or tables of contents so that they can be used efficiently; a master table of contents should be provided for all materials sent. Failure to provide complete, properly bound, and indexed reference materials may prompt the NRC to return the materials to the person at the highest level of corporate management responsible for plant operations. The returned reference materials will be accompanied by a cover letter explaining the deficiencies in the materials and the basis for postponing or canceling the examinations.

Unless otherwise instructed by the NRC's regional office, the facility licensee is expected to provide the following reference materials for each NRC initial licensing examination:

1.   Materials used by the facility licensee to ensure operator competency

   a.   The following types of materials used to train applicants for initial RO and SRO licensing, as necessary to support examination development:

      ·   learning objectives, student handouts, and lesson plans

      ·   system descriptions, drawings, and diagrams of all operationally relevant flow paths, components, controls, and instrumentation

- material used to clarify and strengthen understanding of normal, abnormal, and emergency operating procedures

- complete, operationally useful descriptions of all safety system interactions and, where available, balance-of-plant system interactions under emergency and abnormal conditions, including consequences of anticipated operator errors, maintenance errors, and equipment failures, as well as plant-specific risk insights based on a probabilistic risk analysis (PRA) and individual plant examination (IPE)

These materials should be complete, comprehensive, and of sufficient detail to support the development of accurate and valid examinations without being redundant.

b.   Questions and answers specific to the facility training program that may be used in the written examinations or operating tests

c.   Copies of facility-generated simulator scenarios that expose the applicants to abnormal and emergency conditions, including degraded pressure control, degraded heat removal capability, and containment challenges, during all modes of operation, including low-power conditions (A description of the scenarios used for the training class may also be provided.)

d.   All JPMs used to ascertain the competence of the operators in performing tasks within the control room complex and outside the control room (i.e., local operations) as identified in the facility's job task analysis (JTA) (JPMs should evaluate operator responsibilities during normal, abnormal, and emergency conditions and events, and during all modes of operation including cold shutdown, low power, and full power.)

2.   Complete index of procedures (including all categories sent)

3.   All administrative procedures applicable to reactor operation or safety

4.   All integrated plant procedures (normal or general operating procedures)

5.   All emergency procedures (emergency instructions, abnormal or special procedures)

6.   Standing orders (important orders that are safety-related and may modify the regular procedures)

7.   Surveillance procedures that are run frequently (i.e., weekly) or that can be run on the simulator

8.   Fuel handling and core loading procedures (if SRO applicants will be examined)

9.   All annunciator and alarm procedures

10.   Radiation protection manual (radiation control manual or procedures)

11.   Emergency plan implementing procedures

12.   Technical Specifications or similar technical requirements documents
(and interpretations, if available) for all units for which licenses are sought

13.   System operating procedures

14.   Technical data book and plant curve information used by operators,
as well as the facility precautions, limitations, and set points document

15.   The following information pertaining to the simulation facility:

     a.   list of all initial conditions

     b.   list of all malfunctions with identification numbers and cause-and-effect information,
including a concise description of the expected result or range of results
that will occur upon initiation and an indication of which annunciators
will be actuated as a result of the malfunction

     c.   a description of the simulator's failure capabilities for valves, breakers, indicators,
and alarms

     d.   the range of severity of each variable malfunction (e.g., the size of a reactor
coolant or steam leak, or the rate of a component failure such as a feed pump,
turbine generator, or major valve)

     e.   a list of modeling conditions (e.g., simplifications, assumptions, and limits)
and problems that may affect the examination

     f.   a list of any known performance test discrepancies not yet corrected

     g.   a list of differences between the simulator and the reference plant's control room

     h.   simulator instructor's manual

16.   Any additional plant-specific material that the NRC examiners have requested
to develop examinations that meet the guidelines of these standards and the regulations

(Date)

(Name, Title)
(Name of facility)
(Address)
(City, State, Zip code)

Dear (Name):

In a telephone conversation on (date) between Mr./Ms. (Name, Title) and Mr./Ms. (Name, Title), arrangements were made for the administration of licensing examinations at (facility name) during the week(s) of (date).

As agreed during the telephone conversation, [your staff][[the staff of the U.S. Nuclear Regulatory Commission (NRC)]] will prepare the examinations based on the guidelines in Revision 9, Supplement 1, of NUREG-1021, "Operator Licensing Examination Standards for Power Reactors." [The NRC's regional office will discuss with your staff any changes that might be necessary before the examinations are administered.][[Your staff will be given the opportunity to review the examinations during the week of (date).]]

To meet the above schedule, it will be necessary for your staff to furnish the [examination outlines by (date). The written examinations, operating tests, and supporting] reference materials identified in Attachment 3 to ES-201 [will be due] by (date). [Pursuant to Title 10, Section 55.40(b)(3), of the Code of Federal Regulations (10 CFR 55.40(b)(3)), an authorized representative of the facility licensee shall approve the outlines, examinations, and tests before they are submitted to the NRC for review and approval. All materials shall be complete and ready-to-use.] We request that any personal, proprietary, sensitive unclassified, or safeguards information in your response be contained in a separate enclosure and appropriately marked. Any delay in receiving the required [examination and] reference materials, or the submittal of inadequate or incomplete materials, may cause the examinations to be rescheduled.

In order to conduct the requested written examinations and operating tests, it will be necessary for your staff to provide adequate space and accommodations in accordance with ES-402, and to make the simulation facility available on the dates noted above. In accordance with ES-302, your staff should retain the original simulator performance data (e.g., system pressures, temperatures, and levels) generated during the dynamic operating tests until the examination results are final.

Appendix E to NUREG-1021 contains a number of NRC policies and guidelines that will be in effect while the written examinations and operating tests are being administered.

To permit timely NRC review and evaluation, your staff should submit preliminary reactor operator and senior reactor operator license applications (Office of Management and Budget (OMB) approval number 3150-0090), medical certifications (OMB approval number 3150-0024), and waiver requests (if any)(OMB approval number 3150-0090) at least 30 days before the first examination date. If the applications are not received at least 30 days before the examination date, a postponement may be necessary. Signed applications certifying that all training has been completed should be submitted at least 14 days before the first examination date.

This letter contains information collections that are subject to the *Paperwork Reduction Act of 1995* (44 U.S.C. 3501 et seq.). These information collections were approved by OMB, under approval number 3150-0018, which expires on June 30, 2009. The public reporting burden for this collection of information is estimated to average [500] [[50]] hours per response, including the time for reviewing instructions, gathering and maintaining the data needed, [writing the examinations,]and completing and reviewing the collection of information. Send comments on any aspect of this collection of information, including suggestions for reducing the burden, to the Information and Records Management Branch (T-6 F33), U.S. Nuclear Regulatory Commission, Washington, DC 20555-0001, or by Internet electronic mail to BJS1@nrc.gov; and to the Desk Officer, Office of Information and Regulatory Affairs, NEOB-10202, (3150-0018), Office of Management and Budget, Washington, DC 20503.

The NRC may neither conduct nor sponsor, and a person is not required to respond to, an information collection, unless it displays a currently valid OMB control number.

In accordance with 10 CFR 2.390 of the NRC's "Rules of Practice," a copy of this letter and its enclosures will be available electronically for public inspection in the NRC's Public Document Room or from the Publicly Available Records (PARS) component of the NRC's Agencywide Documents Access and Management System (ADAMS). ADAMS is accessible from the Electronic Reading Room page of the NRC's public Web site at http://www.nrc.gov/reading-rm/adams.html.

Thank you for your cooperation in this matter. (Name) has been advised of the policies and guidelines referenced in this letter. If you have any questions regarding the NRC's examination procedures and guidelines, please contact (name of regional contact) at (telephone number), or (name of responsible regional supervisor) at (telephone number).

Sincerely,

(Appropriate regional representative, Title)

Docket No.:   50-(Number)

Distribution: Public
                 NRC Document Control System
                 Regional Distribution

[ ]   Include only for examinations to be prepared by the facility licensee.
[[ ]]   Include only for examinations to be prepared by the NRC.

(Date)

(Name, Title)
(Name of facility)
(Address)
(City, State, Zip code)

SUBJECT: OPERATOR LICENSING EXAMINATION APPROVAL

Dear (Name):

The purpose of this letter is to confirm the final arrangements for the upcoming operator licensing examinations at (Facility).

The NRC has completed its review of the operator license applications submitted in connection with this examination and separately provided a list of approved applicants to (Name, Title). Note that any examination waivers and application denials have been addressed in separate correspondence.

The NRC has approved the subject examinations and hereby authorizes you to administer the written examinations in accordance with Revision 9, Supplement 1, of NUREG-1021, "Operator Licensing Examination Standards for Power Reactors," on (date). The NRC staff will administer the operating tests during the week of (date). This examination has undergone extensive review by my staff and representatives responsible for licensed operator training at your facility. Based on this review I have concluded that the examination meets the guidelines of NUREG-1021 for content, operational, and discrimination validity. By administering this examination, you also agree that it meets NUREG-1021 guidelines, and is appropriate for measuring the qualifications of licensed operator applicants at your facility. If you determine that this examination is not appropriate for licensing operators at your facility, do not administer the examination and contact me at (phone number).

Please contact your Chief Examiner, (Name), at (phone number), if you have any questions or identify any errors or changes in the license level (RO or SRO) or type of examination (partial or complete written examination and/or operating test) specified for each applicant.

Sincerely,

(Appropriate regional representative, Title)

Docket No.: 50-

cc: Public
       NRC Document Control System
       Regional Distribution

Facility: _____   Date of Examination: _____

Developed by: Written - Facility ☐   NRC ☐   //   Operating - Facility ☐   NRC ☐

Target Date*	Task Description (Reference)	Chief Examiner's Initials
-180	1. Examination administration date confirmed (C.1.a; C.2.a and b)	
-120	2. NRC examiners and facility contact assigned (C.1.d; C.2.e)	
-120	3. Facility contact briefed on security and other requirements (C.2.c)	
-120	4. Corporate notification letter sent (C.2.d)	
[-90]	[5. Reference material due (C.1.e; C.3.c; Attachment 3)]	
{-75}	6. Integrated examination outline(s) due, including Forms ES-201-2, ES-201-3, ES-301-1, ES-301-2, ES-301-5, ES-D-1's, ES-401-1/2, ES-401-3, and ES-401-4, as applicable (C.1.e and f; C.3.d)	
{-70}	{7. Examination outline(s) reviewed by NRC and feedback provided to facility licensee (C.2.h; C.3.e)}	
{-45}	8. Proposed examinations (including written, walk-through JPMs, and scenarios, as applicable), supporting documentation (including Forms ES-301-3, ES-301-4, ES-301-5, ES-301-6, and ES-401-6, and any Form ES-201-3 updates), and reference materials due (C.1.e, f, g and h; C.3.d)	
-30	9. Preliminary license applications (NRC Form 398's) due (C.1.l; C.2.g; ES-202)	
-14	10. Final license applications due and Form ES-201-4 prepared (C.1.l; C.2.i; ES-202)	
-14	11. Examination approved by NRC supervisor for facility licensee review (C.2.h; C.3.f)	
-14	12. Examinations reviewed with facility licensee (C.1.j; C.2.f and h; C.3.g)	
-7	13. Written examinations and operating tests approved by NRC supervisor (C.2.i; C.3.h)	
-7	14. Final applications reviewed; 1 or 2 (if >10) applications audited to confirm qualifications / eligibility; and examination approval and waiver letters sent (C.2.i; Attachment 5; ES-202, C.2.e; ES-204)	
-7	15. Proctoring/written exam administration guidelines reviewed with facility licensee (C.3.k)	
-7	16. Approved scenarios, job performance measures, and questions distributed to NRC examiners (C.3.i)	

* Target dates are generally based on facility-prepared examinations and are keyed to the examination date identified in the corporate notification letter. They are for planning purposes and may be adjusted on a case-by-case basis in coordination with the facility licensee.
[Applies only] {Does not apply} to examinations prepared by the NRC.

Facility:		Date of Examination:			

Item	Task Description	Initials		
		a	b*	c#
1. WRITTEN	a. Verify that the outline(s) fit(s) the appropriate model, in accordance with ES-401.			
	b. Assess whether the outline was systematically and randomly prepared in accordance with Section D.1 of ES-401 and whether all K/A categories are appropriately sampled.			
	c. Assess whether the outline over-emphasizes any systems, evolutions, or generic topics.			
	d. Assess whether the justifications for deselected or rejected K/A statements are appropriate.			
2. SIMULATOR	a. Using Form ES-301-5, verify that the proposed scenario sets cover the required number of normal evolutions, instrument and component failures, technical specifications, and major transients.			
	b. Assess whether there are enough scenario sets (and spares) to test the projected number and mix of applicants in accordance with the expected crew composition and rotation schedule without compromising exam integrity, and ensure that each applicant can be tested using at least one new or significantly modified scenario, that no scenarios are duplicated from the applicants' audit test(s), and that scenarios will not be repeated on subsequent days.			
	c. To the extent possible, assess whether the outline(s) conform(s) with the qualitative and quantitative criteria specified on Form ES-301-4 and described in Appendix D.			
3. W I T	a. Verify that the systems walk-through outline meets the criteria specified on Form ES-301-2: (1) the outline(s) contain(s) the required number of control room and in-plant tasks distributed among the safety functions as specified on the form (2) task repetition from the last two NRC examinations is within the limits specified on the form (3) no tasks are duplicated from the applicants' audit test(s) (4) the number of new or modified tasks meets or exceeds the minimums specified on the form (5) the number of alternate path, low-power, emergency, and RCA tasks meet the criteria on the form.			
	b. Verify that the administrative outline meets the criteria specified on Form ES-301-1: (1) the tasks are distributed among the topics as specified on the form (2) at least one task is new or significantly modified (3) no more than one task is repeated from the last two NRC licensing examinations			
	c. Determine if there are enough different outlines to test the projected number and mix of applicants and ensure that no items are duplicated on subsequent days.			
4. GENERAL	a. Assess whether plant-specific priorities (including PRA and IPE insights) are covered in the appropriate exam sections.			
	b. Assess whether the 10 CFR 55.41/43 and 55.45 sampling is appropriate.			
	c. Ensure that K/A importance ratings (except for plant-specific priorities) are at least 2.5.			
	d. Check for duplication and overlap among exam sections.			
	e. Check the entire exam for balance of coverage.			
	f. Assess whether the exam fits the appropriate job level (RO or SRO).			

	Printed Name/Signature	Date
a. Author	_____	_____
b. Facility Reviewer (*)	_____	_____
c. NRC Chief Examiner (#)	_____	_____
d. NRC Supervisor	_____	_____

Note:	# Independent NRC reviewer initial items in Column "c"; chief examiner concurrence required. * Not applicable for NRC-prepared examination outlines

1. Pre-Examination

I acknowledge that I have acquired specialized knowledge about the NRC licensing examinations scheduled for the week(s) of _____ as of the date of my signature. I agree that I will not knowingly divulge any information about these examinations to any persons who have not been authorized by the NRC chief examiner. I understand that I am not to instruct, evaluate, or provide performance feedback to those applicants scheduled to be administered these licensing examinations from this date until completion of examination administration, except as specifically noted below and authorized by the NRC (e.g., acting as a simulator booth operator or communicator is acceptable if the individual does not select the training content or provide direct or indirect feedback). Furthermore, I am aware of the physical security measures and requirements (as documented in the facility licensee's procedures) and understand that violation of the conditions of this agreement may result in cancellation of the examinations and/or an enforcement action against me or the facility licensee. I will immediately report to facility management or the NRC chief examiner any indications or suggestions that examination security may have been compromised.

2. Post-Examination

To the best of my knowledge, I did not divulge to any unauthorized persons any information concerning the NRC licensing examinations administered during the week(s) of _____. From the date that I entered into this security agreement until the completion of examination administration, I did not instruct, evaluate, or provide performance feedback to those applicants who were administered these licensing examinations, except as specifically noted below and authorized by the NRC.

PRINTED NAME   JOB TITLE / RESPONSIBILITY   SIGNATURE (1)   DATE   SIGNATURE (2)   DATE   NOTE

1.
2.
3.
4.
5.
6.
7.
8.
9.
10.
11.
12.
13.
14.
15.

NOTES:

PRIVACY ACT INFORMATION — FOR OFFICIAL USE ONLY

Facility:	Written Examination Date: Operating Test Dates:						
Applicant Name	Docket No.	Exam Level	Written		Operating Test		
			RO	SRO	Adm.	Sys.	Sim.

Instructions: For each approved applicant, enter the exam level (RO, SRO-I, or SRO-U) and an "X" or "W" to indicate whether each portion of the examination is to be administered or waived.

PRIVACY ACT INFORMATION — FOR OFFICIAL USE ONLY

ES-202
PREPARING AND REVIEWING OPERATOR LICENSING APPLICATIONS

A.   Purpose

This standard provides instructions for facility licensees and applicants to prepare and the NRC to review initial licensing applications. It also discusses the experience, training, education, and certification requirements and guidelines that an applicant should satisfy before being allowed to take an NRC reactor operator (RO), senior reactor operator (SRO), or limited senior reactor operator (LSRO) licensing examination.

B.   Background

In accordance with Title 10, Section 55.31(a)(4), of the *Code of Federal Regulations* (10 CFR 55.31(a)(4)), as amended by a rule change dated March 25, 1987, a license applicant must provide evidence that he or she has successfully completed the facility licensee's requirements to be licensed as an RO or SRO. An authorized representative of the facility licensee shall certify this evidence on the license application; the required certification must include the details of the applicant's qualifications, training, and experience. In lieu of these details, the Commission may accept certification that the applicant has successfully completed a Commission-approved training program that is based on a systems approach to training (SAT) and uses a simulation facility that is acceptable to the Commission.

Revision 2 of Regulatory Guide (RG) 1.8, "Qualification and Training of Personnel for Nuclear Power Plants," which was published in conjunction with the 1987 rule change, provided guidance on an acceptable method of implementing this regulation. However, the NRC staff had reviewed[1] the industry's licensed operator training program experience guidelines in effect at the time of the 1987 rule change and determined that they were equivalent to the baseline experience criteria of RG 1.8, Revision 2. Consequently, as indicated in the statement of consideration for the 1987 rule change, a facility licensee's training program would be considered approved by the NRC when it is accredited by the National Nuclear Accrediting Board (NNAB).

On March 19, 1987, the NRC staff published Generic Letter (GL) 87-07, "Information Transmittal of Final Rulemaking for Revisions to Operator Licensing: 10 CFR Part 55 and Conforming Amendments." Specifically, GL 87-07 informed facility licensees that they have the option to substitute an accredited, SAT-based program in lieu of the operator training program that the NRC staff previously approved for the given facility. The GL also indicated that facility licensees may implement this option upon providing written notification to the NRC and without the need for any staff review. In addition, the GL noted the NRC's expectation that facility licensees would update their licensing-basis documents (e.g., their final safety analysis reports (FSARs) and technical specifications (TSs)), as necessary, to conform with their accredited program status.

---

[1]   The NRC staff conducted this review pursuant to the Commission's continued endorsement of the industry's accreditation process, which the Commission first conferred in its "Final Policy Statement on Training and Qualification of Nuclear Power Plant Personnel" (50 FR 11147), dated March 20, 1985.

In November 1987, the NRC published NUREG-1262, "Answers to Questions at Public Meetings Regarding Implementation of Title 10, *Code of Federal Regulations*, Part 55 on Operators' Licenses," which reiterated and clarified the NRC staff's expectations regarding licensees' compliance with 10 CFR 55.31(a), Revision 2 of RG 1.8, and accredited training programs, as well as the need for facility licensees to update their licensing-basis documents in accordance with 10 CFR 50.71(e). NUREG-1262 also reminded facility licensees that Revision 2 of RG 1.8 would go into effect on March 31, 1988. In addition, this NUREG noted that facilities with NNAB-accredited license training programs do not need to meet the guidance in Revision 2 of RG 1.8.

In summary, the NRC has not changed its requirements or position with regard to license eligibility for ROs and SROs since 1987. RG 1.8 (Revision 2 or 3) and the guidelines for education and experience promulgated by the National Academy for Nuclear Training (NANT)[2] — including those that were in effect in 1987 and those that were issued in January 2000 — outline acceptable methods for implementing the Commission's regulations in this area. In addition, methods different from those set forth in RG 1.8 (Revision 2 or 3) or the NANT guidelines may be acceptable if a facility licensee provides an adequate basis for using such methods.

The staff encourages all facility licensees to review their requirements and commitments related to RO and SRO education and experience and to update their documentation (e.g., FSAR, TSs, and training program descriptions) to enhance consistency and minimize confusion.

When a facility licensee's licensed operator training program description and/or licensing-basis documents contain education and experience requirements that are more restrictive than either Revision 3 of RG 1.8 or the current NANT guidelines, the most restrictive requirements will continue to apply pending the initiation of action by the licensee to amend these requirements; any required TS changes would be considered administrative in nature.

Operator license applicants and facility licensees must provide the NRC with sufficient information to enable the staff to determine whether to grant or deny the applications. However, some facility licensees did not respond to GL 87-07 and/or failed to update their licensing-basis documents to eliminate inconsistencies and contradictions. This has made it difficult for the NRC staff to determine whether some license applicants have successfully completed their facility licensee's requirements to be licensed as an RO or SRO. Nonetheless, the fact that every facility licensee has voluntarily obtained and periodically renewed the accreditation of its licensed operator training program suggests that every facility licensee is implementing the education and experience guidelines endorsed by the NNAB. Specifically, the NRC staff understands that the current version of those guidelines are outlined in the NANT "Guidelines for Initial Training and Qualification of Licensed Operators,"[3] which were issued in January 2000

---

[2]   The NANT operates under the auspices of the Institute of Nuclear Power Operations (INPO). It integrates the training efforts of all U.S. nuclear utilities, the activities of the NNAB, and the training-related activities of INPO.

[3]   The NRC staff has reviewed the NANT guidelines and considers them to be equivalent to the agency's guidelines in Revision 3 of RG 1.8, which was published in May 2000. RG 1.8 now endorses American National Standards Institute/American Nuclear Society (ANSI/ANS) Standard 3.1-1993, "Selection, Qualification, and Training of Personnel for Nuclear Power Plants," with certain clarifications, additions, and exceptions. It replaces Revision 2 of RG 1.8, which was issued in conjunction with the 1987 rule change and endorsed the 1981 revision of ANSI/ANS 3.1.

(NANT 2000 guidelines). Consequently, unless otherwise informed by a facility licensee, the NRC staff believes that the education and experience guidelines described in the NANT 2000 guidelines constitute the facility licensee's education and experience requirements to be licensed as an RO or SRO.

In an effort to clarify the situation, the NRC staff revised NRC Form 398, "Personal Qualifications Statement: Licensee," to clarify that when a facility licensee certifies, pursuant to 10 CFR 55.31(a)(4), that an applicant has successfully completed a Commission-approved, SAT-based training program, it means that the applicant meets or exceeds the minimum education and experience guidelines currently outlined by the NANT (and, by extension, Revision 3 of RG 1.8). Facility licensees can use the revised NRC Form 398 to document any exceptions or waivers that the applicant has taken from the baseline education and experience guidelines outlined by the NANT. In addition, recognizing that the only significant difference between Revision 3 of RG 1.8 and the current accreditation guidelines pertains to certified instructors seeking an SRO license, those applicants can use the revised NRC Form 398 to document the details of their experience. This will minimize the potential for misunderstanding and the need to seek additional information.

C.   Responsibilities

The regulatory requirements associated with the license application process are detailed in Subpart D, "Applications," of 10 CFR Part 55, while the medical requirements for license applicants and licensed operators appear in Subpart C, "Medical Requirements." NRC staff and license applicant should refer to these requirements as necessary when preparing and reviewing license applications.

1.   <u>Applicant/Facility Licensee</u>

a.   To apply for an RO or SRO license, an applicant must submit NRC Form 398, and NRC Form 396, "Certification of Medical Examination by Facility Licensee." (Computer-generated facsimiles are acceptable.) The application is not complete until both forms are filled out, signed by the appropriate personnel, and received by the NRC. Detailed instructions for completing NRC Form 398 are provided with the form. Applicants and facility licensees should pay particular attention to the instructions and note related to Item 12. Additional instructions regarding waivers of training, experience, and examination requirements are provided in ES-204. Instructions for completing NRC Form 396 are also provided with the form. Both Form 396 and 398 are available on the NRC's operator licensing Web page at http://www.nrc.gov/reactors/operator-licensing/licensing-process.html.

If the applicant is reapplying following a license denial, 10 CFR 55.35 applies, and the applicant must complete and submit a new Form 398; however, as discussed below, a new Form 396 may not be required. The applicant may file the second application 2 months after the date of the first final denial, a third application 6 months after the date of the second final denial, and successive applications 2 years after the date of each subsequent denial. Each new Form 398 shall describe the extent of the applicant's additional training

since the denial and shall include a certification by the facility licensee
that the applicant is ready for reexamination.

If the applicant previously passed either the written examination or the operating test,
he or she may request a waiver of that portion of the licensing examination.
Such waivers are limited to the first re-application and must be requested
within 1 year of the date on which the applicant completed the original examination.
The NRC staff will also consider written examination waivers for ROs in good
standing who prefer to take only the 25-question, SRO portion of the written
examination when they apply to upgrade their licenses. Refer to ES-204
for a more detailed discussion of these and other waiver criteria.

Prior to licensing, every applicant must have a complete medical examination
that meets the guidelines in the applicable version of ANSI/ANS 3.4, "Medical
Certification and Monitoring of Personnel Requiring Operator Licenses for
Nuclear Power Plants," as endorsed by RG 1.134, "Medical Evaluation of
Licensed Personnel at Nuclear Power Plants." Although licensed operators
can go up to 24 months between medical examinations, new license applicants
are generally expected to be examined and certified as fit (on NRC Form 396)
no more than 6 months before the anticipated date of licensing. However, if
more than 6 months have passed since the date of an applicant's last medical
examination or fitness certification on NRC Form 396, the applicant/facility
licensee may request a waiver of medical reexamination by checking Item 4.f.4
on NRC Form 398 and certifying in writing, in Item 17, "Comments,"
that the applicant has not developed any physical or mental condition
that would be reportable under 10 CFR 55.25. The NRC staff will consider
such a waiver if an applicant is reapplying for a license (because of withdrawing a
previous application, final license denial on a previous application, or terminating a
previous license at the same facility), or if an examination is delayed from its
originally scheduled date. (Refer to ES-204 for more information on waivers.)
However, if an applicant's physical or mental condition has changed, or the time
since the applicant's last complete medical examination is expected to exceed
24 months before the licensing action is completed, the applicant shall be
reexamined by a physician and the facility licensee shall recertify the applicant's
medical fitness on NRC Form 396. Licensed ROs upgrading to an SRO license
need not have an additional medical examination or waiver request, as long as
their medical status as a licensed RO is up to date at the time of application,
including a complete medical examination within the past 24 months.

In accordance with Section 3.1 of ANSI/ANS 3.4-1996, which the NRC endorsed
in Revision 3 of RG 1.134, the examining physician may delegate portions
of the medical examination to a licensed nurse practitioner or licensed
physician's assistant who is familiar with the ANSI/ANS 3.4-1996 and
the activities required of a nuclear power plant operator or senior operator.
However, the physician has the ultimate responsibility for certifying that
the medical examination was conducted in accordance with the standard
and that the applicant meets the medical requirements. The names and license
numbers of all medical practitioners (but not laboratory technicians) who were
substantially involved in the examination should be entered on NRC Form 396.

b.   Each new applicant (except those applying for an LSRO license or an SRO upgrade license at the same facility) must satisfactorily complete the NRC's generic fundamentals examination (GFE) section of the written operator licensing examination for the applicable reactor type (boiling- or pressurized-water) within 24 months before the date of application. Applicants who passed a GFE on the same reactor type more than 24 months before the date of application may request a waiver of the GFE in accordance with ES-204. Refer to ES-205 for more information on the GFE program.

c.   Pursuant to 10 CFR 55.31(a)(5), new applications must include the number of significant control manipulations affecting reactivity or power level in Item 14, "Significant Control Manipulations." At least five manipulations are required on the facility for which the license is sought or a plant-referenced simulator. Control manipulations performed on the plant-referenced simulator may be chosen from a representative sampling of the control manipulations and plant evolutions described in 10 CFR 55.59(c)(3)(i)(A–F), (R), (T), (W), and (X), as applicable to the design of the plant for which the license application is submitted. Power changes (Items (E) and (F)) performed on the simulator must be 10 percent or greater in magnitude, while those on the plant may be smaller but of sufficient magnitude for the operator to experience appropriate feedback (i.e., clearly observable effects on the plant) as a result of the control manipulation. Every effort should be made to perform at least some of the manipulations on the actual plant and to diversify the reactivity and power changes for each applicant. For ROs applying for an SRO license, certification that the operator has successfully operated the controls of the facility as a licensed operator shall be accepted as evidence of having completed the required manipulations.

Facility licensees who propose to use a plant-referenced simulator to perform the control manipulations required by 10 CFR 55.31(a)(5) must ensure that simulator fidelity has been demonstrated pursuant to 10 CFR 55.46(c).

d.   Neither 10 CFR Part 55 nor Section 107 of the *Atomic Energy Act* requires license applicants to be citizens of the United States; therefore, non-citizens may apply for a license without having to obtain a waiver or exemption. However, all applicants must meet the requirements for unescorted access to a nuclear power facility pursuant to 10 CFR 73.56 – 57, including a criminal history check and background investigation.

e.   As noted in ES-201, the facility licensee should submit preliminary, uncertified license applications and medical certifications for review by the NRC's regional office at least 30 days before the examination date. This will permit the NRC staff to make preliminary eligibility determinations, process the medical certifications, evaluate any waivers that might be appropriate, and obtain additional information, if necessary, while allowing the facility licensee to finish training the applicants before the certified applications are due.

f.   The facility licensee's senior management representative on site (i.e., an authorized representative of the facility licensee, such as the plant manager or site vice-president) must certify when an applicant has completed all of the facility licensee's requirements and commitments for the desired license level (i.e., experience, control manipulations, training, and medical). Such certification involves placing a check in Item 19.b of NRC Form 398, signing the form, and submitting it to the NRC's regional office at least 14 days before the examination date. The senior management representative must also sign Item B, "Certification," on NRC Form 396.

Pursuant to 10 CFR 55.5, "Communications," facility licensees may submit these forms to the NRC by mail, in person, or, where practicable, via electronic information exchange (EIE) or on CD-ROM. Electronic submissions must be made in a manner that enables the NRC to receive, read, authenticate, distribute, and archive the submission, and process and retrieve it one page at a time. Detailed guidance on making electronic submissions can be obtained by visiting the NRC's public Web site at http://www.nrc.gov/site-help/eie.html, calling (301) 415-6030, sending an email message to EIE@nrc.gov, or writing to the Office of the Chief Information Officer, U.S. Nuclear Regulatory Commission, Washington, DC 20555-0001. Forms that have only a single signature, such as NRC Form 396, may be submitted electronically using an electronic digital signature. However, forms with multiple signatures, such as NRC Form 398, must rely on handwritten optically scanned signatures, because of the limited digital signature capability of the EIE system. For any textual documents submitted in an optically scanned format, please note that Searchable Image (Exact) PDF is required, to preclude optical character recognition errors. When sending these forms via EIE, facility licensees are encouraged to follow up with a phone call or e-mail message to the operator licensing assistant in the regional office to ensure the forms are received.

The facility must also submit a written request to administer the written examination and operating test to the applicant.

g. When the NRC's regional office denies a license application, the applicant need not accept the proposed denial. In such instances, the applicant may request that the Director, Division of Inspection and Regional Support, Office of Nuclear Reactor Regulation (NRR), review the application denial or request a hearing in accordance with 10 CFR 2.103(b)(2). Further action will be taken in accordance with ES-502.

h. The facility licensee is expected to inform the NRC's regional office in writing if it wishes to withdraw an application before the licensing process is complete.

2. NRC Regional Office

a. The NRC's regional office shall review preliminary applications as soon as possible after they are received. In that way, the regional office can process the medical certifications, evaluate and resolve any waiver requests in accordance with ES-204, and obtain from the facility licensee any additional information that might be necessary in order to support the final eligibility determinations.

With regard to medical certifications, the regional office shall forward the applicant's NRC Form 396 and supporting medical evidence to the NRC's contract physician for evaluation any time the examining physician recommends that the NRC should issue a restricted license to the applicant, that the NRC should grant the applicant a waiver (exception) of any requirement set forth in the applicable ANSI/ANS standard, or that the NRC should change an existing restriction (by checking any of blocks A.2 to A.10 on Form 396). If, on the date of the licensing examination, the NRC's physician is still reviewing an applicant's medical certification but there is no reason to expect that the physician will disqualify the applicant, the NRC's regional office should allow the applicant to take the examination, with the understanding that the NRC will withhold the license until the medical certification is approved.

The NRC will not process a retake application if the applicant's request for reconsideration or a hearing on the previous license denial is still outstanding. (Refer to ES-502.)

Before entering the applicants' data in the operator licensing tracking system (OLTS), the NRC's regional office shall verify that none of the applicants' names appear on the list of "Escalated Enforcement Actions Issued to Individuals." The regional office shall check with the appropriate contact in the Office of Enforcement by telephone or email to verify that the information on the subject individuals is current before using the information on the list to deny a licensing action.

b. The regional office will verify that the applicant has successfully passed the GFE, if required, and review the data on NRC Form 398 to ensure that it is complete.

Affirmative responses to Items 12.a and 12.b on NRC Form 398, indicate that the applicant has successfully completed a Commission-approved, SAT-based training program that (1) meets the education and experience requirements outlined by the NNAB and (2) uses a simulation facility acceptable to the Commission under 10 CFR 55.45(b). If the facility licensee checks "yes" in response to these items, the licensee need not complete Item 13, "Training," or Item 15, "Experience Details," on NRC Form 398, except as noted below, and the regional office may accept the application without further review.

The regional office will verify that new applications include at least five significant control manipulations affecting reactivity or power level in Item 14 of NRC Form 398 (refer to Section C.1.c).

As noted in the instructions for Item 12 on NRC Form 398, certified instructors (who may not have the requisite responsible nuclear power plant experience, or RNPPE, defined in RG 1.8, Revision 3) seeking an SRO license must complete Item 15. Moreover, any exceptions or waivers from the education and experience requirements outlined in the NANT "Guidelines for Initial Training and Qualification of Licensed Operators" must be explained in Item 17.

If an applicant checks "no" in response to Items 12.a and 12.b, provides information that is not required, or indicates that exceptions or waivers have been taken (in Item 17 on NRC Form 398), the regional office shall review the application against the specific eligibility requirements and commitments applicable to the facility licensee and shall refer any eligibility issues (e.g., any failure to meet the minimum guidelines established by the NNAB or RG 1.8, Revision 3) and questions to the NRR operator licensing program office for resolution.

If the applicant is reapplying after a previous examination failure and license denial, the regional office shall evaluate the applicant's additional training to determine if the facility licensee made a reasonable effort to remediate the deficiencies that caused the applicant to fail the previous examination.

c.   The regional office may determine (1) that the preliminary application is incomplete, (2) more information is necessary to make a waiver determination, or (3) the applicant does not meet the requirements in 10 CFR 55.31. In such instances, the regional office will note the deficiencies and request that the facility licensee supply additional information when it submits the final, certified license application (or sooner if possible).

Conversely, the regional office may determine that the preliminary application is complete, and the applicant meets the eligibility requirements or is expected to meet the requirements pending the receipt of additional information. In such instances, the regional office shall enter the applicant's name, docket number, and examination requirements on the "List of Applicants" in accordance with ES-201.

d.   Upon receiving the final, certified license application, the reviewer shall evaluate any new information to ensure that the eligibility criteria are satisfied. If so, the reviewer shall check the "meets requirements" block at the bottom of NRC Form 398 and shall sign and date the form. If necessary, the reviewer shall add the applicant's name and other data to the "List of Applicants" in accordance with ES-201. The reviewer shall also ensure that the list accurately reflects any examination waivers that may have been granted in accordance with ES-204.

If the regional office determines that the applicant still does not meet the eligibility requirements, the regional licensing authority will (1) discuss its decision with the NRR operator licensing program office, (2) notify the applicant in writing that the application is being denied, and (3) identify the deficiencies on which the denial is based (Attachment 1). The responsible regional supervisor, or designee, shall check the "does not meet requirements" block at the bottom of Form 398, and shall sign and date the form. The applicant's name shall be stricken from the "List of Applicants," and the applicant shall not be permitted to take the licensing examination until the regional office determines that he or she meets the eligibility criteria.

In accordance with ES-204, the region may administer a license examination to an applicant who has not satisfied the applicable training or experience requirements at the time of the examination, but is expected to complete them shortly thereafter. Assuming that the applicant passes the examination, the regional office shall not issue the applicant's license until the facility licensee certifies that all of the requirements have been completed. (Refer to ES-501 for additional guidance.)

e. During either the preparatory site visit or the examination week, the regional office shall audit a sample (approximately 10 percent) of the license applications (i.e., NRC Form 398s) to confirm that they accurately reflect the subject applicants' qualifications. The review should focus primarily on the applicants' experience and on-the-job training, including reactivity manipulations, to ensure that they comply with Part 55 and the facility's licensing-basis documents and licensed operator training program description. The regional office will refer specific eligibility questions and deficiencies to the NRR operator licensing program office for review before making the licensing decisions.

D. NRC License Eligibility Guidelines

Regulatory Guide (RG) 1.8, "Qualification and Training of Personnel for Nuclear Power Plants," describes a method that the NRC staff finds acceptable for complying with the Commission's regulations with regard to the training and qualifications of nuclear power plant personnel. For the positions of shift supervisor, senior operator, and licensed operator, Revision 3 of RG 1.8, which was issued in May 2000, endorses the guidelines contained in ANSI/ANS 3.1-1993; specific clarifications, additions, and exceptions are noted in Section C, "Regulatory Position," of RG 1.8. The license eligibility guidelines in RG 1.8, Revision 3, and ANSI/ANS 3.1-1993 are summarized below; refer to those documents for more detailed information. No backfitting is intended or required in connection with the issuance of the revised RG.

As noted in Section B, above, the NRC has reviewed the current education and experience guidelines outlined in the NANT "Guidelines for Initial Training and Qualification of Licensed Operators," and concluded that they are equivalent to the NRC staff guidelines in RG 1.8, Revision 3.

Except as specifically noted below, experience and training are separate aspects of license eligibility. As stated in NUREG-1262 (in response to Question No. 113), a person should meet the experience guidelines before entering the license training program. Time spent in training before entering the license training program may qualify as experience, but time spent in an NRC-approved training program leading up to license eligibility should normally not be double-counted as experience.

1. Reactor Operator

a. Experience

(1) The applicant should have a minimum of 3 years of power plant experience, at least 1 of which should be spent at the nuclear power plant for which the license is sought (preferably in the performance of non-licensed operator duties) and should not include any of the time spent in the control room as an extra person on shift.

(2)   The applicant should spend at least 6 months performing plant operational duties as a non-licensed operator at the nuclear power plant for which the license is sought.

b.   Training

(1)   Before being assigned RO duties, the applicant should complete at least 3 months as an extra person on shift in training for the RO position. This training should include all phases of day-to-day operations and should be conducted under the supervision of licensed personnel. This time should not count toward the 1-year onsite experience specified in Item D.1(a)(1) above.

(2)   The applicant should complete an RO training program that is established and maintained using a systems approach to training (SAT).

(3)   The applicant must manipulate the controls of the reactor or a plant-referenced simulator that meets the requirements of 10 CFR 55.46(c) during five significant changes in reactivity or power level (refer to 10 CFR 55.31(a)(5) and Section C.1.c above). Every effort should be made to perform at least some of the manipulations on the actual plant and to diversify the reactivity and power changes for each applicant.

3.   Education

The applicant should have a high school diploma or equivalent.

2.   Senior Reactor Operator

a.   Experience

(1)   A non-licensed (i.e., instant SRO) applicant should have a minimum of 3 years of responsible nuclear power plant experience (RNPPE), as defined in RG 1.8. At least 6 months of the RNPPE should be at the plant for which the applicant seeks a license and should not include any of the time spent in the control room as an extra person on shift. A maximum of 1 year of RNPPE may be fulfilled by academic or related technical training on a one-for-one basis.

(2)   Applicants for an SRO license who do not hold a bachelor's degree in engineering or the equivalent should have held an operator's license and should have been actively involved in the performance of licensed duties for at least 1 year or have at least 2 years in a position that is equivalent (or superior) to a licensed RO at a military reactor (e.g., propulsion plant watch officer, reactor operator, engineering officer of the watch, propulsion plant watch supervisor, or engineering watch supervisor). Maintaining a minimally active operator's license pursuant to 10 CFR 55.53(e) is not sufficient to satisfy this experience guideline.

(3) During the years of responsible nuclear power plant experience, the applicant should participate in reactor operator activities at power levels greater than 20 percent for at least 6 weeks.

(4) The eligibility of equipment operators, plant technicians, and non-degreed licensed operator instructors, who do not satisfy the strict definition of RNPPE and might otherwise be disqualified, will be evaluated on a case-by-case basis. The NRR operator licensing program office will assess their experience to determine the degree of equivalence and amount of credit to be granted.

b. Training

(1) Before being assigned SRO duties, the applicant should complete at least 3 months as an extra person on shift in training for the SRO position. This training should include all phases of day-to-day operations and should be conducted under the supervision of licensed personnel. This time does not count toward the 6-month onsite responsible experience guideline in Item D.2(a)(1) above. However, any portion of the 3 months that is spent at or above 20 percent power may also be used to satisfy the experience guideline in Section D.2.a(3).

(2) If the applicant has not held an RO license at the facility for which a license is sought, the applicant must complete the required control manipulations as discussed in Section C.1.c above.

(3) The applicant should complete a SAT-based SRO training program.

c. Education

The applicant should have a high school diploma or equivalent.

3. **Limited Senior Reactor Operator**

a. Experience

The applicant should have 3 years of RNPPE that includes active participation in at least one refueling outage at the site for which the license is sought or at a similar facility. Pursuant to 10 CFR 55.31(a)(5), the applicant must perform five significant control manipulations that affect reactivity (e.g., by loading or unloading fuel into, out of, or within the reactor vessel). Six months of the RNPPE should be at the site for which the LSRO license is sought or at a similar facility owned by the same facility licensee.

b. Training

The applicant is expected to have satisfactorily completed a SAT-based training program.

c. Education

The applicant should have a high school diploma or equivalent.

4.   Cold License Eligibility

Cold examinations are those administered before the unit completes pre-operational testing and the initial startup test program as described in the FSAR.

Each applicant must satisfactorily complete the training programs described in Section 13.2 of the FSAR and approved by the NRC. The NRC's review and approval are based on information contained in Section 13.2.1 of the Standard Review Plan (SRP) (NUREG-0800).

E.   Attachments/Forms

Attachment 1,   "Sample Initial Application Denial from Region"

NRC Letterhead

<u>(date)</u>

(Applicant's name)
(Street address)
(City, State, Zip code)

Dear <u>(Name)</u>:

This is to inform you that your application, dated <u>(date)</u>, for a <u>(reactor operator, senior reactor operator)</u> license, submitted in connection with <u>(facility name)</u>, is hereby denied.

<u>(Region to discuss deficiencies and which part of 10 CFR 55.31, ES-202, NRC-approved facility training program, or Regulatory Guide 1.8 was involved.)</u> When you have met the requirements of Title 10, Section 55.31, of the Code of Federal Regulations (10 CFR 55.31), you may submit another application.

If you do not accept this denial, you may, within 20 days of the date of this letter, take one of the following actions:

- You may request that the NRC reconsider the denial of your application by writing to the Director, Division of Inspection and Regional Support, Office of Nuclear Reactor Regulation, U.S. Nuclear Regulatory Commission, Washington, DC 20555. If submitting via private courier (e.g., FedEx, UPS), send your request to 11555 Rockville Pike, Rockville, Maryland 20852, instead of using the Washington, DC, address. Your request must include specific reasons for your belief that your application was improperly denied. If the NRC determines that the denial of your application remains appropriate, you still have the right to request a hearing pursuant to 10 CFR 2.103(b)(2), as described below.

- You may request a hearing in accordance with 10 CFR 2.103(b)(2). Submit your request, in writing, to the Office of the Secretary, U.S. Nuclear Regulatory Commission, Washington, DC 20555-0001, Attention: Rulemakings and Adjudications Staff, with a copy to the Associate General Counsel for Hearings, Enforcement, and Administration, Office of the General Counsel, at the same address. (Refer to 10 CFR 2.302 for additional filing options and instructions.) If submitting via private courier (e.g., FedEx, UPS), send your request to 11555 Rockville Pike, Rockville, Maryland 20852, instead of using the Washington, DC, address.

If you have any questions, please contact <u>(name)</u> at <u>(telephone number)</u>.

Sincerely,

(Regional branch chief or above)

Docket No. 55-<u>(number)</u>

cc: <u>(Facility representative who signed the applicant's NRC Form 398)</u>

CERTIFIED MAIL — RETURN RECEIPT REQUESTED

ES-204
PROCESSING WAIVERS REQUESTED BY
REACTOR OPERATOR AND SENIOR REACTOR OPERATOR APPLICANTS

A. Purpose

This standard provides guidance concerning the processing of waivers requested by reactor operator (RO) and senior reactor operator (SRO) license applicants at power reactor facilities.

B. Background

In accordance with Title 10, Section 55.35, "Re-applications," of the *Code of Federal Regulations* (10 CFR 55.35), and 10 CFR 55.47, "Waiver of Examination and Test Requirements," an applicant may request to be excused from a written examination or an operating test. The NRC may waive any or all of the examination requirements if it determines that the applicant has presented sufficient justification. In an effort to expedite the resolution of applicant requests, the Office of Nuclear Reactor Regulation (NRR) has delegated the authority to grant routine waivers of certain operator licensing requirements to the NRC regional offices.

C. Responsibilities

1. Applicant/Facility Licensee

a. An applicant may request a waiver of a license requirement by checking the appropriate block in Item 4.f on NRC Form 398, "Personal Qualifications Statement: Licensee." The applicant should also explain the basis for requesting the waiver in Item 17, "Comments."

b. The facility licensee's senior management representative on site must certify the final license application, thereby substantiating the basis for the applicant's waiver request.

c. Facility licensees having units designed by the same nuclear steam supply system vendor and operated at approximately the same power level may request dual licensing for their operators. Similarly, if the units of a multi-unit facility are nearly identical, the facility licensee may request a waiver of the examination requirements for the second and subsequent units.

In either case, the facility licensee must justify to the NRC that the differences between the units are not so significant that they could affect the operator's ability to operate each unit safely and competently. Further, the facility licensee must submit for NRC review the details of the training and certification program. The analysis and summary of the differences on which the applicants must be trained will include the following, as applicable:

- facility design and systems relevant to control room personnel

- technical specifications

- procedures (primarily abnormal and emergency operating)

- control room design and instrument location

- operational characteristics

- administrative procedures related to conduct of operations at a multi-unit site (e.g., shift manning and response to accidents and fires)

- the expected method of rotating personnel between units and the refamiliarization training to be conducted before an operator assumes responsibility on a new unit

2. NRC Regional Office

a. The regional office will evaluate waiver requests on a case-by-case basis against the waiver criteria discussed in Section D of this examination standard.

b. The regional office may grant routine waivers identified in Section D.1 without first obtaining concurrence from the NRR operator licensing program office.

However, waivers of experience requirements, completion of training, or completion of examinations not specifically identified in Section D.1 must be approved by NRR. The regional office should evaluate the waiver request and forward its approval recommendation to the NRR operator licensing program office for concurrence.

The region does not require written concurrence from NRR to deny an applicant's waiver request, but it should discuss its decision with the operator licensing program office before informing the applicant; formal concurrence may be desirable in some cases.

c. If additional information is required to reach a decision on a waiver request, the regional office shall generally request the necessary information from the facility licensee in accordance with ES-202.

d. Upon deciding to grant or deny a waiver, the regional office shall promptly notify the applicant in writing concerning the disposition of the request, and provide an explanation for the denial. If time is too short to notify the applicant in writing before the examination date, the regional office shall notify the facility training representative by telephone concerning the disposition of the waiver request and provide a followup written response to the applicant. The regional office shall include the NRR operator licensing program office on distribution for all waiver disposition letters.

e. The region shall document the disposition of every waiver request, whether granted or denied, by completing the block designated "For NRC Use" on the applicant's NRC Form 398 and by entering the data in the operator licensing tracking system (OLTS).

f.   NRC examiners assigned to a particular examination will be notified of approved waivers by the appropriate regional supervisor and by an entry on the list of applicants (Form ES-201-4).

g.   If the applicant is determined to be ineligible to take the licensing examination, the regional office shall issue a denial letter in accordance with ES-202.

D.   Waiver Criteria

1.   <u>Routine Waivers</u>

a.   If an applicant failed *only* one portion of the site-specific initial licensing examination (i.e., either the written examination overall, the SRO-only section of the written examination, the simulator operating test, the walk-through overall, or the administrative portion of the walk-through), the region may waive those examination areas that were passed. This is only applicable for the first retake examination and only if it takes place within 1 year of the date on which the original examination was completed.

Note that an SRO applicant who passed the operating test, achieved a score of 80 percent on the RO portion of the written examination, 76 percent on the SRO-only questions, and 79 percent overall would ***not*** be eligible for a waiver of the RO portion because the overall 80-percent cut score was not achieved. An SRO-instant applicant who passed everything except the SRO-only portion of the written examination may reapply for an RO license, and a full RO examination waiver, after accepting a final denial of the original SRO application; however, this is ***not*** considered a routine waiver and must be forwarded to NRR for approval as discussed in Section C.2.b. Such a waiver would be contingent upon the applicant's eligibility for an RO license (refer to the training and experience guidelines in ES-202) and the applicant's demonstration of control board competence during the simulator operating test (refer to ES-303).

b.   The region may waive training requirements specified in the final safety analysis report (FSAR) when the FSAR authorizes waiver of those specific requirements and the applicant otherwise meets NRC requirements (e.g., waiver of some training requirements for applicants previously licensed at a comparable facility).

c.   The medical data in support of NRC Form 396 are normally good for 6 months from the date of the medical examination for a person applying for an RO or SRO instant license. For re-applications (e.g., following a license denial or withdrawal of an application, or to request reinstatement of a terminated license) or for an examination which is delayed from its originally scheduled date, the NRC regional office may grant waivers extending the 6-month period, provided that the date of the original medical examination is within 24 months of the anticipated licensing date and Item 17, "Comments," of NRC Form 398 certifies that the applicant has not developed any physical or mental condition that would be reportable under 10 CFR 55.25. For renewal and SRO upgrade applicants, the medical examination documented on NRC Form 396 is good for 2 years from the date of the medical examination.

Waivers/exceptions and license conditions/restrictions that might be requested if an applicant does not meet the medical standards in the applicable version of ANSI/ANS 3.4, "Medical Certification and Monitoring of Personnel Requiring Operator Licenses for Nuclear Power Plants," will be coordinated with the NRC contract physician as discussed in ES-202.

d.   Substitutions allowed by Regulatory Guide (RG) 1.8, Revision 3, are not considered to be waivers and, therefore, do not require approval. For example, substitution of related technical training for up to 1 year of experience for an SRO is not a waiver. However, training for the examination applied for may not be counted as related technical training.

e.   If the facility licensee certifies that the applicant has successfully completed a training program accredited by the Institute of Nuclear Power Operations using an acceptable simulation facility, the region may waive the requirement for 10 startups on an operating reactor, which is typically required by NRC-approved cold license training programs.

f.   For those applicants who are unable to meet the requirement for 6 weeks on shift at greater than 20 percent power (because of extended plant shutdowns or other extraordinary circumstances), the NRC regional office may waive this requirement upon application if the following criteria are satisfied:

   (1)   Facility training objectives for the desired licensed position have been developed using a properly validated job task analysis (JTA).

   (2)   The facility licensee's training program is based on a systems approach to training (SAT) using the five elements defined in 10 CFR 55.4.

   (3)   The facility licensee can accomplish the required training objectives for plant operation at greater than 20 percent power using a plant-referenced or NRC-approved simulation facility.

g.   If an operator was previously licensed at a facility and reapplies for a license at the same facility and the same or lower license level, the regional office may, pursuant to 10 CFR 55.47, waive the requirement for the applicant to pass a written examination (including the generic fundamentals examination (GFE)) and an operating test if it finds that the applicant meets the following criteria:

   (1)   previously discharged his or her responsibilities competently and safely and is capable of continuing to do so

(2)   terminated participation in the facility licensee's requalification program less than 2 years (24 months) before the date of the license application

(3)   successfully completed "Additional Training," pursuant to 10 CFR 55.59(b), and a facility-prepared written examination and operating test, which ensure that the applicant is up-to-date in the licensed operator requalification training program (including GFE topics)

(4)   will successfully complete at least 40 hours of shift functions under the direction of an operator or senior operator, as appropriate, and in the position to which the applicant will be assigned (see 10 CFR 55.53(f)) before being assigned to licensed duties

(5)   complies with the requirements of 10 CFR 55.31

h.   If an applicant is unable to perform the five significant control manipulations required by 10 CFR 55.31(a)(5), the regional office may process the application, administer the examination, and issue a conditional license that is only valid with the reactor in cold shutdown and refueling (or simply delay licensing the applicant until the facility licensee certifies that the required manipulations have been completed; refer to Section D.3.c of ES-501). The regional office will not remove the license condition until the facility licensee supplies the required evidence that the applicant has successfully completed the control manipulations (refer to ES-501).

i.   The region may authorize a facility licensee to defer completion of the following specific experience and training guidelines until after the licensing examination is passed:

(1)   up to 6 months of the 3 years of (responsible nuclear) power plant experience for an RO (or an SRO), but not to exceed 2 months of the year of onsite experience for an RO and 1 month of the 6 for an SRO

(2)   up to 2 months of the year actively performing duties as a licensed RO at the facility for which an SRO upgrade license is sought

(3)   up to 1 month of the 3 spent as an extra RO or SRO on-shift in training

The facility licensee must provide evidence that the deferred items have been completed before the region will issue the license (refer to ES-501).

j.   If an individual is currently licensed as an RO at a facility and applies for an SRO license at the same facility, the regional office may waive the requirement for the applicant to take the RO portion of the SRO written examination if the applicant satisfies the following requirements:

(1)   Pursuant to 10 CFR 55.47(a)(1), which requires extensive actual operating experience within the previous 2 years, the applicant must

have maintained an active license for at least 12 of the 24 months preceding the date of application. This would also satisfy the SRO-upgrade eligibility criteria in Section D.2.a(2) of ES-202 and the similar guidelines established by the National Academy for Nuclear Training.

(2) Pursuant to 10 CFR 55.47(a)(2), the applicant must have discharged his or her responsibilities competently and safely and be capable of continuing to do so. As in 10 CFR 55.57, the NRC will consider the applicant's past performance and certification by the facility licensee when making this determination.

(3) Pursuant to 10 CFR 55.47(a)(3), the applicant must have learned the operating procedures for and be qualified to safely and competently operate the facility. This requirement would be satisfied if the applicant passed his or her most recent requalification examination and was up-to-date in the facility licensee's requalification training program at the time that he or she entered the upgrade training program.

Applications who do not satisfy these requirements shall be referred to the NRR operator licensing program office in accordance with Section C.2.b, above.

k. If an applicant passed the GFE more than 24 months before the date of license application, the regional office may waive the requirement to pass another GFE if the applicant meets any one of the following criteria (as explained in Item 17 on NRC Form 398):

(1) The applicant terminated an RO or SRO license at a comparable (boiling- or pressurized-water) facility less than 24 months before the date of application and was up-to-date in the requalification program at the time of license termination.

(2) Within the 24 months preceding the date of application, the applicant completed self-study or classroom instruction, as deemed necessary by the facility licensee, and passed a prior GFE that was randomly selected from among those contained on the NRC's GFE Web page and administered, under controlled conditions, by the facility licensee.

(3) Within the 24 months preceding the date of application, the applicant completed self-study or classroom instruction, as deemed necessary by the facility licensee, and passed a GFE prepared by the facility licensee in accordance with Section D of ES-205 and administered under controlled conditions.

2. Examination Waivers for Operators Previously Licensed at Comparable Facilities

Depending on the justification provided by the applicant and the facility licensee, NRR will consider examination waivers for operators who were previously licensed at a comparable facility. Pursuant to 10 CFR 55.47, the Commission may waive any or all requirements for a written examination and operating test.

3. **Multi-Unit Examination Waivers**

   a. Generally, personnel will *not* be examined on or allowed to hold licenses for "different units" simultaneously. For purposes of this standard, "different units" owned or managed by a single facility licensee are defined as follows:

   - units having the same vendor, but significantly different age and/or power level (e.g., Nine Mile Point Units 1 and 2)

   - units having the same vendor and similar design, but different locations (e.g., Sequoyah and Watts Bar, Byron and Braidwood)

   - units having different vendors (pressurized-water reactors only), but located on the same site (e.g., Arkansas Units 1 and 2, Millstone Units 2 and 3)

   NRR may authorize a limited senior reactor operator (LSRO) to be licensed at multiple sites, provided that the units are manufactured by the same vendor and are of similar design. The applicant must pass an examination that addresses the differences in the designs, procedures, technical data, and administrative controls of the separate facilities for which the license is being sought.

   b. With regard to the examination requirements for "identical" second or subsequent units at the same site, NRR may waive any or all requirements for a written examination and operating test if the staff finds that the applicant meets the criteria specified in 10 CFR 55.47, as noted in Item D.2 above. If the situation warrants, the Commission may impose other examination requirements, such as NRC-administered operating tests and written examinations concerning the plant differences.

ES-205
PROCEDURE FOR ADMINISTERING
THE GENERIC FUNDAMENTALS EXAMINATION PROGRAM

A. Purpose

This standard describes the procedures and policies pertaining to administration of the generic fundamentals examination (GFE) section of the written operator licensing examination at power reactor facilities. It describes how the examinations are scheduled and constructed, how to solicit facility licensees for applicants to take the examinations, and how to promulgate the examination results.

B. Background

Title 10, Sections 55.41 and 55.43, of the *Code of Federal Regulations* (10 CFR 55.41 and 55.43) require that the written operator licensing examinations for reactor operators (ROs) and senior reactor operators (SROs) must include questions concerning various mechanical components, principles of heat transfer, thermodynamics, and fluid mechanics. These regulations also require that the written examinations must address fundamentals of reactor theory, including the fission process, neutron multiplication, source effects, control rod effects, criticality indications, reactivity coefficients, and poison effects.

The fundamental knowledge and abilities (K/As) required of an operator do not vary significantly between license levels or among facilities of the same vendor type. As a result, the NRC implemented the GFE program to standardize the fundamental examination coverage for all applicants at pressurized- and boiling-water reactors (PWRs and BWRs). Having passed a GFE as an RO or an SRO applicant, an operator will not have to take another GFE unless he or she transfers to a facility of the other vendor type or discontinues, for a period exceeding 2 years (24 months), participation in an accredited licensed operator requalification training program that maintains proficiency in the GFE topics. Refer to Section D.1.k of ES-204 for guidance regarding waivers of the GFE. The GFE program does not pertain to limited senior reactor operator (LSRO) license applicants.

The GFE examinations for BWRs and PWRs are typically administered four times a year, on the Wednesday following the first Sunday in March, June, September, and December.

C. Responsibilities

1. Facility Licensee

a. The facility licensee must certify that all individuals who plan to take the GFE are enrolled in a facility-sponsored training program that will satisfy the eligibility requirements for an RO or SRO license. The operator trainees need not complete all of the training required for the license before they take the GFE.

The facility licensee may use the sample registration letter enclosed with the NRC notification letter (Attachment 1) or any similar format that contains the required information and certification. If the facility licensee must add or delete an individual after submitting its registration letter, the facility licensee should inform the Office of Nuclear Reactor Regulation (NRR) operator licensing program office of the change, as specified in the examination cover letter, *before* the examinations are administered.

b.   Upon receiving the examinations from the GFE contractor, the facility licensee shall reproduce and safeguard the examinations as described in the examination cover letter.

c.   On the designated examination day, the facility licensee shall administer and proctor the GFE in accordance with the instructions contained in the examination package.

The facility licensee will start and stop the GFE in accordance with the time zone map contained within the examination package. Late arrivals will be allowed to take the examination; however, all examinees must hand in their examinations at the completion time designated in the proctor instructions enclosed with the examination cover letter (refer to Section C.2.d).

d.   No later than the day after the GFE is administered, the facility licensee shall send the following items via overnight mail to the name and address designated in the examination package:

- the original answer sheets
- the signed exam cover sheets
- the signed security statements

2.   NRR Operator Licensing Program Office and GFE Contractor

a.   The NRR operator licensing program office will designate a coordinator to oversee the GFE activities of the regional offices, the GFE contractor, and the facility licensees.

b.   Beginning in 2005, the NRC will send a notification letter (Attachment 1) to each facility licensee at the beginning of each calendar year. The letter will announce the GFE administration dates for the entire year and inform facility licensees when the registration letters are due to the NRC.

c.   The GFE contractor will prepare the examinations as described in Section D of this examination standard. The examiner assigned responsibility for developing the GFE shall submit the examinations to the NRR GFE coordinator and any other designated reviewers at least 20 calendar days before the scheduled administration date. The NRR operator licensing program office will provide comments and recommended changes to the examination author as soon as possible. The final examinations should be ready at least 14 days before the GFE administration date.

d.   The GFE contractor will assemble the approved examination packages
      as described below, and mail the packages to the names and addresses
      designated by the participating facility licensees. The examinations should
      normally be mailed 1 week before the examinations are scheduled to be
      administered.

      The examination packet will contain the following information, enclosures, and
      attachments:

      - cover letter (Attachment 2 is a sample letter)
      - proctor instructions
      - security agreement
      - single copies of appropriate exam, forms A and B
      - exam time zone map
      - sample answer sheet
      - facility docket number sheet
      - applicant docket number sheet
      - appropriate number of answer sheets
      - applicant answer sheet instructions

e.   On the day that the GFE is administered, the NRR GFE coordinator
      and GFE contractor shall be available to answer questions from facility proctors
      if the need arises.

f.   Upon receiving the examination answer sheets from the facility licensees,
      the GFE contractor shall score, grade, and tabulate the overall item statistics,
      and generate facility and regional grade reports for each GFE examination.
      The contractor shall forward the regional and facility grade reports, including
      individual scores and copies of individual answer sheets, and corrected answer keys
      to the applicable regional office for distribution.

      The GFE contractor shall develop individual item statistics on all questions used
      on the GFE examinations. Questions with acceptable statistical characteristics
      shall be moved into the "validated" GFE question bank.

      The contractor will provide copies of all grade reports to the NRR GFE coordinator,
      along with the following additional items:

      - exam-wide item statistics (PWR and BWR)
      - analysis reports of specific items deleted or answers changed
      - corrected answer keys
      - original answer sheets
      - original signed exam cover sheets
      - signed security statements

g.   The NRR operator licensing program licensing assistant will ensure that copies
      of the final master BWR and PWR examinations are placed in the NRC's Public
      Document Room.

3. **NRC Regional Office**

   a. Regional management should assign an individual to coordinate GFE administration in the region.

   b. Beginning in 2005, the NRC will issue a single notification letter each year, and the regional office will informally remind facility licensees (by electronic mail or telephone) to submit their registration letters for the June, September, and December examinations.

   c. The regional operator licensing assistant (OLA) shall assign a docket number to each individual identified in the facility licensee's registration letter. The OLA shall then forward the list of names and docket numbers for each facility to the GFE contractor, with a copy to the NRR GFE coordinator, no later than 20 days before the examination administration date.

   d. The regional GFE coordinator should keep the NRR GFE coordinator informed of any changes in the number of applicants scheduled to take the GFE at any facility.

   e. The regional office shall distribute the GFE examination results to their participating facility licensees. A sample cover letter is provided in Attachment 3 to this examination standard.

   f. The regional OLA shall update the applicants' status (pass or fail) in the operator licensing tracking system (OLTS) and ensure that a hard copy of the GFE results is placed in each applicant's docket file.

4. **Industry**

   The industry may make arrangements to review and comment on the GFEs before they are administered by contacting the NRR operator licensing program office at least 2 months before the scheduled examination date. The review will be limited to one BWR and one PWR instructor (provided by the facility) who will not be fielding applicants during the subject examination. These reviewers will be required to sign security agreements, in accordance with Section D.2.b of ES-201, before and after seeing the examinations. The reviewers must complete the review (including the new, modified, and previously validated questions, as desired) and provide feedback to the NRC staff within 3 working days from the date of receipt. If the NRR operator licensing program office does not receive the reviewers' comments within the allotted time, the examinations will proceed on schedule. Otherwise, the NRR operator licensing program office and GFE contractor will evaluate the reviewers' comments and make changes as deemed appropriate.

D. Examination Scope and Structure

Each GFE shall contain 50 questions covering the "Components" and "Theory" (including reactor theory and thermodynamics) sections of NUREG-1122, "Knowledge and Abilities Catalog for Nuclear Power Plant Operators: Pressurized-Water Reactors," or NUREG-1123, "Knowledge and Abilities Catalog for Nuclear Power Plant Operators: Boiling-Water Reactors." The passing grade for the GFE is 80 percent.

The K/A topics applicable to the GFE for PWRs and BWRs have been categorized into various component, reactor theory, and thermodynamics groups as shown in Attachment 4 to this examination standard. That attachment also identifies the number of test questions required to evaluate each topic.

The questions used on the GFE examination shall conform with the applicable construction and style guidelines in Appendix B. The examination shall include 40 questions taken directly from the NRC's GFE question bank for the applicable vendor type, 5 questions that are derived from existing bank questions by making one or more significant modifications, and 5 questions that are newly developed.

E. Attachments/Forms

Attachment 1,   "Sample Notification Letter"
Attachment 2,   "Sample Examination Cover Letter"
Attachment 3,   "Sample Results Letter"
Attachment 4,   "GFE Test Item Distribution"

NRC Letterhead

<u>(Date)</u>

<u>(Name, Title)</u>
<u>(Facility name)</u>
<u>(Street address)</u>
<u>(City, State Zip code)</u>

Dear <u>(Name)</u>:

The U.S. Nuclear Regulatory Commission (NRC) plans to administer the generic fundamentals examination (GFE) section of the written operator licensing examination on the following dates during this calendar year:

March ##, 200#
June ##, 200#
September ##, 200#
December ##, 200#

To register personnel to take the GFE, an authorized representative of your facility must submit a letter to the appropriate regional administrator with a copy addressed as follows:

> Chief, Operator Licensing and Human Performance Branch
> Mail Stop OWFN 12D19
> U.S. Nuclear Regulatory Commission
> Washington, DC. 20555-0001

Your letter should identify the individuals who will take the examination, and it should certify that they are enrolled in a facility licensee-sponsored program leading to NRC operator or senior operator licensing and that they will have completed their fundamentals training by the date of the examination. The letter should also identify the personnel who will have access to the examinations before they are administered (e.g., proctors) and the address to which the examinations are to be sent. To allow the NRC to assign docket numbers, your letter should be received by both the NRC regional administrator and the Chief, Operator Licensing and Human Performance Branch, **30 days before each desired examination date** shown above. A sample registration letter is enclosed.

Copies of the administered GFEs and their answer keys will be available for review in the NRC's Public Document Room approximately 45 days following each examination. The NRC's GFE Web page (which is available at http://www.nrc.gov/reactors/operator -licensing/generic-fundamentals-examinations.html) will be updated semi-annually, approximately 60 days following the June and December examinations.

Sincerely,

<u>(Appropriate regional representative)</u>

Docket No. 50-<u>(Number)</u>
Enclosure: As stated

Enclosure

(Name)
Regional Administrator
U.S. Nuclear Regulatory Commission
Region (Number)
(Street address)
(City, State Zip code)

Dear (Name),

(Facility name) requests approval from the U.S. Nuclear Regulatory Commission (NRC) to have the following (number) individuals take the (BWR or PWR) generic fundamentals examination (GFE) section of the written operator licensing examination to be administered on (date):

| Name | Date of Birth | Previous Docket No. |

(Insert the name, date of birth, and previous 10 CFR Part 55 Docket Number (if applicable) for each person.)

All of the listed personnel are enrolled in the (facility name) operator licensing training program and will have completed the generic fundamentals portion of the program by the examination date.

The following personnel will have access to the examinations before they are administered:

| Name | Title |

(Insert the name and title of each person who will have access to the examinations before they are administered (e.g., proctors).

Please address the examinations to the **overnight mail** address, as follows (note that home addresses are not acceptable):

Name, Title
Street address
City, State Zip code

If you have any questions, please contact (facility contact name) at (telephone number).

Sincerely,

(Name, title)

cc:
Chief, Operator Licensing and Human Performance Branch, NRR

(Date)

(Name, Title of designated addressee)
(Facility name)
(Street address)
(City, State Zip code)

Dear (Name):

The U.S. Nuclear Regulatory Commission (NRC) has scheduled your facility to administer the generic fundamentals examination (GFE) section of the NRC's written operator licensing examination on (date). (Name of contractor) is authorized to support the NRC under contract in the administration of GFE-related activities.

Note: For security reasons, please open the sealed envelope now and page-check the examination using the enclosed checklist. Then, immediately and no later than (date), contact one of the persons listed below informing (him or her) that you have received this package and noting any discrepancies:

(Name), (Telephone Number)
(Name), (Telephone Number)

This letter and its enclosures provide the instructions and guidelines for administering the GFE and returning the completed exams and related materials to (Name of contractor). Please read this letter **now**, and follow the directions in the accompanying enclosures.

Enclosure 1. Security Agreement. Please refer to the enclosed NRC Security Agreement. A copy of this agreement must be completed by each and every exam administrator and/or proctor who sees or has knowledge of the GFE contents. For security reasons, the number of persons who see or have knowledge of this exam's contents before the exam must be limited to **three** persons who **have a need to know**.

The top portion of the security agreement is expected to be completed **now**, and the bottom portion immediately **after** the exam has been completed. Fill in the spaces for each individual's **name** and the **name of the facility** for both portions, and have the individual(s) sign the form(s).

Please note: The signed security agreements **must** be returned to (Name of contractor) along with the completed exam answer sheets before any scoring will be performed.

Enclosure 2. Exam Copies. Two single copies of Forms A and B of the exam are provided. These alternative forms are identical in content; however, for security purposes, the test item sequence on each form is different to reduce the possibility of an applicant copying any answers from a nearby test answer sheet. (See the separate Proctor Instructions in Enclosure 3 for further exam administration instructions.)

You are responsible for reproducing the number of exam copies required for the number of individuals taking the exam. Prior to the exam, store the original copies in a locked cabinet or safe and reproduce the necessary number of copies **only** on the day immediately preceding the exam; in this case, copies should be made on (date). Please note that your total number of copies should consist of one half Form A and one half Form B. After making the necessary number of copies, secure the original and all copies from view of unauthorized persons, storing them in a locked cabinet or safe until the exam date.

Each individual who takes the exam must sign the security statement on the exam cover page. This page must be removed from the exam copy and mailed to (Name of contractor) along with the answer sheets and administrator/proctor security agreements.

After the exam has been given, the exam copies become public knowledge and no longer need security. Therefore, exam copies may subsequently be kept or disposed of as desired.

Enclosure 3. Proctor Instructions. The proctor instructions detail the guidelines for administering the exam. Please note that the specific instructions presented are designed to be adhered to and followed identically by each proctor at **all** facilities. This process will ensure uniform administration and equity of results nationwide. As noted in the Proctor Instructions, all GFE exams will be administered at the same time in accordance with the local time zone in which the facility is located.

Enclosure 4. Exam Answer Sheets. The appropriate number of answer sheets (extra copies included) is enclosed for the number of applicants you identified to take the exam. All applicants must use the original enclosed answer sheets for recording answers during the exam.

Summary of Items to be Returned to (Name of contractor)

The following items must be mailed via ***Overnight Delivery Service*** to (Name of contractor) and postmarked no later than (date).

- completed answer sheets
- applicant-signed exam cover sheets
- administrator/proctor-signed security statement(s)

Mail all of the above exam-related materials addressed as follows:

(Name)
(Name of contractor)
(Street address)
(City, State Zip code)

For further questions regarding the specifics of this exam, please contact (Name) at (telephone number). For questions regarding the GFE in general, please contact (Name), NRC, at (telephone number).

For matters regarding candidate withdrawals or cancellations, contact either (Name) or (Name) at (telephone numbers) for specific guidance.

> (Name), Chief
> Operator Licensing and Human Performance Branch
> Division of Inspection and Regional Support
> Office of Nuclear Reactor Regulation

Enclosures:
  As stated

Distribution: w/o enclosures
  Director, DIRS
  Chief, Operator Licensing and Human Performance Branch
  NRR GFE Coordinator
  Project Manager
  Public

NRC Letterhead

(Date)

(Name, Title)
(Facility name)
(Street address)
(City, State Zip code)

Dear (Name):

On (date), the U.S. Nuclear Regulatory Commission (NRC) administered the generic fundamentals examination (GFE) section of the written operator licensing examination to employees of your facility. Enclosed with this letter are copies of both forms of the examination, including answer keys, the grading results for your facility, and copies of the individual answer sheets for each of your employees. Please forward the results to the individuals along with the copies of their respective answer sheets. A "P" in the RESULTS column indicates that the individual achieved a passing grade of 80 percent or better on the GFE, while an "F" indicates that the individual failed the examination.

If you have any questions concerning this examination, please contact (Name of the NRR GFE coordinator) at (phone number).

Sincerely,

(Appropriate regional representative)

Docket No. 50-(Number)

Enclosures:
1. Examination Form "A" and "B" with answers
2. Examination Results Summary for (Facility Name)
3. Individual Answer Sheets

K/A	Pressurized-Water Reactors Topic	No. of Items
	Components	
191001	Valves	2
191002	Sensors and Detectors	4
191003	Controllers and Positioners	3
191004	Pumps	4
191005	Motors and Generators	2
191006	Heat Exchangers and Condensers	2
191007	Demineralizers and Ion Exchangers	2
191008	Breakers, Relays, and Disconnects	3
	Reactor Theory	
192001	Neutrons	1
192002	Neutron Life Cycle	1
192003	Reactor Kinetics and Neutron Sources	1
192004	Reactivity Coefficients	2
192005	Control Rods	2
192006	Fission Product Poisons	2
192007	Fuel Depletion and Burnable Poisons	1
192008	Reactor Operational Physics	4
	Thermodynamics	
193001	Thermodynamic Units and Properties	1
193002	Not Applicable	0
193003	Steam	2
193004	Thermodynamic Processes	1
193005	Thermodynamic Cycles	1
193006	Fluid Statics and Dynamics	2
193007	Heat Transfer	1
193008	Thermal Hydraulics	4
193009	Core Thermal Limits	1
193010	Brittle Fracture and Vessel Thermal Stress	1
Total Items		50

K/A	Boiling-Water Reactors Topic	No. of Items
	Components	
291001	Valves	3
291002	Sensors and Detectors	4
291003	Controllers and Positioners	2
291004	Pumps	4
291005	Motors and Generators	2
291006	Heat Exchangers and Condensers	3
291007	Demineralizers and Ion Exchangers	2
291008	Breakers, Relays, and Disconnects	2
	Reactor Theory	
292001	Neutrons	1
292002	Neutron Life Cycle	1
292003	Reactor Kinetics and Neutron Sources	1
292004	Reactivity Coefficients	2
292005	Control Rods	2
292006	Fission Product Poisons	2
292007	Fuel Depletion and Burnable Poisons	1
292008	Reactor Operational Physics	4
	Group I Thermodynamics	
293001	Thermodynamic Units and Properties	1
293002	Basic Energy Concepts	0
293003	Steam	1
293004	Thermodynamic Processes	1
293005	Thermodynamic Cycles	1
293006	Fluid Statics	2
293007	Heat Transfer and Heat Exchangers	1
293008	Thermal Hydraulics	3
293009	Core Thermal Limits	3
293010	Brittle Fracture and Vessel Thermal Stress	1
Total Items		50

ES-301
PREPARING INITIAL OPERATING TESTS

A.   Purpose

All applicants for reactor operator (RO) and senior reactor operator (SRO) licenses at power reactor facilities are required to take an operating test, unless it has been waived in accordance with Title 10, Section 55.47, of the Code of Federal Regulations (10 CFR 55.47). (Refer to ES-204, "Processing Waivers Requested by Reactor Operator and Senior Reactor Operator Applicants.") The specific content of the operating test depends on the type of license for which the applicant has applied.

This standard describes the procedure for developing operating tests that meet the requirements of 10 CFR 55.45, including the use of reactor plant simulation facilities and the conduct of multi-unit evaluations.

B.   Background

To the extent applicable, the operating test will require the applicant to demonstrate an understanding of, and the ability to perform, the actions necessary to accomplish a representative sampling of the 13 items identified in 10 CFR 55.45(a). (All 13 items do not need to be sampled on every operating test). In addition, the content of the operating test will be identified, in part, from learning objectives contained in the facility licensee's training program and information in the final safety analysis report, system description manuals and operating procedures, the facility license and amendments thereto, licensee event reports, and other materials that the Commission requests from the facility licensee.

The structure of the operating test is dictated, in part, by 10 CFR 55.45(b). Specifically, that requirement states that the test will be administered in a plant walk-through and in either a simulation facility that the Commission has approved pursuant to 10 CFR 55.46(b), a plant-referenced simulator that conforms with 10 CFR 55.46(c), or the plant, if approved by the Commission under 10 CFR 55.46(b).

The walk-through portion of the operating test consists of two parts ("Administrative Topics" and "Control Room/In-Plant Systems"), each focusing on specific knowledge and abilities (K/As) required for licensed operators to safely discharge their assigned duties and responsibilities. The second major portion of the operating test (the "Simulator Test") is administered on an NRC-approved or plant-referenced simulator. Unless specifically waived in accordance with ES-204 and documented on the "List of Applicants" (Form ES-201-4), each license applicant must complete the entire operating test.

Each part of the operating test is briefly described below. Section D of this standard provides detailed instructions for developing each part. Procedures for administering and grading the operating test are contained in ES-302, "Administering Operating Tests to Initial License Applicants," and ES-303, "Documenting and Grading Initial Operating Tests," respectively.

1.   **"Administrative Topics"**

This part of the walk-through operating test covers K/As that are generally associated with administrative control of the plant. It implements items 9–12 of 10 CFR 55.45(a) and is divided into four administrative topics, as described below. The scope and depth of coverage required in each topic is based on the applicant's license level.
The applicant's competence in each topic is evaluated by administering job performance measures (JPMs) and asking specific "for cause" followup questions, as necessary, based on the applicant's performance (refer to ES-302).

The first topic, "Conduct of Operations," evaluates the applicant's knowledge of the daily operation of the facility. The following subjects are examples of the types of information that could be evaluated under this topic:

- shift turnover
- shift staffing requirements
- access controls for vital/controlled plant areas
- operator responsibilities and procedure usage
- purpose, function, and controls for plant systems
- fuel handling and refueling

The second topic, "Equipment Control," addresses the administrative requirements associated with managing and controlling plant systems and equipment. The following subjects are examples of the types of information that could be evaluated under this topic:

- surveillance testing
- pre-startup activities
- maintenance
- tagging and clearances
- temporary modification of systems
- changes to procedures and plant design
- technical specifications, including plant mode
- familiarity with and use of piping and instrument drawings

The third topic, "Radiation Control," evaluates the applicant's knowledge and abilities with respect to radiation hazards and protection (of plant personnel and the public). The following subjects are examples of the types of information that could be evaluated under this topic:

- use and function of portable radiation and contamination survey instruments and personnel monitoring equipment
- knowledge of significant radiation hazards
- radiological safety principles and procedures
- radiation exposure limits under normal or emergency conditions
- radiation work permits
- control of radiation releases

The fourth topic, "Emergency Procedures/Plan," evaluates the applicant's knowledge of the facility's emergency plan, including, as appropriate, the responsibility of the RO or SRO to decide whether the plan should be executed and duties assigned under the plan. The following subjects are examples of the types of information that could be evaluated under this topic:

- lines of authority during an emergency
- operator responsibilities during an emergency
- emergency operating procedures
- emergency action levels and classifications
- emergency facilities
- emergency communications
- emergency protective action recommendations
- security event procedures (non-safeguards information)

The "Administrative Topics" are administered in a one-on-one, walk-through format in accordance with ES-302 and graded in accordance with ES-303.

2.   **"Control Room/In-Plant Systems"**

This part of the walk-through operating test is used to determine whether the applicant has an adequate knowledge of plant system design and is able to safely operate those systems. This part implements the requirements of items 3, 4, 7, 8, and 9 identified in 10 CFR 55.45(a) and encompasses several types of systems, including primary coolant, emergency coolant, decay heat removal, auxiliary, radiation monitoring, and instrumentation and control.

This part of the walk-through focuses primarily on those systems with which licensed operators are most involved (i.e., those having controls and indications in the main control room). To a lesser extent, it also ensures that the applicant is familiar with the design and operation of systems located outside the main control room. The applicant's knowledge and abilities relative to each system are evaluated by administering JPMs and, when necessary, specific followup questions based on the applicant's performance of each JPM.

This part of the operating test is administered in a one-on-one, walk-through format in accordance with ES-302 and graded in accordance with ES-303.

3. **"Simulator Operating Test"**

This part of the operating test implements items 1–8 and 11–13 of 10 CFR 55.45(a). This is the most performance-based aspect of the operating test and is used to evaluate the applicant's ability to safely operate the plant's systems under dynamic, integrated conditions.

The simulator test is administered in a team format with up to three applicants (or surrogates) filling the RO and SRO license positions (as appropriate) on an operating crew. (Refer to ES-201, "Initial Operator Licensing Examination Process," for additional guidance on crew composition and ES-302 for test administration instructions.) This format enables the examiner to evaluate each applicant's ability to function within the control room team as appropriate to the assigned position, in such a way that the facility licensee's procedures are adhered to and that the limitations in its license and amendments are not violated. [Refer to 10 CFR 55.45(a)(13).]

Each team or crew of applicants is administered a set of scenarios designed so that the examiners can individually evaluate each applicant on a range of competencies applicable to the applicant's license level. Appendix D describes those competencies, and Forms ES-303-3 and ES-303-4, the "Simulator Competency Grading Worksheets" for ROs and SROs, break down each competency into a number of specific rating factors to be considered during the grading process (refer to ES-303).

Each applicant must demonstrate proficiency on every competency applicable to his or her license level. The only exception is that SRO Competency Number 3, "Control Board Operations," is optional for SRO-upgrade applicants (i.e., SRO-upgrade applicants do not have to fill a position that requires control board operations; however, if they do rotate into such a position, they will be graded on this competency even though they may not be individually observed by an NRC examiner, as discussed in ES-302).

C. Responsibilities

1. **Facility Licensee**

The facility licensee is responsible for the following activities, as applicable, depending upon the examination arrangements confirmed with the NRC's regional office in accordance with ES-201 approximately 4 months before the scheduled examination date:

a. Prepare proposed examination outlines in accordance with Section D and submit them to the NRC's regional office for review and approval in accordance with ES-201.

b. Submit the reference materials necessary for the NRC regional office to prepare and/or review the requested examination(s). (Refer to ES-201, Attachment 3.)

c.   Prepare and review the final operating tests in accordance with the previously approved examination outline(s) and the instructions in Sections D and E, and submit the tests to the NRC's regional office in accordance with ES-201.

d.   Make the simulation facility available, as necessary, for NRC examiners to prepare for the operating tests.

e.   Meet with the NRC examination team in the regional office or at the facility, when and as necessary, to review the proposed operating tests and discuss potential changes. (Refer to ES-201.)

f.   Revise the operating test outlines and the final tests as applicable and as agreed upon by the NRC regional office. (Refer to ES-201.) The NRC retains final authority to approve the operating tests.

2.   NRC Regional Office

The NRC's regional office is responsible for the following activities:

a.   Ensure that the operating tests are developed in accordance with Section D.

b.   Ensure that the operating tests are reviewed for quality in accordance with Section E.

c.   Meet with the facility licensee, when and as appropriate, to pre-review the operating tests in accordance with ES-201.

D.   Instructions

Prepare each category of the operating test in accordance with the following general guidelines and specific instructions:

1.   General Guidelines

a.   In an effort to reduce examination preparation effort, the same operating test may be used to examine multiple applicants and simulator crews. Depending on the number and license level of the applicants being examined, it might be possible to use the same set of JPMs and scenarios to examine all of the applicants if the operating test is administered in multiple segments (e.g., single scenarios or two to four JPMs), each of which can be given to all of the applicants in a single day. The facility licensee and the NRC's chief examiner shall discuss the options and reach agreement on the process before developing the operating tests.

To minimize predictability and maintain test integrity, varied subjects, systems, and operations shall be evaluated with applicants who are not being examined at the same time, unless measures are taken to preclude interaction among the applicants. The same JPMs and simulator scenarios shall not be repeated on subsequent days.

Operating tests may not duplicate test items (simulator scenarios or JPMs) from the applicants' audit test (or tests if the applicant is retaking the examination) given at or near the end of the license training class. Simulator events and JPMs that are similar to those that were tested on the audit examination are permitted provided that the actions required to mitigate the transient or complete the task (e.g., using an alternative path as discussed in Appendix C) are significantly different from those required during the audit examination. The facility licensee shall identify for the NRC chief examiner those simulator events and JPMs that are similar to those that were tested on the audit examination.

Sufficient operating test materials shall be developed to ensure that all applicants can be tested with the available personnel according to the schedule agreed upon by the NRC's regional office and the facility licensee (refer to ES-201).

b.   To the extent permitted for each part of the operating test, select and modify testing materials (i.e., JPMs and simulator scenarios) from the facility's examination banks. Every selected test item must satisfy the qualitative and quantitative criteria specified for the applicable section of the operating test or be modified accordingly.

c.   Consider the K/As associated with normal, abnormal, and emergency tasks and evolutions as a source of topics for use in evaluating applicant competency in each part of the operating test.

The K/As associated with the tasks and questions planned for the operating test should have importance factors of at least 2.5. Tasks with importance factors of less than 2.5 may be used if there is a substantive reason for including them (e.g., a recent licensee event or a significant system modification). Failure to train the applicants on a particular K/A is not an acceptable basis for rejecting that K/A.

The K/As should be appropriate to the plant-specific requirements for the applicant's license level. Refer to the facility's job and task analysis (if available), learning objectives, and other reference material to confirm that the operating test is correctly oriented to the facility and the applicant's license level.

The facility licensee's site-specific task list may be used to supplement or override, on a case-by-case basis, selected individual items in the NRC's K/A catalogs. In order to maintain examination consistency, the site-specific task list shall not be used in place of the entire K/A catalog.

d.   When selecting and developing JPMs and scenarios for the operating test, ensure that the materials contribute to the test's overall capacity to differentiate between those applicants who are competent to safely operate the plant and those who are not. Additionally, all of the test items should include the three facets of test validity (i.e., content, operational, and discrimination) discussed in Appendix A. Any test items that, when missed, would raise questions regarding adequate justification for denying the applicant's license should not be included on the operating test.

e.   SRO applicants, whether upgrade or instant, will be examined for the highest on-shift position for which the SRO's license is applicable (e.g., shift supervisor), regardless of the position to be assigned when licensed. SRO applicants should demonstrate their supervisory abilities and an attitude of responsibility for safe operation, and are expected to assume a management role during plant transients and upset conditions while taking the simulator operating test. The operating test briefing, discussed in Appendix E, ensures that the applicants are advised of this policy.

Differences in administrative controls and facility design will affect the SRO's responsibilities; however, in general, the following guidelines should be used to differentiate the SRO operating test from that of an RO:

- In directing licensed activities, the SRO must evaluate plant performance and make operational judgments accordingly. SRO applicants should, therefore, be more knowledgeable in areas such as operating characteristics, reactor behavior, and instrument interpretation.

- In directing licensed activities, the SRO must have a broader and more thorough knowledge of facility administrative controls and methods, including limitations imposed by the regulations and the facility's technical specifications and their bases.

- The SRO may be assigned responsibilities for auxiliary systems that are outside the control room (e.g., waste disposal and fuel handling systems) and are not normally operated by licensed operators. Because the SRO may have these additional responsibilities, the SRO license applicant should demonstrate knowledge of the designs of such systems as they relate to maximum permissible concentrations, effluent release rates, and other radiological considerations.

f.   Incorporate facility-specific and industry-generic operating experience into the operating test whenever possible. Documentation such as licensee event reports, significant event reports, and service information letters are readily available sources of operationally oriented plant anomalies.

Evaluate the dominant accident sequences (DASs) for the facility to determine whether they are suitable for testing, on a sampling basis, during the dynamic simulator or walk-through tests. DASs are those sequences that contribute significantly to the frequency of core damage as determined by the facility licensee's probabilistic risk assessment (PRA) or individual plant examination (IPE).

The PRA/IPE should also be used to identify risk-important operator actions. Chapter 13, "Operational Perspectives," of NUREG-1560, "Individual Plant Examination Program: Perspectives on Reactor Safety and Plant Performance," identifies a number of important human actions that may be appropriate for evaluation on the operating test. In determining what actions to evaluate, do not overlook actions that are relied upon or result in specific events being driven to low risk contribution. This will help identify those human actions that are assumed to be very reliable, but might otherwise not show up in a list of risk-dominant actions.

g.   If the applicants at a facility qualify for dual or multi-unit licenses, the operating tests should evaluate their knowledge of the design, procedural, and operational differences between the units.

Divide the operating test coverage among the units and do not become predictable by conducting the walk-through tests on only one unit. Different applicants may be examined on different units, or each applicant may be asked to explain or demonstrate his or her understanding of variations in control board layouts, systems, instrumentation, and procedural actions between the units at the facility.

Most dual- or multi-unit stations have a simulator that is modeled after only one of the units. Therefore, ensure that the applicants are properly tested on the different systems, control board layouts, and any other differences between the units during the walk-through portion of the operating test. For example, after administering the simulator operating test on Browns Ferry Unit 1, the control room systems portion of the walk-through operating test could be administered on Unit 2 or Unit 3 or both.

h.   The operating test should examine a broad range of knowledge and abilities, systems and components, and operations and events. The walk-through and simulator tests should not be redundant, nor should they duplicate material that is covered on the written examination. It is particularly important that the simulator and control room systems walk-through be developed and reviewed as a package to preclude the same tasks and events from appearing on both parts of the test.

i.   Every facet of the operating test, including the walk-through JPMs and simulator scenarios, should be planned, researched, validated, and documented to the maximum extent possible before the test is administered.

j.   Examiners who will be administering the operating tests but were not involved in their development are expected to research and study the topics and systems to be examined on the operating test so that they are prepared to ask whatever performance-based followup questions might be necessary to determine whether the applicant is competent in those areas. As stated in 10 CFR 55.45(a), the operating test requires the applicant to demonstrate an understanding of and the ability to perform the actions necessary to accomplish a representative sample from among 13 items listed in the rule. If the applicant correctly performs a JPM (including both critical and noncritical steps) and demonstrates familiarity with the administrative topic, equipment, and procedures, it is not necessary to ask any followup questions. However, if the applicant fails to accomplish the task standard for the JPM or demonstrates a lack of understanding regarding the administrative topic, equipment, and procedures such as having difficulty locating information, control board indications, or controls, the examiner must be prepared to ask performance-based followup questions, as necessary, to clarify or confirm the applicant's understanding of the administrative topic or system as it relates to the task that was performed.

Examination team members are strongly encouraged to meet as a group with the chief examiner to review the examination materials after they have been approved for administration by the responsible supervisor. The discussions should focus on those test items that might require extensive cuing by the examiner and those that are unique to the facility and require a response different from what the examiner might expect based on past experience.

k.   Performance-based followup questions during any part of the operating test may include a combination of open- and closed-reference items. Open-reference items that require applicants to apply their knowledge of the plant to postulated normal, abnormal, and emergency situations are preferred. Closed-reference items may be used to evaluate the immediate actions of emergency and other procedures, certain automatic actions, operating characteristics, interlocks, set points, and routine administrative activities, as appropriate to the facility.

Refer to Attachment 1 for more guidance regarding the development and use of open reference questions. To the extent possible, the concepts in the attachment should be applied to performance-based followup questions.

l.   If it becomes necessary to deviate from a test outline that has been approved by the NRC's chief examiner in accordance with ES-201, discuss the proposed deviation with the chief examiner and obtain concurrence before proceeding with the changes. Be prepared to explain why the original proposal could not be implemented and why the proposed replacement is considered an acceptable substitute.

2. **Walk-Through Guidelines**

   a. In order to protect the integrity and security of the examination process, the examination author must limit how much of the examination is taken directly from the facility's testing materials without significant modification and how much of the walk-through test is repeated from the last two NRC licensing examinations at the facility. A significant modification means that at least one condition has been substantively changed in a manner that alters the course of action of the JPM. If JPMs are repeated from the past two NRC examinations, they must be randomly selected from all the JPMs used on the past 2 examinations. Refer to Forms ES-301-1 and ES-301-2 for specific limits on JPM bank use and repetition from the previous 2 NRC examinations.

   b. JPMs should include the elements identified in Appendix C (e.g., initiating and terminating cues, critical steps, and performance criteria). The guidelines and forms (or equivalents) in that appendix should be used when developing new JPMs. Facility procedures may be adapted for use as JPMs by identifying critical steps and entering comments on how to execute particular steps.

   c. The JPMs should, individually and as a group, have meaningful performance requirements that will provide a legitimate basis for evaluating the applicant's understanding of and ability to safely operate the plant (as required by 10 CFR 55.45).

3. **Specific Instructions for the "Administrative Topics" Walk-Through**

   Although the administrative topics may be examined separately, it is preferable, whenever possible, to link, associate, or integrate them with tasks and events conducted during the systems and simulator portions of the operating test. However, it is important to keep in mind that the applicant's proficiency in the administrative topics should be deliberately evaluated and not inferred solely from observations made during the other portions of the operating test.

   a. For each of the administrative topics listed below, select the required number of subjects to be evaluated during the operating test based on the applicant's license level.

Topic	Number of Subjects	
	RO	**SRO and RO Retakes**
"Conduct of Operations"	1 (or 2)	2
"Equipment Control"	1 (or 0)	1
"Radiation Control"	1 (or 0)	1
"Emergency Procedures/Plan"	1 (or 0)	1
Total	4	5

RO applicants need not be evaluated on every topic (as indicated above, "Equipment Control," "Radiation Control," or "Emergency Procedures/Plan" can be omitted by doubling-up on "Conduct of Operations"), unless the applicant is retaking only the "Administrative Topics" (with a waiver of the systems walk-through and simulator test pursuant to ES-204).

K/As associated with each administrative topic shall be selected from Section 2 of the applicable NRC K/A catalog for pressurized- or boiling-water reactors (i.e., NUREG-1122 and 1123, respectively). For the "Emergency Procedures/Plan" topic, only those K/As related to the emergency plan and implementing procedures [not those associated with the emergency operating procedures (EOPs)] are applicable to this category of the operating test.

b.   For each administrative subject, select a performance-based activity for which an administrative JPM can be developed. The administrative JPMs may require the applicant to identify and respond to one or more postulated administrative errors in a manner similar to the alternate path methodology discussed in Appendix C.

c.   In general, SROs have more administrative responsibilities than ROs, so SRO applicants should be evaluated in greater depth on the administrative topics. RO applicants need only understand the mechanics and intent of the related subjects, as they pertain to tasks at the facility.

d.   The following specific guidelines should be applied when selecting or developing JPMs to confirm the applicant's competence with regard to each topic:

"Conduct of Operations"

Many of these subjects can be covered within the framework of a shift turnover or by integrating them into other discussions, as they apply, throughout the examination.

The applicant's awareness of access controls for vital/controlled plant areas should be evaluated by observing his or her behavior during the operating test. However, passive observations, in and of themselves, are insufficient to justify an evaluation in that subject area.

The subject of fuel handling can be covered in the control room, but attempt to cover this subject in the fuel handling areas of the plant whenever possible. The RO applicant should be aware of his or her duties in the control room during fuel handling. These duties include monitoring instrumentation and responding to alarms from the fuel handling area, communicating with the fuel handling and storage facility, and operating systems from the control room in support of (re)fueling operations. For the SRO applicant, evaluate topics such as core alterations, new and spent fuel storage and movement, the design of the fuel handling area, use of the fuel handling tools, and fuel handling casualties.

"Equipment Control"

These subjects can be evaluated within the framework of a normal maintenance evolution. For example, have the applicant demonstrate how he or she would take a failed system or component out of service, initiate maintenance on the system, and test the system before placing it back in service. During the maintenance evolution, have the applicant demonstrate the use of piping and instrument drawings and technical specifications.

"Radiation Control"

This topic is best covered in conjunction with the JPMs prepared for the in-plant systems walk-through. It is most appropriate to evaluate these subjects during the required entry into the radiologically controlled area (RCA).

The levels of knowledge expected of RO and SRO applicants in some radiation control subjects are significantly different. The RO's duties generally require knowledge of radiation worker responsibilities and operation of plant systems associated with liquid and gaseous waste releases. Therefore, the depth to which RO applicants are evaluated should be limited to their responsibilities and the monitoring requirements before, during, and after the release. The SRO, however, may be involved in reviewing and approving release permits and should be cognizant of the requirements associated with those releases, as well as their potential effect on the health and safety of the public. The SRO applicants may be asked to simulate a planned release (e.g., liquid, gaseous, or containment purge) when examining these topics.

"Emergency Procedures/Plan"

There are significant differences between the knowledge required of RO and SRO applicants in this area. RO applicants should be familiar with the emergency plan and with their plant-specific responsibilities under the emergency plan implementing procedures (EPIPs). By contrast, SRO applicants must demonstrate additional knowledge based upon their responsibility to direct and manage the implementation of the EPIPs during the initial phases of an emergency. As a result, SRO applicants should have a more detailed understanding of the EPIPs, in general, and should be familiar with event classification procedures, protective action recommendations, and communication requirements and methods. As discussed in Section D.1, ensure that the test does not become predictable by always performing a different variation of the same activity (e.g., repetitive emergency classifications with different events).

This topic is best evaluated by linking a JPM to a simulator transient that requires implementation of the emergency plan. Such a JPM can be conducted immediately following a simulator scenario or during the walk-through examination.

e. The planned administrative subjects should normally take no more than 1 hour and 1.5 hours to administer to RO and SRO applicants, respectively.

f. On Form ES-301-1, "Administrative Topics Outline," briefly describe the specific administrative activities selected for evaluation.

g. Forward the completed outline to the NRC's chief examiner so that it is *received* by the date agreed upon with the NRC regional office at the time the examination arrangements were confirmed; the outline is normally due approximately 75 days before the scheduled examination date. Refer to ES-201 for additional instructions regarding the review and submittal of the examination outline.

   The NRC's chief examiner and responsible supervisor shall review the test outline coverage as soon as possible in accordance with ES-201 and forward any comments to the originator for resolution.

h. After the NRC's chief examiner approves the operating test outline, prepare the final administrative JPMs in accordance with the general operating test guidelines in Sections D.1 and 2 and the JPM guidelines in Appendix C.

i. When the materials are complete, review the quality of the final administrative walk-through test using Form ES-301-3, "Operating Test Quality Checklist." This review shall be performed in conjunction with the associated systems walk-through and the dynamic simulator operating test as noted in Sections D.4 and D.5.

   Submit the entire operating test package to the designated facility reviewer or the NRC's chief examiner, as appropriate, for review and approval in accordance with Section E. The NRC's chief examiner must receive the test approximately 45 days before the scheduled administration date, unless other arrangements have been made.

4. **Specific Instructions for the "Control Room/In-Plant Systems" Walk-Through**

This part of the operating test evaluates the applicant on systems-related K/As by having the applicant perform selected tasks and, when necessary, based on the applicant's performance, probing his or her knowledge of the task and its associated system with specific followup questions. The selected tasks are *in addition to* and shall be *different from* the events and evolutions conducted during the simulator operating test. A task that is similar to a scenario event may be acceptable if the actions required to complete the task are significantly different from those required in response to the scenario event.

a. Refer to Section 1.9 of the K/A catalog applicable to the type of reactor for which the applicant is seeking a license (i.e., NUREG-1122 for PWRs and NUREG-1123 for BWRs). From the nine safety function groupings identified in the catalog, select the appropriate number of systems (see the table below) to be evaluated based on the applicant's license level. The emergency and abnormal plant evolutions (E/APEs) listed in Section 1.10 of the appropriate NUREG may also be used to evaluate the applicable safety function (as specified for each E/APE in the first tier of the written examination outlines attached to ES-401, "Preparing Initial Site-Specific Written Examinations").

License Level	Control Room	In-Plant	Total
RO	8	3	11
SRO-instant (I)	7	3	10
SRO-upgrade (U)	2 or 3	3 or 2	5

Each of the control room systems and evolutions (and separately each of the in-plant systems and evolutions) selected for RO and SRO-I applicants should evaluate a different safety function, and the same system or evolution should not be used to evaluate more than one safety function in each location. For PWR operating tests, the primary and secondary systems listed under Safety Function 4, "Heat Removal From Reactor Core," in Section 1.9 of NUREG-1122 may be treated as separate safety functions (i.e., two systems, one primary and one secondary, may be selected from Safety Function 4).

The five systems and evolutions selected for an SRO-U applicant should evaluate at least five different safety functions. One of the control room systems or evolutions must be an engineered safety feature, and the same system or evolution should not be used to evaluate more than one safety function.

b. For each system selected for evaluation, select from the applicable K/A catalog or the facility licensee's site-specific task list *one* task for which a JPM exists or can be developed. Review the associated simulator outline if it has already been prepared (refer to Section D.5), and avoid those tasks that have already been selected for evaluation on the dynamic simulator test.

At least one of the tasks shall be related to a shutdown or low-power[1] condition, and four to six of the tasks for ROs and instant SROs and two to three of the tasks for upgrade SROs shall require the applicant to execute alternative paths within the facility's operating procedures. In addition, at least one of the tasks conducted in the plant shall evaluate the applicant's ability to implement actions required during an emergency or abnormal condition, and another shall require the applicant to enter the RCA. This provides an excellent opportunity for the applicant to discuss or demonstrate the radiation control administrative subjects.

---

[1] NUREG-1449, "NRC Staff Evaluation of Shutdown and Low-Power Operation," defines "low power" to include the range from criticality to 5 percent power.

If it is not possible to develop or locate a suitable task and/or JPM for each of the selected systems, return to Step (a), above, and select a different system or evolution. After identifying a JPM for each system, list the JPM and its associated safety function number on Form ES-301-2, "Control Room/In-Plant Systems Outline." Also indicate the type of JPM by entering the applicable code(s) identified at the bottom of the form.

c.   Forward the completed walk-through test outline to the NRC's chief examiner so that it is *received* by the date agreed upon with the NRC's regional office at the time the examination arrangements were confirmed; the outlines are normally due approximately 75 days before the scheduled examination date. Refer to ES-201 for additional instructions regarding the review and submittal of examination outlines.

The NRC's chief examiner and responsible supervisor shall review the test outline in accordance with ES-201 and forward any comments to the originator for resolution.

d.   After the NRC's chief examiner approves the operating test outline, prepare the final JPMs in accordance with the general guidance in Sections D.1 and D.2 and the JPM guidelines in Appendix C.

e.   When the materials are complete, review the completed walk-through test for quality using Form ES-301-3, "Operating Test Quality Checklist," and make any changes that might be necessary. To minimize duplication, this review shall be performed in conjunction with the associated administrative topics and the simulator operating test (refer to Sections D.3 and D.5).

Submit the entire operating test package to the designated facility reviewer or the NRC's chief examiner, as appropriate, for review and approval in accordance with Section E. The NRC's chief examiner must receive the test approximately 45 days before the scheduled review date, unless other arrangements have been made.

5.   Specific Instructions for the "Simulator Operating Test"

a.   Based on the anticipated crew compositions, determine the number of scenarios and scenario sets necessary to rotate each RO and SRO-I applicant into the lead reactor operator (i.e., the "at-the-controls") position. For example, a crew consisting of two ROs and one SRO-I will normally require three scenarios to evaluate each applicant's performance on the reactor controls; however, a surrogate SRO will have to fill the supervisory role while the SRO-I applicant is in the lead operator position. Additionally, the crews and scenarios will have to be planned so that every SRO applicant (U and I) fills the supervisory role and every RO applicant rotates through the balance-of-plant (BOP) position for at least one scenario.

SRO-U applicants are given credit for their previous RO license evaluation and experience and are normally not required to manipulate the controls.

It may be possible to significantly reduce the number of simulator scenario sets required to examine a large group of applicants by administering the same set of scenarios on the same day to two (or more) different crews of applicants. However, provisions must be made to ensure that the crews remain out of contact until all crews have completed the set of scenarios (refer to ES-302).

Additional or replacement scenarios should also be prepared and available while administering the operating tests in accordance with ES-302, in case one of the planned scenarios does not work as intended.

b. The simulator operating tests (i.e., scenario sets) will be constructed by selecting and modifying scenarios from existing facility licensee or NRC scenario banks and by developing new scenarios.

In order to maintain test integrity, every applicant shall be tested on at least one new or significantly modified scenario that he or she has not had the opportunity to rehearse or practice. A significant modification means that at least one condition or event has been substantively changed to alter the course of action in the scenario. Furthermore, any other scenarios that are extracted from the facility licensee's bank must be altered to the degree necessary to prevent the applicants from immediately recognizing the scenarios based on the initial conditions or other cues.

c. The initial conditions, normal operations, malfunctions, and major transients should be varied among the scenarios and should include startup, low-power[2], and full-power situations. Review the associated walk-through outline if it has already been prepared (refer to Section D.4), and take care not to duplicate operations that will be tested during the systems walk-through portion of the operating test.

d. In order to maximize the quality and consistency of the operating tests, develop new scenarios in accordance with the instructions in Appendix D. Modify existing scenarios, as necessary, to make them conform with the qualitative and quantitative attributes described in that appendix and enumerated on Form ES-301-4, "Simulator Scenario Quality Checklist." The quantitative attribute target ranges that are specified on the form are not absolute limitations; some scenarios may be an excellent evaluation tool, but may not fit within the ranges. A scenario that does not fit into these ranges shall be evaluated to ensure that the level of difficulty is appropriate. Whenever possible, the critical tasks should be distributed so that each applicant is required to respond.

---

[2] NUREG-1449, "NRC Staff Evaluation of Shutdown and Low-Power Operation," defines "low power" to include the range from criticality to 5 percent power.

At a minimum, each scenario set must require each applicant to respond to the types of evolutions, failures, technical specification (TS) evaluations, and transients in the quantities identified for the applicant's license level on Form ES-301-5, "Transient and Event Checklist." An applicant should only be given credit for those events that require the applicant to perform verifiable actions that provide insight to the applicant's competence. The required instrument and component failures should normally be completed before starting the major transient; those that are initiated after the major transient should be carefully reviewed because they may require little applicant action and provide little insight regarding their performance. With the exception of the SRO TS evaluations, each event should only be counted once per applicant; for example, a power change can be counted as a normal evolution *or* as a reactivity manipulation and, similarly, a component failure that immediately results in a major transient counts as one or the other, but not both.

Any normal evolution, component failure, or abnormal event (other than a reactor trip or other automatic power reduction) that requires the operator to perform a *controlled* power or reactivity change will qualify as a reactivity manipulation. This includes events such as an emergency borating, a dropped rod recovery, a significant rod bank realignment, or a manual reactor power reduction in response to a secondary system upset. Such events may produce a more timely operator and plant response than a normal power change.

Furthermore, each scenario set must also allow the examiner to evaluate the applicant's performance on each competency and rating factor that is germane to the applicant's license level. Use Form ES-301-6, "Competencies Checklist," to verify that the competencies are adequately evaluated by entering the scenario and event numbers that are intended to assess each competency. To minimize the need to run an additional scenario if an applicant makes a single, uncompensated error related to a rating factor (refer to Section D.3.n of ES-302), it is recommended that each applicant be given multiple opportunities to demonstrate competence in any particular area.

If the facility licensee normally operates with and is required by its technical specifications to have more than two ROs in the control room, the chief examiner may authorize the use of additional surrogates to fill out the crews. In such cases, take care in planning the scenarios to ensure that the additional operators do not reduce the examiners' ability to evaluate each applicant on the required number of events and on every competency and rating factor.

Appendix D provides detailed instructions for completing Form ES-D-1, the "Scenario Outline," and Form ES-D-2, the "Required Operator Actions," that examiners will use to administer the simulator operating tests. In order to minimize the amount of rework that might be required as a result of changes in the planned scenario events, Form ES-D-2 should be completed after the NRC's chief examiner has had the opportunity to review and comment on the proposed simulator operating test outlines (i.e., Form ES-D-1) in accordance with ES-201.

e.	When the proposed simulator operating test outlines are complete, forward them to the NRC's chief examiner so they are *received* by the date agreed upon with the NRC's regional office at the time the examination arrangements were confirmed; the outlines are normally due approximately 75 days before the scheduled examination date. Refer to ES-201 for additional instructions regarding the review and submittal of the examination outlines.

The NRC's chief examiner shall review the operating test outlines in accordance with ES-201, and forward any comments to the originator for resolution.

f.	After the NRC's chief examiner approves the operating test outlines, prepare the final simulator test materials by revising Form(s) ES-D-1 as requested by the NRC's chief examiner and completing a detailed operator action form (ES-D-2) for each event. All required operator actions (e.g., opening, closing, and throttling valves; starting and stopping equipment; raising and lowering level, flow, and pressure; making decisions and giving directions; acknowledging or verifying key alarms and automatic actions) shall be documented, and critical tasks shall be identified. Events that do not require an operator to take one or more substantive actions will not count toward the minimum number of events required for each operator per Form ES-301-5.

g.	Review the completed simulator operating test for quality using Form ES-301-4, "Simulator Scenario Quality Checklist," and make any changes that might be necessary. This review shall be performed in conjunction with the associated walk-through test (refer to Sections D.3 and D.4) to minimize duplication.

Submit the entire operating test package to the designated facility reviewer or the NRC's chief examiner, as appropriate, for review and approval in accordance with Section E. The NRC's chief examiner must receive the test approximately 45 days before the scheduled administration date, unless other arrangements have been made.

## E.	Quality Reviews

1.	<u>Facility Management Review</u>

If the operating test was prepared by the facility licensee, the preliminary outline and the proposed test shall be independently reviewed by a supervisor or manager before they are submitted to the NRC's regional office for review and approval in accordance with ES-201. The reviewer should evaluate the outline and test using the criteria on Forms ES-201-2, ES-301-3, and ES-301-4 and include the signed forms (for each different operating test) in the examination package submitted to the NRC in accordance with ES-201.

2. NRC Examiner Review

a. The NRC's chief examiner shall ensure that each operating test is independently reviewed for content, wording, operational validity, and level of difficulty. As a minimum, the examiner shall check the items listed on Forms ES-301-3 and ES-301-4, as applicable. The examiner should keep in mind that counting the number of scenario quantitative attributes is not always indicative of the scenario's level of difficulty. Although there are no definitive minimum or maximum attribute values that can be used to identify scenarios that will not discriminate because they are too easy or difficult, scenarios that fall outside the target ranges specified on Form ES-301-4 should be carefully evaluated to ensure they are appropriate. Refer to Section C.3 of ES-201 for additional guidance regarding examination reviews.

b. The NRC examiner should review the operating tests as soon as possible after receipt so that supervisory approval can be obtained before the final review with the facility licensee, which is normally scheduled about 2 weeks before the administration date. It is especially important that the examiner promptly review tests prepared by a facility licensee because of the extra time that may be required if extensive changes are necessary. The chief examiner shall consolidate the comments from other regional reviewers and submit one set of comments to the author.

c. If the facility licensee developed the operating test, the facility licensee is primarily responsible for technical accuracy and compliance with the restrictions concerning the use of examination banks. However, the chief examiner is expected to use his or her best judgment and take reasonable measures, including selective review of reference materials and past tests, to verify these attributes.

d. The chief examiner will note/review any changes that need to be made and forward the tests to the responsible supervisor for review and comment in accordance with Section E.3 before reviewing the examinations with the author or facility contact. There are no minimum or maximum limits on the number or scope of changes the chief examiner may direct the author or facility contact to make to the proposed tests, provided that they are necessary to make the tests conform with established acceptance criteria. Refer to ES-201 for additional guidance regarding NRC response to facility-developed examinations that are significantly deficient.

e. Upon supervisory approval, and generally at least 14 days before the operating tests are scheduled to be given, the chief examiner will review the tests with the facility licensee in accordance with ES-201.

Tests that were developed by the NRC should be clean, properly formatted, and "ready-to-give" before they are reviewed with the facility licensee. The regional office should not rely on the facility licensee to ensure that the tests are of acceptable quality to administer.

f.   After reviewing the tests with the facility licensee, the chief examiner will ensure that any comments and recommendations are resolved and the tests are revised as necessary. If the facility licensee developed the tests, it will generally be expected to make whatever changes the NRC recommends.

g.   After the necessary changes have been made and the chief examiner is satisfied with the test, he or she will sign Form(s) ES-301-3 and forward the test package to the responsible supervisor for final approval.

3.   **NRC Supervisory Review**

a.   In accordance with ES-201, the responsible supervisor shall review the operating tests before authorizing the chief examiner to proceed with the facility pre-review. The supervisory review is not intended to be another detailed review, but rather a general assessment of test quality, including a review of the changes recommended by the chief examiner, and a check to ensure that all of the applicable administrative requirements have been implemented.

b.   The responsible supervisor should ensure that any significant deficiencies in the original operating tests submitted by a facility licensee are evaluated in accordance with ES-201 to determine the appropriate course of action. At a minimum, the supervisor should ensure that they are addressed in the final examination report in accordance with ES-501, "Initial Post-Examination Process."

c.   Following the facility review, the responsible supervisor should again review the tests to ensure that the concerns expressed by the facility licensee and the chief examiner have been appropriately addressed. The supervisor shall not sign Form(s) ES-301-3 until he or she is satisfied that the examination is acceptable to be administered.

F.   Attachments/Forms

Attachment 1,	"Open-Reference Question Guidelines"
Form ES-301-1,	"Administrative Topics Outline"
Form ES-301-2,	"Control Room/In-Plant Systems Outline"
Form ES-301-3,	"Operating Test Quality Checklist"
Form ES-301-4,	"Simulator Scenario Quality Checklist"
Form ES-301-5,	"Transient and Event Checklist"
Form ES-301-6,	"Competencies Checklist"

1. The most appropriate format is the short-answer question, which requires the applicant to compose a response rather than select from among a set of alternative responses, as is the case with multiple-choice, matching, and true/false questions.

2. Provide clear, explicit directions and/or guidelines for answering the question so that the applicant understands what constitutes a fully correct response. Choose words carefully to ensure that the stipulations and requirements of the question are appropriately conveyed. Words such as "evaluate," "outline," and "explain," can invite a lot of detail that is not necessarily relevant.

3. Make sure that the expected response matches (and is limited to) the requirements posed in the question. Consider the amount of partial credit to be granted for an incomplete answer. For questions requiring computation, specify the degree of precision expected. Try to make the answer turn out to be whole numbers.

4. Avoid giving away part or all of the answer by the way the question is worded. For example, "If the letdown line became obstructed, could borating of the plant be accomplished shortly after a reactor trip to put the plant in cold shutdown? If so, how?"

   A test-wise applicant can realize that the answer has to be yes, or else the second part of the question would have read something like "If not, why not?"

5. Avoid what could be considered "trick" questions, in which the expected answer does not precisely match the question. For example, asking "How do the SI termination criteria change following an SI re-initiation?" implies that the termination criteria will change, when in actuality they do not.

6. Do not use direct look-up questions that only require the applicant to recall where to find the answer to the question. The operational orientation required of questions on the walk-through test and the applicant's access to reference documents, argue against the use of questions that test recall and memorization. Any questions that do not require any analysis, synthesis, or application of information by the applicant should be answerable without the aid of reference materials. Refer to ES-602, Attachment 1, for a more detailed discussion of direct look-up questions.

7. Questions should also adhere to the generic item construction principles and guidelines in Appendix B. Moreover, Form ES-602-1, "NRC Checklist for Open-Reference Test Items," contains a list of questions that can be used to evaluate the suitability of the questions for the walk-through portion of the operating test. Although the checklist was developed for use in evaluating requalification written examinations, all of the criteria except 9–11, and the K/A rating on item 7 are generically applicable.

Facility: _____ Date of Examination: _____

Examination Level: RO ☐ SRO ☐ Operating Test Number: _____

Administrative Topic (see Note)	Type Code*	Describe activity to be performed
Conduct of Operations		
Conduct of Operations		
Equipment Control		
Radiation Control		
Emergency Procedures/Plan		

NOTE: All items (5 total) are required for SROs. RO applicants require only 4 items unless they are retaking only the administrative topics, when all 5 are required.

* Type Codes & Criteria: (C)ontrol room, (S)imulator, or Class(R)oom
                                          (D)irect from bank ($\leq$ 3 for ROs; $\leq$ 4 for SROs & RO retakes)
                                          (N)ew or (M)odified from bank ($\geq$ 1)
                                          (P)revious 2 exams ($\leq$ 1; randomly selected)

Facility: _____	Date of Examination: _____
Exam Level: RO ☐ SRO-I ☐ SRO-U ☐	Operating Test No.: _____

Control Room Systems@ (8 for RO); (7 for SRO-I); (2 or 3 for SRO-U, including 1 ESF)

System / JPM Title	Type Code*	Safety Function
a.		
b.		
c.		
d.		
e.		
f.		
g.		
h.		

In-Plant Systems@ (3 for RO); (3 for SRO-I); (3 or 2 for SRO-U)

i.		
j.		
k.		

@   All RO and SRO-I control room (and in-plant) systems must be different and serve different safety functions; all 5 SRO-U systems must serve different safety functions; in-plant systems and functions may overlap those tested in the control room.

* Type Codes	Criteria for RO / SRO-I / SRO-U
(A)lternate path	4-6 / 4-6 / 2-3
(C)ontrol room	
(D)irect from bank	≤ 9 / ≤ 8 / ≤ 4
(E)mergency or abnormal in-plant	≥ 1 / ≥ 1 / ≥ 1
(EN)gineered safety feature	- / - / ≥1 (control room system)
(L)ow-Power / Shutdown	≥ 1 / ≥ 1 / ≥ 1
(N)ew or (M)odified from bank including 1(A)	≥ 2 / ≥ 2 / ≥ 1
(P)revious 2 exams	≤ 3 / ≤ 3 / ≤ 2 (randomly selected)
(R)CA	≥ 1 / ≥ 1 / ≥ 1
(S)imulator	

Facility:	Date of Examination:	Operating Test Number:			

			Initials		
1. General Criteria			a	b*	c#
a.	The operating test conforms with the previously approved outline; changes are consistent with sampling requirements (e.g., 10 CFR 55.45, operational importance, safety function distribution).				
b.	There is no day-to-day repetition between this and other operating tests to be administered during this examination.				
c.	The opera ing test shall not duplicate items from the applicants' audit test(s). (see Section D.1.a.)				
d.	Overlap with the written examination and between different parts of the operating test is within acceptable limits.				
e.	It appears that the operating test will differentiate between competent and less-than-competent applicants at the designated license level.				
2. Walk-Through Criteria			--	--	--

a.	Each JPM includes the following, as applicable:
	• initial conditions
	• initiating cues
	• references and tools, including associated procedures
	• reasonable and validated time limits (average time allowed for completion) and specific designation if deemed to be time-critical by the facility licensee
	• operationally important specific performance criteria that include:
	  – detailed expected actions with exact criteria and nomenclature
	  – system response and other examiner cues
	  – statements describing important observations to be made by the applicant
	  – criteria for successful completion of the task
	  – identification of critical steps and their associated performance standards
	  – restrictions on the sequence of steps, if applicable				

b.	Ensure hat any changes from the previously approved systems and administrative walk-through outlines (Forms ES-301-1 and 2) have not caused the test to deviate from any of the acceptance criteria (e.g., item distribution, bank use, repetition from the last 2 NRC examinations) specified on those forms and Form ES-201-2.				
3. Simulator Criteria			--	--	--
The associated simulator operating tests (scenario sets) have been reviewed in accordance with Form ES-301-4 and a copy is attached.					

	Printed Name / Signature	Date
a.	Author _____	_____
b.	Facility Reviewer(*) _____	_____
c.	NRC Chief Examiner (#) _____	_____
d.	NRC Supervisor _____	_____

NOTE:	*	The facility signature is not applicable for NRC-developed tests.
	#	Independent NRC reviewer initial items in Column "c"; chief examiner concurrence required.

Facility:	Date of Exam:	Scenario Numbers: / / Operating Test No.:		

QUALITATIVE ATTRIBUTES		Initials		
		a	b*	c#
1.	The initial conditions are realistic, in that some equipment and/or instrumentation may be out of service, but it does not cue the operators into expected events.			
2.	The scenarios consist mostly of related events.			
3.	Each event description consists of • the point in the scenario when it is to be initiated • the malfunction(s) that are entered to initiate the event • the symptoms/cues that will be visible to the crew • the expected operator actions (by shift position) • the event termination point (if applicable)			
4.	No more than one non-mechanistic failure (e.g., pipe break) is incorporated into the scenario without a credible preceding incident such as a seismic event.			
5.	The events are valid with regard to physics and thermodynamics.			
6.	Sequencing and timing of events is reasonable, and allows the examination team to obtain complete evaluation results commensurate with the scenario objectives.			
7.	If time compression techniques are used, the scenario summary clearly so indicates. Operators have sufficient time to carry out expected activities without undue time constraints. Cues are given.			
8.	The simulator modeling is not altered.			
9.	The scenarios have been validated. Pursuant to 10 CFR 55.46(d), any open simulator performance deficiencies or deviations from the referenced plant have been evaluated to ensure that functional fidelity is maintained while running the planned scenarios.			
10.	Every operator will be evaluated using at least one new or significantly modified scenario. All other scenarios have been altered in accordance with Section D.5 of ES-301.			
11.	All individual operator competencies can be evaluated, as verified using Form ES-301-6 (submit the form along with the simulator scenarios).			
12.	Each applicant will be significantly involved in the minimum number of transients and events specified on Form ES-301-5 (submit the form with the simulator scenarios).			
13.	The level of difficulty is appropriate to support licensing decisions for each crew position.			
Target Quantitative Attributes (Per Scenario; See Section D.5.d)	Actual Attributes	--	--	--
1.	Total malfunctions (5–8)	/ /			
2.	Malfunctions after EOP entry (1–2)	/ /			
3.	Abnormal events (2–4)	/ /			
4.	Major transients (1–2)	/ /			
5.	EOPs entered/requiring substantive actions (1–2)	/ /			
6.	EOP contingencies requiring substantive actions (0–2)	/ /			
7.	Critical tasks (2–3)	/ /			

Facility:		Date of Exam:											Operating Test No.:				

APPLICANT	EVENT TYPE	Scenarios												TOTAL	MINIMUM(*)		
		1			2			3			4						
		CREW POSITION			CREW POSITION			CREW POSITION			CREW POSITION						
		SRO	ATC	BOP	SRO	ATC	BOP	SRO	ATC	BOP	SRO	ATC	BOP		R	I	U
RO ☐ SRO-I ☐ SRO-U ☐	RX														1	1	0
	NOR														1	1	1
	I/C														4	4	2
	MAJ														2	2	1
	TS														0	2	2
RO ☐ SRO-I ☐ SRO-U ☐	RX														1	1	0
	NOR														1	1	1
	I/C														4	4	2
	MAJ														2	2	1
	TS														0	2	2
RO ☐ SRO-I ☐ SRO-U ☐	RX														1	1	0
	NOR														1	1	1
	I/C														4	4	2
	MAJ														2	2	1
	TS														0	2	2
RO ☐ SRO-I ☐ SRO-U ☐	RX														1	1	0
	NOR														1	1	1
	I/C														4	4	2
	MAJ														2	2	1
	TS														0	2	2

Instructions:

1. Check the applicant level and enter the operating test number and Form ES-D-1 event numbers for each event type; TS are not applicable for RO applicants. ROs must serve in both the "at-the-controls (ATC)" and "balance-of-plant (BOP)" positions; Instant SROs must serve in both the SRO and the ATC positions, including at least two instrument or component (I/C) malfunctions and one major transient, in the ATC position. If an Instant SRO *additionally* serves in the BOP position, one I/C malfunction can be credited toward the two I/C malfunctions required for the ATC position.

2. Reactivity manipulations may be conducted under normal or *controlled* abnormal conditions (refer to Section D.5.d) but must be significant per Section C.2.a of Appendix D. (*) Reactivity and normal evolutions may be replaced with additional instrument or component malfunctions on a 1-for-1 basis.

3. Whenever practical, both instrument and component malfunctions should be included; only those that require verifiable actions that provide insight to the applicant's competence count toward the minimum requirements specified for the applicant's license level in the right-hand columns.

Facility:	Date of Examination:							Operating Test No.:							
	APPLICANTS														
	RO ☐ SRO-I ☐ SRO-U ☐				RO ☐ SRO-I ☐ SRO-U ☐				RO ☐ SRO-I ☐ SRO-U ☐				RO ☐ SRO-I ☐ SRO-U ☐			
Competencies	SCENARIO				SCENARIO				SCENARIO				SCENARIO			
	1	2	3	4	1	2	3	4	1	2	3	4	1	2	3	4
Interpret/Diagnose Events and Conditions																
Comply With and Use Procedures (1)																
Operate Control Boards (2)																
Communicate and Interact																
Demonstrate Supervisory Ability (3)																
Comply With and Use Tech. Specs. (3)																

Notes:
(1)   Includes Technical Specification compliance for an RO.
(2)   Optional for an SRO-U.
(3)   Only applicable to SROs.

Instructions:

Check the applicants' license type and enter one or more event numbers that will allow the examiners to evaluate every applicable competency for every applicant.

ES-302
ADMINISTERING OPERATING TESTS TO INITIAL LICENSE APPLICANTS

A. Purpose

This standard describes how to administer operating tests to initial license applicants in accordance with the requirements of Title 10, Section 55.45, of the *Code of Federal Regulations* (10 CFR 55.45). It includes policies and guidelines for administering both the walk-through and integrated plant operations portions of the operating test. This standard presumes that the operating test was prepared in accordance with ES-301, "Preparing Initial Operating Tests."

B. Background

As noted in ES-201, "Initial Operator Licensing Examination Process," facility licensees will generally prepare proposed operating tests in accordance with ES-301 and submit them to the responsible NRC regional office for review and approval. Regardless of whether the facility licensee or the NRC prepared a given operating test, an NRC licensing examiner will independently administer and grade every test in accordance with the instructions contained herein and in ES-303, "Documenting and Grading Initial Operating Tests."

C. Responsibilities

1. Facility Licensee

The facility licensee is responsible for the following activities:

a. Make the plant and simulation facility available, as necessary, for validating and administering the operating tests.

b. Safeguard the integrity and security of the operating tests in accordance with facility procedures established pursuant to 10 CFR 55.40(b)(2) and the guidelines discussed in Attachment 1 to ES-201.

c. Provide administrative and logistics support (e.g., personnel to operate the simulation facility, surrogate operators, copies of the approved operating test materials as arranged with the chief examiner, etc.) to facilitate the administration of the operating tests in accordance with Section D.

d. Inform the NRC's regional office in writing if an applicant withdraws from the examination process before it is complete.

2. NRC Regional Office

The NRC regional office is responsible for the following activities:

a. Work with the facility contact to coordinate the operating test administration schedule in a manner that maximizes efficiency and maintains security. Normally, the operating tests should be administered within 30 days before or after the written examinations. The regional office shall obtain concurrence from the NRR operator licensing program office if the examination dates diverge by more than 30 days. (Refer to ES-201 for additional guidance regarding examinations that have to be rescheduled to achieve an acceptable product.)

b. Administer the operating tests in accordance with Section D.

D. Test Administration Instructions and Policies

1. General

a. Before beginning the operating test, an examiner shall brief the applicant(s) using Parts A, C, D, and E of Appendix E. To save time, the examiner(s) may brief the applicants as a group.

b. If an applicant requests to withdraw during any part of the examination process, the examiner shall inform the applicant that this will result in automatic license denial and that he or she may reapply in accordance with 10 CFR 55.35. The chief examiner will then ask the facility licensee to document the applicant's withdrawal in a letter to the NRC's regional administrator.

c. Each applicant identified on the "List of Applicants" (Form ES-201-4) shall be administered an operating test as indicated on the form.

d. For purposes of test integration and continuity, the chief examiner should generally schedule the same examiner to administer both the walk-through and simulator portions of the operating test to an applicant. However, under certain circumstances, such as when a licensee's simulation facility is not located near the plant or if a licensee requests examinations for an unusually large group of applicants, the responsible regional supervisor may authorize the chief examiner to divide the operating test among different examiners. (However, simulator operating tests consisting of multiple scenarios shall not be divided among examiners.) The chief examiner will be responsible for ensuring that each applicant receives a complete operating test and that the tests are thoroughly and accurately documented.

Normally, an NRC examiner will be assigned to individually evaluate each applicant during the simulator operating test. However, if a three-person operating crew consists entirely of senior reactor operator (SRO) upgrade applicants (who do not have to be evaluated on the control boards), the chief examiner may assign only two examiners to observe the crew. In addition, although applicants in the reactor operator (RO) and balance-of-plant positions may not be individually evaluated, they will be held accountable for any errors that occur as a result of their action(s) or inaction(s), and they will be graded on their ability to "operate the control boards" (i.e., SRO Competency 3). By contrast, SRO-instant applicants will always be individually evaluated by an NRC examiner, regardless of what operating position they fill during a given scenario.

e.   The examiner is expected to administer the planned operating test in accordance with the prepared and approved walk-through test outlines (Forms ES-301-1, "Administrative Topics Outline," and ES-301-2, "Control Room/In-Plant Systems Outline") and simulator scenarios (Forms ES-D-1, "Scenario Outline," and ES-D-2, "Required Operator Actions"). Examiners shall document every significant aspect of each applicant's performance for later evaluation, but they shall *not* use the applicant's unplanned actions and statements to displace any part of the planned operating test.

Normally, examiners should substitute or replace planned operating test materials only if an item is determined to be invalid or impossible to perform or simulate because of unanticipated access restrictions or equipment failures.

f.   Examiners may administer the same operating test (walk-through and simulator) to consecutive applicants and crews on the same day, but they must ensure that the security of the operating test is maintained. The same simulator scenarios and job performance measures (JPMs) shall not be repeated during subsequent days.

If previously agreed upon by the facility licensee, examiners may also administer the same operating test (walk-through and simulator) by dividing the test into segments that can be administered to all of the applicants on the same day. This will minimize the amount of effort required to develop different operating tests, but will complicate the scheduling process.

g.   The examiner should normally administer the systems walk-through and the simulator operating test first and attempt to concurrently evaluate as many of the planned administrative subjects as possible. The examiner should then evaluate the remaining administrative subjects in accordance with the approved outline.

h.   The examiner must take sufficient notes to facilitate thorough documentation of any and all applicant deficiencies in accordance with ES-303. The examiner must be able to cross-reference each comment to a specific JPM, simulator event, or for-cause followup question.

i.   The making of videotapes during the administration of operating tests is not authorized.

j. The number of persons present during an operating test should be limited to ensure the integrity of the test and to minimize distractions to the applicants:

- Except for the simulation facility operators, no other member of the facility's staff shall be allowed to observe an operating test without the chief examiner's permission. Facility management and other personnel deemed necessary by the facility licensee should generally be allowed access to the examination (under security agreements, as appropriate), provided that the simulation facility can accommodate them and there is no impact on the applicants.

  Although the simulation facility operator will normally assume the role of the other personnel that the applicants direct or notify regarding plant operations, the chief examiner may permit other members of the facility training or operations staff (e.g., a shift technical advisor (STA)) to augment the operating shift team if necessary. In such instances, the chief examiner shall fully brief those individuals regarding their responsibilities, reporting requirements, duties, and level of participation before the operating test begins. All participants in the testing process must also be mindful of their responsibilities with regard to examination integrity pursuant to 10 CFR 55.49.

  Although the applicants will generally be expected to perform "peer checks" in accordance with the facility licensee's operations and training procedures and practices, additional personnel may not be stationed or called upon for this purpose.

  Surrogate operators should be used only when they are necessary to complete an operating crew. A facility licensee may not replace license applicants with surrogates solely because the applicants have performed the minimum number of events or scenarios. If an applicant would be exposed to only *one* additional scenario above the minimum required, a surrogate operator should not be used in place of a license applicant. However, no applicant will be required to participate in *more* than one scenario above the minimum required, in which case, a surrogate operator should be used. If, at the discretion of the chief examiner, it is desired to use surrogate operators contrary to the above guidance, the operator licensing program office should be consulted prior to implementation.

  When surrogate operators are required to complete the operating crew (e.g., during retake tests or for a class consisting entirely of ROs), the chief examiner shall ensure that the surrogate operator(s) are briefed regarding the content of the scenario(s) and their expected actions in response to every event. The examiners must not restrict the surrogate operators' activities to such an extent that the applicants being evaluated are required to assume responsibilities beyond the scope of their respective positions. The surrogate operators do not need to be licensed at the facility, but they must have the knowledge and abilities required to assume the full responsibilities of the roles they take in the operating test. Consultations with an STA shall be conducted in accordance with the facility licensee's normal control room practice (e.g., an STA shall not be stationed in the simulator if they are on-call at the site). If used, the STA shall also be briefed regarding the content of the scenario(s) and their expected actions in response to every event. Surrogates and STAs should not take a proactive role in assisting or coaching the applicants because such interventions would hinder the examiners' ability to evaluate the applicants' competence. Examiners shall run additional scenarios if necessary to make a licensing decision.

If the facility licensee normally operates with and is required by its technical specifications to have more than two ROs in the control room, the chief examiner may authorize the use of additional surrogates to fill out the crews. In such cases, examiners must take care that the presence of additional operators does not dilute the examiners' ability to evaluate each applicant during the required number of events and on every applicable competency and rating factor. Examiners shall not hesitate to run additional scenarios, as necessary, to ensure that every applicant has the opportunity to demonstrate his or her competence. Only one individual (applicant or surrogate) is allowed to fill a shift supervisor or manager position during the simulator operating test.

- Under *no* circumstances will another applicant be allowed to observe an operating test. Operating tests are *not* to be used as training vehicles for future applicants.

- Other examiners may observe an operating test as part of their training or to audit the performance of the examiner administering the operating test.

- The chief examiner may permit other NRC employees, such as resident inspectors, regional personnel, researchers, or NRC supervisors, to observe an operating test. Personnel who are not NRC employees (e.g., representatives from the Institute of Nuclear Power Operations (INPO)) may observe the operating tests with prior approval from the NRR operator licensing program office. The chief examiner will control the observers' activities in accordance with guidance provided by NRR. The examiner should also give the applicants the opportunity to object to the presence of observers.

k.   The chief examiner should confirm with the facility licensee that the simulator instructor's station, programmers' tools, and external interconnections do not compromise operating test security while conducting examinations (refer to Section F of Appendix D). The primary objective is to ensure that the exam material cannot be read or recorded at other unsecured consoles and is either physically secured or electronically protected when not in use by individuals listed on the security agreement.

Examiners should also take reasonable measures to ensure that any notes documenting the applicants' performance on the operating test are not accessible to the facility staff. Notwithstanding the fact that the facility staff has signed the security agreement, such notes are predecisional and should not be left unattended or unsecured in the simulator or examination room to which the facility staff has access.

l. Pursuant to 10 CFR 55.46(d), the chief examiner should confirm that any uncorrected simulator performance deficiencies do not interfere with the conduct of the planned operating tests.

m. The chief examiner should arrange for any NRC examiners who are not familiar with the facility to obtain a tour before they administer any operating tests. Such tours shall not be conducted or observed by any of the applicants. In addition, the tours should concentrate on areas of the plant that will be used during the examination process, such as the control room, the simulation facility, and planned walk-through locations.

n. The chief examiner will conduct an exit briefing with the facility licensee after the operating tests are complete. The briefing should address any generic weaknesses noted during the operating tests, as well as any other significant issues (e.g., problems with the reference materials, the simulation facility, or the plant) that might be addressed in the examination report. The individual operating test results are predecisional until approved by NRC management in accordance with ES-501, "Initial Post-Examination Activities," and shall *not* be shared with the facility licensee during the exit briefing.

2. Walk-Through

a. The examiner should validate any JPMs that were not previously validated by the facility licensee or the NRC during a preparatory site visit. This is particularly important for complex JPMs and those that require the applicant to implement an alternative method directed by plant procedures.

b. To the extent possible, the examiner should have the applicant perform the control room JPMs on the simulator, rather than asking the applicant to describe how he or she would accomplish the task.

   If the examiner observes a discrepancy between the simulator setup and the conditions specified in a JPM, the examiner shall stop the JPM and correct the situation, as necessary. If the task can be completed with different values (e.g., wind direction when determining a protective action recommendation during an emergency), the examiner shall document the differences and coordinate with the facility contact and the NRC chief examiner to validate the applicant's response under the actual conditions.

   The chief examiner is expected to coordinate the administration of the JPMs to maximize the use of the simulator. To increase efficiency, different JPMs may be administered simultaneously to multiple applicants, but the examiners must ensure that mutual interference is minimized and test integrity is not compromised.

Under certain circumstances, it may be more efficient to administer some or all of the JPMs in "station-keeping" mode, in which the examiners remain in position at designated operating stations and the applicants, under escort, rotate through the various stations. Such arrangements would have to be agreed to by and coordinated with the facility licensee; moreover, the guidelines in Sections D.1.d and D.1.f would apply.

When JPMs or followup discussions are conducted in the control room, the examiners shall make every effort to accommodate and not interfere with normal shift operations. The chief examiner should ask the facility training manager to notify the shift supervisor when the NRC will be conducting examination activities in the control room. If the number of persons or the noise level in the control room is excessive, the examiner should, if possible, move to a quieter location, modify the sequence of the JPMs and return when the level of activity in the control room has abated, or ask the facility training manager to address the issue.

c.   The examiner should encourage the applicants to sketch diagrams, flow paths, or other illustrations to aid in answering any followup questions that might be necessary. In all cases, the examiner shall collect the supporting material because it provides additional documentation to support a pass or fail decision (refer to ES-303). To facilitate photocopying, the applicant's drawings should be restricted to one side of separate sheets of 8.5-inch by 11-inch paper.

d.   The examiner should encourage the applicants to use such materials as facility forms, schedules, and procedures if they are relevant to the tasks to be performed or the followup questions to be asked.

e.   The examiner should keep in mind that the applicant's proficiency in every administrative topic and each control room and in-plant system should be deliberately evaluated in a manner that is consistent with the operating test that was prepared in accordance with ES-301.

f.   As stated in 10 CFR 55.45(a), the operating test requires applicants to demonstrate an understanding of and the ability to perform the actions necessary to accomplish a representative sample from among 13 items listed in the rule. If an applicant correctly performs a JPM (including both critical and noncritical steps) and demonstrates familiarity with the equipment and procedures, the examiner should infer that the applicant has an adequate understanding of the system/task and should refrain from asking followup questions. However, if the applicant fails to accomplish the task standard for the JPM; exhibits behavior that demonstrates a lack of familiarity with the equipment and procedure; or is unable to locate information, control board indications, or controls, the examiner should ask performance-based followup questions (as necessary) to clarify or confirm the applicant's understanding of the system as it relates to the task performed.

The examiner shall document all performance-based questions and answers for later evaluation.

If the applicant exceeds twice the validated time estimate for any JPM (including time-critical) because he or she has selected an incorrect procedure or operated the wrong equipment (despite being presented with sufficient plant feedback to correct the error), the examiner should stop the JPM, document the circumstances, and proceed with the next JPM. However, if the applicant is on the correct path but has simply stopped making progress toward completing a non-time-critical JPM, the examiner should ask the applicant to describe the work to be done and how long it should take to complete the JPM. If the applicant does not then make timely progress toward completing the described actions, the examiner should inform the applicant that the allowed time for the JPM has elapsed and the applicant will be evaluated on the work completed. The examiner should then proceed with the next JPM.

If an applicant volunteers additional or corrected information after completing a task, the examiner shall offer the applicant the opportunity to take whatever actions would be required in a similar situation in the plant. The examiner will record any revisions to previously performed tasks or answers for consideration when grading the operating test in accordance with ES-303.

g.   If an applicant requests a "peer check," the examiner will simply acknowledge the applicant's request and grade any errors in accordance with ES-303. Similarly, the examiner will not permit an applicant to obtain assistance from a "procedure reader" when performing JPMs.

h.   The examiner should practice other good walk-through evaluation techniques, as discussed in Section D of Appendix C.

3.   **Simulator Operating Test**

a.   Before administering the test(s), the examiners will validate each scenario on the simulator to ensure that it will run as intended. Scenarios that were adapted from previous NRC examinations at the facility or from the facility licensee's bank may not require real-time validation. At a minimum, the examiners will "dry run" those events that have variable inputs and questionable outcomes and discuss the remainder of the scenario with the facility's simulator instructor to ensure that it will run as planned.

In some cases, the scenarios can be validated while the applicants are taking the written examination. However, it may be beneficial to validate the scenarios during a preparatory site visit as determined by NRC regional management (refer to ES-201).

b.   The examiners will take precautions to prevent the scenarios from being revealed to the applicants before the tests begin. If significant portions of the scenarios are dry run or otherwise reviewed with the simulator instructor(s), the chief examiner shall ask the instructor(s) to sign a security agreement (Form ES-201-3) to protect the integrity of the simulator test.

c.   The examiners should revise all copies of Forms ES-D-1 and ES-D-2 to reflect any changes made to the scenario events or the expected operator actions as a result of the scenario validation runs and reviews. These revisions should be neatly written in ink so that the forms can be used in the final write-up of the simulator test, as discussed in ES-303.

d.   The examiners should review the scenarios together and discuss the required procedures, technical specifications, special circumstances, and so forth, related to the scenarios.

e.   Immediately before beginning the simulator tests, the examiners should review the scenario events with the simulator operator and provide him or her with a copy of Form ES-D-1. This review should familiarize the operator with the sequence of events to ensure that they will proceed as planned. This is particularly important if the simulator operator during the test is not the same individual who assisted in validating the scenarios.

f.   The examiners should identify important plant parameters to be monitored during each simulator scenario. The chief examiner should ask the simulator operator to record selected parameters, if possible, on the facility's safety parameter display system(s). Parameter readings should be collected at meaningful intervals, depending on the parameter, the nature of the event, and the capability of the simulation facility. The chief examiner should retain the recordings as backup documentation to augment the notes taken by the examiners during the simulator test.

g.   The examiner in charge of each scenario should arrange a suitable communication system with the simulator operator so that he or she can be prompted to insert the malfunctions without cuing the applicants. Malfunctions may be planned for a predetermined time or power level so that the examiners and the facility operator are aware of the event that is occurring or pending.

   If necessary, the examiners may use time compression to speed up the response of key parameters so that the scenario can proceed to the next event within a reasonable time. Time compression is acceptable as long as it is used judiciously and the operators are given sufficient time to perform the tasks that they would typically perform in real time. If the examiners intend to use time compression, they should inform the applicants of that fact during the operating test briefing (refer to Section D.1.a). The examiners should also mitigate the potential for negative training by debriefing the applicants after any scenario in which time compression was used.

h.   Before beginning each scenario set, the examiners should have the simulator operator advance any control room strip chart recorders that may prove useful in recreating the sequence of events. The charts should be clearly marked with the date, time, and examiner's initials so that they can be accurately matched with the correct operating crew.

i.   The chief examiner should ensure that the simulator operator (or examiner) playing the role of other plant personnel is aware of the time scale for responding to the applicants' requests for information. For example, fast-time could be specified for auxiliary operator checks or lineups to prevent long delays in simulated operations, while maintenance and chemistry sample information can be provided with normal time delays to present the applicants with the same analysis problems that they will face as operators.

j.   Before the simulator test begins, the examiners shall caution the simulator operator to provide only information that is specifically requested by the applicants and does not compromise the integrity of the examination. When the simulator operator is briefing the applicants or communicating with them on the telephone, the examiners should monitor the conversations to ensure that the information provided is appropriate and does not cue the applicants.

k.   Before the simulator test begins, the facility instructor (or examiner) will provide a shift turnover briefing. The briefing will cover present plant conditions, power history, equipment out of service, abnormal conditions, surveillances due, and instructions for the shift, and the applicants will be given time to familiarize themselves with the plant status.

l.   The operating team or crew (including license applicants and surrogates, if applicable) should perform peer checks in accordance with the facility licensee's operations and training procedures and practices. NRC examiners will not perform this function. If an applicant begins to make an error that is corrected by a peer checker, the applicant will be held accountable for the consequences of the potential error without regard to mitigation by the crew.

m.   Each examiner should use the expected actions and behaviors listed on Form ES-D-2 as a guide while administering the simulator tests. If an applicant performs as expected, the examiner may simply note in the left-hand column of the form the time when the expected actions occurred. However, if an applicant does not perform as expected, the examiner should note the applicant's actions (or lack thereof) next to or below the expected action and follow up with appropriate questions after the simulator scenario is completed (refer to Section D.3.n).

Each examiner must determine the best way to document the applicant's actions. Some examiners record a minute-by-minute account of all key plant events and applicant actions as they occur; other examiners record only the applicant's significant actions. Each individual examiner should develop his or her own examination documentation technique; however, the documentation must provide an adequate basis for a licensing decision. In addition, the examiner's notes must provide sufficient information to allow the examiner to confidently assess the applicant's performance on the competencies described in Appendix D.

n. Examiners shall limit discussions with the applicants during the scenarios both to maintain realism and to avoid distracting the applicants from operating the plant. The examiners' questions during the scenarios should be limited to those that are necessary to assess the applicants' understanding of plant conditions and the required operator actions. Whenever possible, the examiner shall defer questioning the applicant until a time when the applicant is not operating or closely monitoring the plant (preferably after the simulator has been placed in "freeze"). The examiner's followup questions or concerns can generally be addressed during a brief question-and-answer period after each scenario or during the control room systems and facility walk-through portion of the operating test if it is performed after the simulator test.

o. The examiners who administer the simulator test shall confer immediately after completing the scenario set to compare notes and verify that each examiner observed his or her applicant performing the required number of transients and events in a manner sufficient to justify a proper evaluation of all required competencies. If necessary, the examiners shall run an additional scenario to ensure that all required evolutions and competencies are covered. For example, if an applicant has only one opportunity to demonstrate competence on a particular rating factor, but makes an error that does not affect his or her performance of a critical task, the examiners shall give the applicant another opportunity to demonstrate competence or to make a second error that would justify an unsatisfactory score for the subject rating factor (refer Section D.2.b of ES-303 for detailed simulator grading instructions). All scenarios will be planned and documented in accordance with Section D of ES-301.

The chief examiner shall ensure that the examiners' observations are consistent and their findings are mutually supportive. If a performance deficiency is "shared" by more than one applicant, both evaluating examiners should note the deficiency. Ideally, this cross-check should be accomplished as soon as possible after running the scenarios while still at the facility. The cross-check must be accomplished before finalizing the examination results in accordance with ES-303.

p. If the applicants did not perform as expected, the examiner shall ask the simulator operator to provide copies of the logs, charts, and other materials that may be required after leaving the facility to evaluate and document the applicants' performance. The examiner of record shall retain all documentation related to any operating test failure until the proposed denial becomes final or a license is issued.

The chief examiner should also ask the simulator operator to retain copies of the same materials until all applicants are licensed or all appeals are settled, as suggested in the sample corporate notification letter shown in Attachment 4 to ES-201.

q. If the simulation facility should become inoperable and cause excessive delay of the operating tests, the chief examiner should discuss the situation with the facility licensee and the responsible regional supervisor so that management can make a decision regarding the conduct of the operating tests; it may be necessary to reschedule the simulator examinations for a later date.

The simulator should be considered inoperable under any of the following conditions:

- The simulator exhibits a mass/energy imbalance, erratic logic, or inexplicable panel indications during model execution.

- The simulator exhibits unplanned and unexplained events or malfunctions that cause the applicants to divert from the expected responses and success path of the planned scenario.

- The simulator automatically goes to the "freeze" state during a scenario, or a "beyond simulated limits" alarm is received on the instructor's station.

- The simulator instructor informs the examination team that a software module has halted or "kicked out."

Occurrence of any of these abnormal simulator operating conditions during an examination constitutes sufficient cause to stop the scenario. Evaluations of the applicants' performance during any of these simulator malfunctions may be unreliable.

When the simulator has been restored to full operability, the chief examiner will determine whether the scenario requires replacement, may be resumed in progress, or may be restarted from the beginning. Examiners will not use the "backtrack" function when restarting a scenario; the simulator must be in a stable plant condition, at a definitive procedural step, before conducting a turnover as discussed in Item D.3(k), above.

ES-303
DOCUMENTING AND GRADING INITIAL OPERATING TESTS

A.   Purpose

This standard describes the procedures for documenting all categories of the operating test, collating the data to arrive at a pass or fail recommendation, and reviewing the documentation to ensure quality.

B.   Background

This standard assumes that the operating test was prepared and administered in accordance with ES-301, "Preparing Initial Operating Tests," and ES-302, "Administering Operating Tests to Initial License Applicants," respectively. The procedures contained herein require the examiner to evaluate each applicant's performance on the operating test and make a judgement as to whether the applicant's level of knowledge and understanding meet the minimum requirements to safely operate the facility for which the license is sought. The examiner evaluates each noted deficiency in light of the total breadth of knowledge and ability (K/A) demonstrated by the applicant in that subject area.

C.   Responsibilities

1.   Facility Licensee

The facility licensee's responsibilities are limited to providing the NRC examiners with whatever additional reference materials and information the examiner might require to evaluate the applicants' performance on the operating tests. Such materials might include simulator strip chart recordings that document plant status during the simulator scenarios, and procedures that document the expected operator actions.

2.   NRC Examiner of Record

As soon as possible after administering the test, the examiner of record shall review, evaluate, and finalize each applicant's operating test documentation in accordance with the instructions in Section D.

If an applicant made an error with *serious safety consequences,* the examiner may recommend an operating test failure even if the grading instructions in Section D would normally result in a passing grade. Conversely, if an applicant made a number of errors with minimal or no safety consequences, the examiner may recommend that the applicant be granted a license even if the grading instructions in Section D would normally result in a failing grade. However, in either case, the examiner shall thoroughly justify and document the basis for the recommendation in accordance with Section D.3. Moreover, the NRC's regional office shall obtain written concurrence from the NRR operator licensing program office before completing the licensing action.

3. **NRC Chief Examiner**

a. The chief examiner shall arrange a meeting of the NRC's examination team members after the simulator scenarios are complete. Such meetings enable the examiners to compare notes to ensure that the documentation for applicants on the same operating crew is consistent and mutually supportive.

b. The chief examiner shall work with the other examiners on the team to resolve any technical questions that might arise during the grading process, and communicate any additional reference material requirements to the facility contact.

c. The chief examiner or a management-approved designee will review the grading of each operating test to verify that the examiner's comments appropriately support his or her recommendation and to ensure that the operating test meets the requirements of ES-301. If the chief examiner or designee does not agree with any of the examiner's recommendations, he or she shall confer with the examiner before overturning the recommendation. Such disagreements are not common and usually arise because an unsatisfactory grade is not adequately justified. It is, therefore, very important for examiners to be complete and accurate in their grading and documentation.

d. The chief examiner or designee shall make an independent pass or fail recommendation, sign the "Final Recommendation" block on Form ES-303-1, "Individual Examination Report," and forward the package to the responsible supervisor for review in accordance with ES-501, "Initial Post-Examination Activities." The supervisor must concur in any recommendation to overturn the examiner's results, and the specific reasons for this action must be explained on Form ES-303-2, "Operating Test Comments" (or equivalent).

D. Grading and Documentation Instructions

1. Review and Categorize Rough Notes and Documentation

a. Review the walk-through job performance measures (JPMs) and simulator scenarios that were performed and the performance-based followup questions that were asked. Evaluate all rough notes and documentation generated while administering the operating test to determine the areas in which the applicant was deficient. If the applicant generated or used any material (such as figures, drawings, flowcharts, or forms) during the operating test, the material may be used to aid in documenting the applicant's performance. If it contributes to an unsatisfactory performance evaluation, the material shall be appropriately marked and cross-referenced to the applicable deficiency and attached to the examination package for retention.

b.   Verify the validity and technical accuracy of any performance-based questions that were asked during the operating test, as well as any unexpected events or actions that occurred during the simulator operating test. If necessary, work through the chief examiner to obtain any additional reference material that might be required to resolve any technical questions.

c.   On the rough notes and documentation, label or highlight *every* action, response, note, or comment that may constitute a performance deficiency.

d.   Review each simulator operating test performance deficiency. Using as a guide the competency and rating factor descriptions in Appendix D and on Form ES-303-3 (RO) or Form ES-303-4 (SRO), code each deficiency with the number and letter of the rating factor(s) it most accurately reflects (e.g., 4.a). Whenever possible, attempt to identify the root cause of the applicant's deficiencies and code each deficiency with no more than two different rating factors. However, one significant deficiency may be coded with additional rating factors if the error can be shown, consistent with the criteria in Section D.3.b, to be relevant to each of the cited rating factors.

As stated in ES-302, it is essential that the simulator operating test documentation is consistent and mutually supportive for all applicants in an operating crew. Operating errors that involved more than one applicant should be noted by each applicant's evaluating examiner. If the examination team members do not have the opportunity to discuss and compare their observations before leaving the site, the chief examiner shall schedule a conference call after the examiners return to their respective offices.

2.   **Evaluate the Applicant's Performance**

After categorizing and coding the rough notes, review, evaluate, and grade the applicant's performance, as follows:

a.   *The "Walk-Through"*

On page 2 of the applicant's Form ES-303-1, enter the titles of the JPMs examined during the "Administrative Topics" and "Control Room and In-Plant Systems" portions of the walk-through test.

To determine a grade for each administrative and systems JPM listed on Form ES-303-1, evaluate each deficiency highlighted in the rough notes. If the following criteria are met, assign a satisfactory grade by placing an "S" in the "Evaluation" column for that JPM; otherwise enter a "U":

•   Time-critical JPMs must be completed within the allotted time. All other JPMs should normally be completed within twice the validated time estimate (refer to Section D.2.f of ES-302). The reason for terminating any JPM shall be documented in accordance with Section D.3, below.

- The task standard for the JPM must be accomplished by correctly completing all of the critical steps.

  If the applicant initially missed a critical step, but later performed it correctly and accomplished the task standard without degrading the condition of the system or the plant, the applicant's performance on that JPM should be graded as satisfactory. However, the applicant's error shall be documented in accordance with Section D.3, below.

- The responses to any performance-based followup questions asked pursuant to Section D.2.f of ES-302 must confirm that the applicant's understanding of the administrative topic/system/JPM is satisfactory.

  If the responses to any of the followup questions reveal that the applicant's understanding of the administrative topic/system/JPM is seriously deficient, the examiner may recommend an unsatisfactory grade for the administrative topic/system even though the applicant successfully completed the task standard for the JPM. The basis for the recommendation shall be thoroughly justified and documented in accordance with Section D.3, below.

  Conversely, if the applicant did not accomplish the task standard and followup questioning revealed that the failure was caused by a deficiency in the procedure or some other factor beyond the applicant's control, the examiner may still recommend a satisfactory grade for the administrative topic/system/JPM. Once again, the basis for the recommendation shall be thoroughly justified and documented in accordance with Section D.3, below.

After grading the applicant's performance on each of the administrative topics/systems, determine an overall grade for the "walk-through" by calculating the percentage of satisfactory topic/system grades. If the applicant has an "S" on fewer than 80 percent of the topics/systems (i.e., 12/15 for RO and SRO-I applicants and 8/10 for SRO-U applicants), the applicant fails the "walk-through" and receives a "U" overall.

Additionally, in order to ensure minimal competence in the administrative area, determine a separate "Administrative Topics" grade by calculating the percentage of satisfactory grades for the administrative JPMs. If an SRO applicant has an "S" on fewer than 60 percent (i.e., 3/5) or an RO applicant has an "S" on fewer than 50 percent (2/4) of the administrative topics/JPMs, the applicant fails this portion of the walk-through. Retake applicants who were granted a waiver of the systems walk-through pursuant to ES-204, "Processing Waivers Requested by Reactor Operator and Senior Reactor Operator Applicants," must achieve a satisfactory grade on at least 80 percent of the topics/JPMs (i.e., 4/5 for RO and SRO applicants) to pass.

Document the applicant's grades by placing an "S" or a "U" in the appropriate blocks in the "Operating Test Summary" on page 1 of Form ES-303-1. Enter "W" if any part of the walk-through was waived in accordance with ES-204. Document and justify every deficiency in accordance with Section D.3, below.

b.   *The "Simulator Operating Test"*

Using Form ES-303-3 or ES-303-4, depending on the applicant's license level, and the following generic guidance, evaluate any deficiencies coded for the simulator test to determine a grade for every applicable rating factor (RF) and competency. Keep in mind that the simulator test is generally graded based on competencies rather than consequences; every error that reflects on an operator's competence is considered equal unless it is related to the performance of a critical task (as determined in accordance with ES-301 and Appendix D).

- If there is no basis upon which to grade a rating factor (i.e., it is "not observed"), circle the "0" "Weighting Factor," enter an "RF Grade" of "N/O," and explain in accordance with Section D.3, below. Depending upon which RF is "N/O," circle the appropriate "Weighting Factor" for each remaining RF applicable to that competency; the "Weighting Factors" for each competency must always add up to "1." If more than one rating factor per competency or more than two rating factors overall are not observed, inform the NRC's regional office management and consult the NRR operator licensing program office to determine whether the test supports a licensing decision. As discussed in ES-301, Competency 3 is optional for SRO upgrade applicants and may be scored as "N/O." However, the examiner shall evaluate Competency 3 if the applicant rotated into an operating crew position that required the applicant to manipulate the controls.

- If an applicant performs activities related to a rating factor and makes no errors, circle an "RF Score" of "3" for that rating factor.

- If an applicant makes a single error related to a rating factor, circle an "RF Score" of "2" for that rating factor, unless the error related to a critical task, in which case a score of "1" would be required. Missing a critical task does not necessarily mean that the applicant will fail the simulator test, nor does success on every critical task prevent the examiner from recommending a failure if the applicant had other deficiencies that, in the aggregate, justify the failure based on the competency evaluations.

- If an applicant makes two errors related to a rating factor, circle an "RF Score" of "1" for that rating factor unless a score of "2" can be justified (and documented as discussed in Section D.3, below) based on correctly performing another activity (or activities) related to the same rating factor; three or more errors generally require a score of "1," regardless of the applicant's compensatory actions.

Multiply each "RF Score" by its associated "Weighting Factor" to obtain a numerical measure ("RF Grade") for the applicant's performance on each rating factor. Then sum the RF Grades to obtain a "Competency Grade" for each competency and enter the corresponding numbers (or "N/O," as appropriate) on page 3 of the RO or SRO applicant's Form ES-303-1.

For each competency on page 3 of Form ES-303-1, sum the rating factor grades and enter the resulting competency grade in the designated column. (The grades should range between 1 and 3.)

Using the following evaluation criteria, determine whether the applicant's overall performance on the simulator test is satisfactory or unsatisfactory, and document the grade by placing an "S" or a "U" in the "Simulator Operating Test" block of the "Operating Test Summary" on page 1 of Form ES-303-1. Enter "W" if this part of the operating test was waived in accordance with ES-204.

- If the grade for *all* competencies is greater than 1.8, the applicant's performance is generally satisfactory.

- If the grade for Competency 4, "Communications and Crew Interactions," is less than or equal to 1.8 but greater than 1.0, *and* the individual grades for *all* other competencies are 2.0 or greater, the applicant's performance is satisfactory.

- If the grade for Competency 4 is 1.0, *or* the grade for any other competency is 1.8 or less, the applicant's performance is unsatisfactory.

Note that Competency 3, "Control Board Operations," is optional for SRO upgrade applicants. However, if it is evaluated, it shall be factored into the applicant's final grade.

Document and justify every deficiency in accordance with Section D.3, below.

3.   Finalize the Documentation

a.   Review and finalize the simulator scenarios that were run during the operating test.

Complete Form ES-D-1, "Scenario Outline," by entering the applicants' names, the positions they occupied during the scenario, and the facility's name on the top of the form. Also enter on Form ES-D-1 any scenario revisions that were made during the test, so that each form accurately shows all of the events that actually occurred during each scenario. Change the event numbers, malfunction numbers, malfunction types, and descriptions, as necessary, to reflect the "as run" conditions. These changes may be made using pen-and-ink or by retyping the scenario, provided that the final form is clear and legible.

Update each Form ES-D-2, "Required Operator Actions," to reflect the "as run" conditions. Discard or mark as "not used" any events that were not run, and fill out new forms for any events that were run but not originally planned. Neatly enter notes, comments, and additional actions in the spaces between the expected operator actions.

The final Forms ES-D-1 and ES-D-2 must be a clear, legible, and sequential record of the actual events and actions that occurred during the simulator operating test. The forms sent to the applicant shall not contain any rough notes or irrelevant comments.

Any events or malfunctions that did not function as expected or were not useful in evaluating the applicants (e.g., a surveillance test that required a long time to perform) should be noted on the master copy of the scenarios to aid in future scenario preparation.

b. Review the applicant's Form ES-303-1 and the rough documentation. Justify *in detail* on Form ES-303-2, "Operating Test Comments" (or equivalent), every knowledge or ability deficiency that contributed to a failure in any part of the operating test. Provide the following specific information, as applicable:

* the task administered (i.e., describe the JPM or simulator scenario and event, as well as the applicant's position on the operating crew)

* the applicant's incorrect action and an indication of whether the action was a JPM critical step or a simulator critical task

* the lack of knowledge or ability that the applicant demonstrated

* the potential or actual consequences of the applicant's incorrect action (particularly if the examiner recommends a failure based on a serious error that would not normally result in a failing grade)

* any for-cause followup questions asked and the applicant's responses

* the correct answer or action, with an appropriate facility reference (e.g., lesson plan, system description, procedure name and number)

* the K/A number and its importance rating (as given in NUREG-1122 or NUREG-1123) and the facility's learning objective

* the item from Title 10, Section 55.45(a), of the *Code of Federal Regulations* [10 CFR 55.45(a)] that the applicant did not understand or was unable to perform

General statements (such as "did not know decay heat removal system") are not adequate.

Whenever possible, substantiate comments with printouts or strip chart recordings generated during the simulator operating test and drawings and illustrations generated by the applicant.

c.   Deficiencies that do not contribute to an operating test failure shall also be documented; however, a brief statement describing the error and the expected action or response is generally sufficient. Examiners should keep in mind that their licensing recommendation and associated documentation are subject to review by the NRC's chief examiner and regional office management. Therefore, the documentation should contain sufficient detail so that the independent reviewer, responsible supervisor, and licensing official can make a logical decision in support of the examiner's recommendation to deny or issue the license.

d.   As noted in Section D.2, above, deviations from the nominal grading criteria must be explained in detail. For example, an examiner may conclude that an applicant's performance is acceptable despite exhibiting deficiencies that would normally result in an unsatisfactory grade (e.g., committing two or more errors related to the same simulator rating factor or failing to accomplish the task standard for a JPM). Conversely, an examiner may conclude that an applicant's performance is unacceptable even though the documented deficiencies would normally result in a passing grade. In either case, the examiner shall document the basis for concluding that the applicant is, in fact, (un)acceptably proficient in that area, why the nominal grading criteria might be too (lenient) severe, and/or how a flaw in the test item might have contributed to the applicant's deficient performance. Moreover, as noted in Section C.2, the NRC's regional office shall obtain written concurrence from the NRR operator licensing program office before completing the licensing action.

Any simulator rating factor that is graded as "not observed" must also be explained in the documentation (e.g., did the simulator malfunction, did an event not take place as planned, or did another applicant intercede?).

e.   Retain rough documentation until the NRC's chief examiner and regional office management have reviewed the examiner's recommendations and concurred in the results (refer to ES-501). Examiners shall retain all applicable notes and documentation associated with proposed denials until the denials become final. Examiners are advised that such notes would be subject to disclosure if requested under the Freedom of Information Act.

f.   Cross-reference each comment on Form ES-303-2 with the specific task, subject, or competency rating factor to which it applies on the applicant's Form ES-303-1. Do this by entering the applicable reference from Form ES-303-1 (e.g., Admin-a, Systems-d, or Simulator-1.c) in the left-hand column of Form ES-303-2, and entering the page number on which the comment is found in the appropriate block on Form ES-303-1.

4.   Make a Final Recommendation

    a.   After grading and documenting the operating test, make an overall recommendation by checking the "Pass" or "Fail" (or "Waive" if the entire operating test was waived in accordance with ES-204) block, and signing and dating the "Examiner Recommendations" section on the applicant's Form ES-303-1. Make a "Pass" recommendation only if *all* summary blocks of the operating test contain satisfactory ("S") grades or the letter "W," indicating that the applicant was not examined in that area.

    b.   Assemble the operating test package (including Forms ES-303-1, ES-303-2, ES-D-1, and ES-D-2 and all supporting documentation such as strip chart recordings and applicant notes and drawings) for each applicant and forward the package to the chief examiner for review in accordance with ES-501.

E.   Attachments/Forms

Form ES-303-1,   "Individual Examination Report"
Form ES-303-2,   "Operating Test Comments"
Form ES-303-3,   "RO Competency Grading Worksheet for the Simulator Test"
Form ES-303-4,   "SRO Competency Grading Worksheet for the Simulator Test"

PRIVACY ACT INFORMATION — FOR OFFICIAL USE ONLY

U.S. Nuclear Regulatory Commission
Individual Examination Report

Applicant's Name				Docket Number 55-			
I	R	Examination Type (Initial or Retake)		Facility Name			
		Reactor Operator		Facility Description			Hot
		Senior Reactor Operator (SRO) Instant					Cold
		SRO Upgrade					BWR
		SRO Limited to Fuel Handling					PWR

Written Examination Summary		
NRC Author/Reviewer	RO/SRO/Total Exam Points	___ / ___ / ___
NRC Grader/Reviewer	Applicant Points	___ / ___ / ___
Date Administered	Applicant Grade (%)	___ / ___ / ___
Operating Test Summary		
Administered by	Date Administered	
Walk-Through (Overall)		
Administrative Topics		
Simulator Operating Test		
Examiner Recommendations		

Check Blocks	Pass	Fail	Waive	Signature	Date
Written Examination					
Operating Test					
Final Recommendation					

License Recommendation			
	Issue License	Supervisor's Signature	Date
	Deny License		

PRIVACY ACT INFORMATION — FOR OFFICIAL USE ONLY

PRIVACY ACT INFORMATION — FOR OFFICIAL USE ONLY

Applicant Docket Number: 55-		Page of
Walk-Through Grading Details	**Evaluation (S or U)**	**Comment Page Number**
Administrative Topics		
a.		
b.		
c.		
d.		
e.		
Systems — Control Room		
a.		
b.		
c.		
d.		
e.		
f.		
g.		
h.		
Systems — In-Plant		
i.		
j.		
k.		

PRIVACY ACT INFORMATION — FOR OFFICIAL USE ONLY

PRIVACY ACT INFORMATION — FOR OFFICIAL USE ONLY

Applicant Docket Number: 55-				Page of	
Reactor Operator Simulator Operating Test Grading Details					
Competencies/ Rating Factors (RFs)	RF Weights	RF Scores	RF Grades	Comp. Grades	Comment Page No.

1. Interpretation/Diagnosis
  a. Recognize & Verify Status
  b. Interpret & Diagnose Conditions
  c. Prioritize Response
	___
	___
	___
		___
		___
		___
			___
			___
			___
				___	___
					___
					___

2. Procedures/Tech Specs
  a. Reference
  b. Procedure Compliance
  c. Tech Spec Entry
	___
	___
	___
		___
		___
		___
			___
			___
			___
				___	___
					___
					___

3. Control Board Operations
  a. Locate & Manipulate
  b. Understanding
  c. Manual Control
	___
	___
	___
		___
		___
		___
			___
			___
			___
				___	___
					___
					___

4. Communications
  a. Provide Information
  b. Receive Information
  c. Carry Out Instructions
	___
	___
	___
		___
		___
		___
			___
			___
			___
				___	___
					___
					___

[Note: Enter RF Weights (nominal, adjusted, or "0" if not observed (N/O)), RF Scores (1, 2, 3, or N/O), and RF Grades from Form ES-303-3 and sum to obtain Competency Grades.]

PRIVACY ACT INFORMATION — FOR OFFICIAL USE ONLY

PRIVACY ACT INFORMATION — FOR OFFICIAL USE ONLY

Applicant Docket Number: 55-					Page of
Senior Reactor Operator Simulator Operating Test Grading Details					
Competencies/ Rating Factors (RFs)	RF Weights	RF Scores	RF Grades	Comp. Grades	Comment Page No.
1. Interpretation/Diagnosis a. Recognize & Attend b. Ensure Accuracy c. Understanding d. Diagnose	_____ _____ _____ _____	_____ _____ _____ _____	_____ _____ _____ _____	____	_____ _____ _____ _____
2. Procedures a. Reference b. EOP Entry c. Correct Use	_____ _____ _____	_____ _____ _____	_____ _____ _____	____	_____ _____ _____
3. Control Board Operations a. Locate & Manipulate b. Understanding c. Manual Control	_____ _____ _____	_____ _____ _____	_____ _____ _____	____	_____ _____ _____
4. Communications a. Clarity b. Crew & Others Informed c. Receive Information	_____ _____ _____	_____ _____ _____	_____ _____ _____	____	_____ _____ _____
5. Directing Operations a. Timely & Decisive Action b. Oversight c. Solicit Crew Feedback d. Monitor Crew Activities	_____ _____ _____ _____	_____ _____ _____ _____	_____ _____ _____ _____	____	_____ _____ _____ _____
6. Technical Specifications a. Recognize and Locate b. Compliance	_____ _____	_____ _____	_____ _____	____	_____ _____

[Note: Enter RF Weights (nominal, adjusted, or "0" if not observed (N/O)), RF Scores (1, 2, 3, or N/O), and RF Grades from Form ES-303-4 and sum to obtain Competency Grades.]

PRIVACY ACT INFORMATION — FOR OFFICIAL USE ONLY

PRIVACY ACT INFORMATION — FOR OFFICIAL USE ONLY

Applicant Docket Number: 55-	Page of
Form ES-303-1 Cross-Reference	Comments

PRIVACY ACT INFORMATION — FOR OFFICIAL USE ONLY

1. Interpret/Diagnose Events and Conditions Based on Alarms, Signals, and Readings

Rating Factors	Weighting Factors		RF Scores	RF Grades	Comp. Grade
(a) Did the applicant RECOGNIZE and VERIFY off-normal trends and status?	N/O	= 0	3		
	Nominal	**= 0.40**	2		
	(b) or (c) N/O	= 0.57	1		
(b) Did the applicant correctly INTERPRET/DIAGNOSE plant conditions based on control room indications?	N/O	= 0	3		
	Nominal	**= 0.30**	2		———
	(c) N/O	= 0.43	1		
	(a) N/O	= 0.50			
(c) Did the applicant ATTEND TO annunciators, alarm signals, and instrument readings in order of importance and severity?	N/O	= 0	3		
	Nominal	**= 0.30**	2		
	(b) N/O	= 0.43	1		
	(a) N/O	= 0.50			

2. Comply with and Use Procedures, References, and Technical Specifications

Rating Factors	Weighting Factors		RF Scores	RF Grades	Comp. Grade
(a) Did the applicant REFER TO the appropriate procedure or reference in a timely manner?	N/O	= 0	3		
	Nominal	**= 0.30**	2		
	(c) N/O	= 0.43	1		
	(b) N/O	= 0.50			
(b) Did the applicant COMPLY WITH procedures (including precautions and limitations) and references in an accurate and timely manner?	N/O	= 0	3		
	Nominal	**= 0.40**	2		———
	(a) or (c) N/O	= 0.57	1		
(c) Did the applicant RECOGNIZE plant conditions that are addressed in technical specifications?	N/O	= 0	3		
	Nominal	**= 0.30**	2		
	(a) N/O	= 0.43	1		
	(b) N/O	= 0.50			

3. Operate the Control Boards

Rating Factors	Weighting Factors		RF Scores	RF Grades	Comp. Grade
(a) Did the applicant LOCATE AND MANIPULATE controls in an accurate and timely manner?	N/O	= 0	3		
	Nominal	**= 0.40**	2		
	(b) or (c) N/O	= 0.57	1		
(b) Did the applicant's actions demonstrate UNDERSTANDING OF SYSTEM OPERATION, including set points, interlocks, and automatic actions?	N/O	= 0	3		
	Nominal	**= 0.30**	2		————
	(c) N/O	= 0.43	1		
	(a) N/O	= 0.50			
(c) Did the applicant demonstrate the ability to take MANUAL CONTROL of automatic functions?	N/O	= 0	3		
	Nominal	**= 0.30**	2		
	(b) N/O	= 0.43	1		
	(a) N/O	= 0.50			

4. Communicate and Interact with Other Crew Members

Rating Factors	Weighting Factors		RF Scores	RF Grades	Comp. Grade
(a) Did the applicant PROVIDE clear and accurate INFORMATION on system status to others for the performance of their jobs?	N/O	= 0	3		
	Nominal	**= 0.34**	2		
	(b) or (c) N/O	= 0.50	1		
(b) Did the applicant effectively RECEIVE INFORMATION from others (including requesting, acknowledging, and attending to information)?	N/O	= 0	3		
	Nominal	**= 0.33**	2		————
	(a) or (c) N/O	= 0.50	1		
(c) Did the applicant successfully CARRY OUT THE INSTRUCTIONS of the supervisor?	N/O	= 0	3		
	Nominal	**= 0.33**	2		
	(a) or (b) N/O	= 0.50	1		

1. Interpret/Diagnose Events and Conditions Based on Alarms, Signals, and Readings

Rating Factors	Weighting Factors		RF Scores	RF Grades	Comp. Grade
(a) Did the applicant RECOGNIZE AND ATTEND TO off-normal trends and status in order of their importance and severity?	N/O	= 0	3		
	Nominal	**= 0.20**	2		
	(b) N/O	= 0.25	1		
	(c) or (d) N/O	= 0.29			
(b) Did the applicant ensure the collection of CORRECT, ACCURATE, and COMPLETE information and reference material on which to base diagnoses?	N/O	= 0	3		
	Nominal	**= 0.20**	2		
	(a) N/O	= 0.25	1		
	(c) or (d) N/O	= 0.28			_____
(c) Did the applicant's directives and actions demonstrate an UNDERSTANDING of how the PLANT, SYSTEMS, and COMPONENTS OPERATE AND INTERACT (including set points, interlocks, and automatic actions)?	N/O	= 0	3		
	Nominal	**= 0.30**	2		
	(a) or (b) N/O	= 0.38	1		
	(d) N/O	= 0.43			
(d) Did the applicant correctly INTERPRET/DIAGNOSE plant conditions based on control room indications?	N/O	= 0	3		
	Nominal	**= 0.30**	2		
	(a) or (b) N/O	= 0.37	1		
	(c) N/O	= 0.43			

2. Comply with and Use Procedures and References

Rating Factors	Weighting Factors		RF Scores	RF Grades	Comp. Grade
(a) Did the applicant REFER to correct procedures, procedural steps, and references when appropriate?	N/O	= 0	3		
	Nominal	**= 0.30**	2		
	(b) N/O	= 0.43	1		
	(c) N/O	= 0.50			
(b) Did the applicant RECOGNIZE EOP ENTRY CONDITIONS?	N/O	= 0	3		
	Nominal	**= 0.30**	2		
	(a) N/O	= 0.43	1		_____
	(c) N/O	= 0.50			
(c) Did the applicant USE PROCEDURES CORRECTLY, including following procedural steps in correct sequence, abiding by procedural cautions and limitations, selecting correct paths on decisions blocks, and correctly transitioning between procedures?	N/O	= 0	3		
	Nominal	**= 0.40**	2		
	(a) or (b) N/O	= 0.57	1		

3. Operate the Control Boards
[NOTE: This competency is optional for SRO-upgrade applicants; refer to Section D.2.b.]

	Rating Factors	Weighting Factors		RF Scores	RF Grades	Comp. Grade
(a)	Did the applicant LOCATE AND MANIPULATE CONTROLS in an accurate and timely manner?	N/O	= 0	3		
		Nominal	**= 0.34**	2		
		(b) or (c) N/O	= 0.5	1		
(b)	Did the applicant's control manipulations demonstrate an UNDERSTANDING OF SYSTEM OPERATION, including set points, interlocks, and automatic actions?	N/O	= 0	3		———
		Nominal	**= 0.33**	2		
		(a) or (c) N/O	= 0.5	1		
(c)	Did the applicant demonstrate the ability to take MANUAL CONTROL of automatic functions?	N/O	= 0	3		
		Nominal	**= 0.33**	2		
		(a) or (b) N/O	= 0.5	1		

4. Communicate and Interact with the Crew and Other Personnel

	Rating Factors	Weighting Factors		RF Scores	RF Grades	Comp. Grade
(a)	Did the applicant communicate in a clear, easily understood manner?	N/O	= 0	3		
		Nominal	**= 0.4**	2		
		(c) N/O	= 0.5	1		
		(b) N/O	= 0.67			
(b)	Did the applicant keep crew members and those outside the control room informed of plant status?	N/O	= 0	3		———
		Nominal	**= 0.4**	2		
		(c) N/O	= 0.5	1		
		(a) N/O	= 0.67			
(c)	Did the applicant ENSURE RECEIPT of clear, easily-understood communications from crew and others?	N/O	= 0	3		
		Nominal	**= 0.2**	2		
		(a) or (b) N/O	= 0.33	1		

5.   Direct Shift Operations

Rating Factors		Weighting Factors		RF Scores	RF Grades	Comp. Grade
(a)	Did the applicant take TIMELY AND DECISIVE ACTION that demonstrated appropriate CONCERN for the SAFETY of the plant, staff, and public?	N/O	= 0	3		
		Nominal	**= 0.30**	2		
		(c) or (d) N/O	= 0.38	1		
		(b) N/O	= 0.43			
(b)	Did the applicant remain ATTENTIVE to control room indications, stay in a position of OVERSIGHT, and provide an APPROPRIATE AMOUNT of DIRECTION and GUIDANCE that facilitated CREW PERFORMANCE?	N/O	= 0	3		
		Nominal	**= 0.30**	2		
		(c) or (d) N/O	= 0.37	1		
		(a) N/O	= 0.43			
(c)	Did the applicant SOLICIT and INCORPORATE FEEDBACK from the crew to foster an effective, team-oriented approach to problem solving and decision making?	N/O	= 0	3		
		Nominal	**= 0.20**	2		
		(d) N/O	= 0.25	1		
		(a) or (b) N/O	= 0.29			
(d)	Did the applicant ensure that CORRECT AND TIMELY ACTIVITIES (including diagnosis, procedural implementation, and control board operations) were carried out BY THE CREW?	N/O	= 0	3		
		Nominal	**= 0.20**	2		
		(c) N/O	= 0.25	1		
		(a) or (b) N/O	= 0.28			

6.   Comply with and Use Technical Specifications (TS)

Rating Factors		Weighting Factors		RF Scores	RF Grades	Comp. Grade
(a)	Did the applicant RECOGNIZE when conditions were covered by the TS and LOCATE the appropriate TS?	N/O	= 0	3		
		Nominal	**= 0.4**	2		
		(b) N/O	= 1.0	1		
(b)	Did the applicant ensure correct COMPLIANCE with TS and LCO action statements?	N/O	= 0	3		
		Nominal	**= 0.6**	2		
		(a) N/O	= 1.0	1		

ES-401
PREPARING INITIAL SITE-SPECIFIC WRITTEN EXAMINATIONS

A. Purpose

This standard specifies the requirements, procedures, and guidelines for preparing site-specific written examinations for the initial licensing of reactor operator (RO) and senior reactor operator (SRO) applicants at power reactor facilities.

B. Background

The content of the written licensing examinations for ROs and SROs is dictated by Title 10, Sections 55.41 and 55.43, of the *Code of Federal Regulations* (10 CFR 55.41 and 55.43), respectively. Each examination shall contain a representative selection of questions concerning the knowledge, skills, and abilities (K/As) needed to perform duties at the desired license level. Both the RO and SRO examinations will sample the 14 items specified in 10 CFR 55.41(b), and the SRO examination will sample the 7 additional items specified in 10 CFR 55.43(b). Given that SRO-U (upgrade) applicants previously passed an RO licensing examination covering the topics specified in 10 CFR 55.41(b), they may apply for a waiver of the RO portion of the SRO written examination pursuant to 10 CFR 55.47. (Refer to ES-204, "Processing Waivers Requested by Reactor Operator and Senior Reactor Operator Applicants.")

The written operator licensing examination is administered in two sections, including a generic fundamentals examination (GFE) and a site-specific examination. The GFE covers those K/As that do not vary significantly among reactors of the same type (i.e., pressurized- or boiling-water) and is generally administered early in the license training process. (For a description of the program, refer to ES-205, "Procedure for Administering the Generic Fundamentals Examination Program.") The instructions in this standard apply only to the site-specific examination.

Except as noted in Section D.1.b of this examination standard (ES), the "Knowledge and Abilities Catalog[s] for Nuclear Power Plant Operators: Pressurized- [and Boiling-] Water Reactors" (NUREG-1122 and 1123, respectively) provide the basis for developing content-valid licensing examinations. Each K/A stem statement has been linked to the applicable item number in 10 CFR 55.41 and/or 55.43. Preparing the license examination using the appropriate K/A catalog, in conjunction with the instructions in this NUREG-series report, will ensure that the examination includes a representative sample of the items specified in the regulations.

C. Responsibilities

1. Facility Licensee

The facility licensee will perform the following activities, as applicable, depending upon the examination arrangements confirmed with the NRC's regional office (in accordance with ES-201, "Initial Operator Licensing Examination Process") approximately 4 months before the scheduled examination date:

a. Prepare the proposed examination outline(s) in accordance with Section D.1, and submit the outline(s) to the NRC's regional office for review and approval in accordance with ES-201.

b. Submit the reference materials necessary for the NRC's regional office to prepare and/or validate the requested examination(s). (Refer to ES-201, Attachment 3.)

c. Prepare the proposed examination(s) in accordance with Sections D.2 through D.4, review the examination(s) in accordance with Section E, and submit the examination(s) to the NRC's regional office in accordance with ES-201.

d. Meet with the NRC staff in the regional office or at the facility, when and as necessary, to review the proposed examination(s) and discuss potential changes. (Refer to ES-201.)

e. Revise the proposed examination outline(s) and examination(s) as agreed upon with the NRC's regional office; however, the NRC retains final authority to approve the examination.

f. Facility licensees that prepare the examination shall ensure that appropriate controls are implemented to keep the comprehensive audit or screening examination that is given at or near the end of the license training class (as well as any practice exams and quizzes that are developed after beginning work on the licensing examination) from compromising the integrity of the licensing examination. Examples of acceptable control measures are as follows (other methods may also be acceptable, but will have to be reviewed and approved on a case-by-case basis):

    • The facility licensee could prepare the audit examination using a systematic and random sampling process that is similar to that used to prepare the NRC's licensing examination as discussed in Section D.

    • The facility licensee could prepare and finalize the audit examination (and any practice exams and quizzes) before it begins developing the NRC's licensing examination outline as discussed in Section D.

    • The facility licensee could develop the audit (as well as any practice exams and quizzes) and the licensing examinations using independent examination teams.

    • The facility licensee could certify as part of the examination submittal that there is no question duplication between the facility licensee's audit and the NRC's licensing examinations.

2. **NRC Regional Office**

The NRC's regional office will perform the following activities:

a. Ensure that the examinations are prepared in accordance with Section D.

b. Ensure that the examinations are reviewed for quality as described in Section E.

c. Meet with the facility licensee, when and as appropriate, to pre-review the examination(s) in accordance with ES-201.

D.   Examination Preparation

1.   Develop the Outline

Develop each written examination outline in accordance with the following general instructions:

a.   Select the appropriate examination outline model for the licensing examination being developed:

- For RO applicants, use only the left side of Form ES-401-1 (BWR) or ES-401-2 (PWR), depending upon the facility design.

- For SRO-I (instant) applicants, use both the RO and SRO portions of Form ES-401-1 (BWR) or ES-401-2 (PWR), depending upon the facility design.

- For SRO-U applicants, use both sides of Form ES-401-1 (BWR) or ES-401-2 (PWR) unless the RO portion is waived in accordance with ES-204.

b.   Systematically and randomly select specific K/A statements (e.g., K1.03 or A2.11) from NUREG-1122 (for PWRs) or NUREG-1123 (for BWRs) to complete each of the three tiers (i.e., Tier 1, Emergency and Abnormal Plant Evolutions; Tier 2, Plant Systems; and Tier 3, Generic Knowledge and Abilities) of the applicable examination outline. In order to maintain examination consistency, the facility licensee's site-specific K/A list shall not be used in place of the NRC's K/A catalog. Attachment 1 provides an example of an acceptable methodology for randomly selecting K/As within the defined structure of the examination outline to achieve as broad a sample as possible. Other methodologies may be used, provided that they are reproducible and scrutable and yield an examination outline that is free of bias, adheres to the applicable examination model, minimizes the number of K/As related to any particular system or evolution (i.e., every system or evolution in the group should be sampled once before selecting a second K/A for any system or evolution), and samples at the specific K/A statement level.

When submitting its examination outline to the NRC, the facility licensee shall describe the process that was used to develop the examination outline (in sufficient detail for the NRC to confirm that it meets the systematic and random selection criteria). Examples of adequate documentation include (1) a statement that the facility licensee used the sampling process described in Attachment 1; (2) identification of the industry standard or widely-available commercial product that was used; or (3) a description or copy of the facility licensee's process document.

Because the NRC's K/A catalogs are based on generic job and task analyses and not all facilities are the same, examination authors can eliminate inapplicable or inappropriate K/A statements by (1) discarding randomly selected K/As during the outline development process and/or (2) pre-screening the entire K/A catalog to eliminate inappropriate K/As before beginning the random selection process.

Refer to the remainder of this section for specific requirements and guidance regarding K/A elimination.

The topics for the generic K/A category in Tiers 1 and 2 (i.e., Column "G" on Forms ES-401-1 and ES-401-2) shall be selected from Section 2, "Generic Knowledge and Abilities," of the applicable K/A catalog.) However, only those topics that are relevant to the selected evolution or system shall be included; therefore, generic K/As for Tiers 1 and 2 for both RO and SRO examinations should be randomly selected from the following: 2.1.7, 2.1.19, 2.1.20, 2.1.23, 2.1.25, 2.1.27, 2.1.28, 2.1.30, 2.1.31, 2.1.32, 2.2.3, 2.2.4, 2.2.12, 2.2.22, 2.2.25, 2.2.36, 2.2.37, 2.2.38, 2.2.39, 2.2.40, 2.2.42, 2.2.44, 2.4.1, 2.4.2, 2.4.3, 2.4.4, 2.4.6, 2.4.8, 2.4.9, 2.4.11, 2.4.18, 2.4.20, 2.4.21, 2.4.30, 2.4.31, 2.4.34, 2.4.35, 2.4.41, 2.4.45, 2.4.46, 2.4.47, 2.4.49, and 2.4.50. All other generic K/As for Tiers 1 and 2 may be eliminated before or after the random selection process, and single-unit facilities may also eliminate K/As 2.2.3 and 2.2.4.

Examination authors and reviewers should ask themselves the following questions to help determine whether or not any K/A statement is appropriate for testing:

- Is the subject K/A relevant (e.g., is the system, component, process, procedure, or event installed, in use, or possible) at the subject facility?

- Is the importance rating of the K/A equal to or greater than 2.5 for the license level of the proposed examination, or is there a site-specific priority that justifies keeping the K/A if its importance rating is below 2.5?

- Is it possible to prepare a psychometrically sound question related to the subject K/A?

- Is it possible to prepare a question at the correct license level related to the subject K/A? A question at the RO level should test one (or more) of the 14 items listed under 10 CFR 55.41(b) that the K/A is linked to, or test at a RO level as determined from the facility's learning objectives. A question at the SRO-only level should test one (or more) of the 7 items listed under 10 CFR 55.43(b) that the K/A is linked to, or test at a level that is unique to the SRO job position as determined from the facility's learning objectives.

If these questions can all be answered in the affirmative, then the subject K/A is probably appropriate for testing. The fact that a K/A does not have a corresponding facility learning objective, was not covered in training, or is subject to selection in multiple tiers, are not sufficient bases for eliminating the K/A from any tier of the outline.

Facility licensees that elect to pre-screen and eliminate any K/A statements from the random selection process should make arrangements for their NRC regional office to review their screening process and results before they submit their examination outline. Any subsequent changes to the list of K/As from which the examination outline is generated would also have to be documented, justified, and reviewed by the NRC. All K/A statements that are eliminated after they have been randomly selected to fill an examination outline shall be documented on Form ES-401-4, "Record of Rejected K/As," or equivalent, and submitted to the NRC regional office for review in conjunction with the proposed outline.

Enter the K/A statement numbers, a brief description of each topic, the topics' importance ratings for the license level of the exam (use the RO and SRO ratings for the RO and SRO-only portions, respectively), and the point totals (system, category, group, and tier) on the examination outline. The proposed point totals for each group and tier must match the number specified on Forms ES-401-1 and ES-401-2, as applicable.

If a facility licensee proposes to use an outline that was previously used at the subject or another facility, the licensee shall identify the source of the outline and explain what effect its reuse is expected to have on examination integrity.

c.   Special attention is required to ensure that the SRO examination tests at the appropriate license level. The SRO outline (refer to the right-hand portion of Forms ES-401-1 or -2, as applicable) shall include 25 K/A statements that relate to the topics in 10 CFR 55.43(b).

A number of the generic K/As in Section 2 of the catalogs are specifically linked to one or more topics specified in 10 CFR 55.43(b), and all of the Category A2, AA2, and EA2 K/A statements are (or, in the case of NUREG-1123, should be) similarly linked. Consequently, the K/As for the SRO examination will be drawn from those K/A categories (denoted by Columns "A2" and "G" in the SRO-only section of the applicable examination outline) and from all K/A categories related to the fuel handling facilities, which are specifically identified for sampling in 10 CFR 55.43(b)(7). The fact that a K/A is linked to both 10 CFR 55.41 and 10 CFR 55.43 does not mean that the K/A cannot be used to develop an SRO-only question, nor does it exclude the K/A from sampling on the RO examination. However, to be used on the SRO-only section of an examination, a question developed from a K/A linked to both 10 CFR 55.41 and 10 CFR 55.43 should test at the level of the 10 CFR 55.43(b) item number(s) that the K/A is linked to, or test a level that is unique to the SRO job position as determined from the facility's learning objectives. K/A topics linked to 10 CFR 55.41(b) may also be appropriate for developing SRO-level questions, if the questions developed evaluate knowledge and abilities at a 10 CFR 55.43(b) level, or at a level that is unique to the SRO job position as determined from the facility's learning objectives.

d.   After completing the outline, check the selected K/As for balance of coverage within and across the three tiers. Ensure that every applicable K/A category is sampled at least twice within each of the three tiers so that a valid sample will likely be maintained in the event that some questions are deleted as a result of post-examination comments. Similarly, ensure that no emergency/abnormal plant evolution (E/APE), system, or K/A category is over-sampled (e.g., avoid selecting more than two K/A topics from a given system unless they relate to plant-specific priorities. Make any adjustments that might be necessary by systematically and randomly selecting replacement K/A statements. Also check the overall balance of the entire licensing examination, including the walk-through and the dynamic simulator test, and make any necessary adjustments. Document and justify all changes on Form ES-401-4 and submit the documentation with the completed outline.

e .   Review and submit the completed outline to the NRC's chief examiner for review and approval in accordance with ES-201. Facility-developed outlines shall be independently reviewed by a facility supervisor or manager before being submitted to the NRC's regional office in accordance with ES-201. Facility licensees are responsible for ensuring that contractor-prepared outlines meet the guidelines herein. The NRC must receive the outlines by the date agreed upon when the examination arrangements were confirmed (normally approximately 75 days before the scheduled examination date).

f.   The NRC's chief examiner will ensure that the outline is independently reviewed within 5 working days (or as otherwise agreed with the facility licensee) and provide comments and recommended changes, as appropriate. The NRC's examiner shall review the sampling methodology, including all K/A rejections and changes, to ensure it is unbiased. The examiner shall also review and approve the site-specific item or topic substitutions. Refer to Section C.3 of ES-201 for additional guidance regarding outline reviews.

2.   Select and Develop Questions

a.   Prepare the site-specific written operator licensing examination using a combination of existing, modified, and new questions that match the specific K/A statements in the previously approved examination outline (refer to Section D.1 and ES-201) and the criteria summarized below. Ensure that the questions selected for Tier 3 maintain their focus on plant-wide generic knowledge and abilities and do not become an extension of Tier 2, "Plant Systems."

When selecting or writing questions for K/As that test coupled knowledge or abilities (e.g., the A.2 K/A statements in Tiers 1 and 2 and a number of generic K/A statements, such as 2.4.1, in Tier 3), try to test both aspects of the K/A statement. If that is not possible without expending an inordinate amount of resources, limit the scope of the question to that aspect of the K/A statement requiring the highest cognitive level (e.g., the (b) portion of the A.2 K/A statements) or substitute another randomly selected K/A.

Any time it becomes necessary to deviate from the previously approved examination outline, discuss the proposed deviations with the NRC's chief examiner and obtain concurrence. Also explain on Form ES-401-4 why the original proposal could not be implemented and why the proposed replacement is considered an acceptable substitute.

b. Ensure that each question is technically accurate *and free of the following psychometric flaws* that could diminish the validity of the examination:

- implausible distractors (C.2.g, h, k; D)
- confusing or ambiguous language (C.1.c; E)
- confusing or inappropriate negatives (C.2.e; E.3)
- collection of true/false statements (C.2.c; F)
- backward logic (C.1.h; G)
- specific determiners (C.2.m)

Appendix B provides a detailed discussion and examples of questions containing each of these and other errors; the parenthetic references (above) identify the applicable sections of Appendix B and its Attachment 2. Appendices A and B contain more detailed instructions and guidelines for preparing and formatting content-valid examinations and should be referred to as necessary while preparing the examination.

c. Ensure that the questions will differentiate between competent and less-than-competent applicants, that they are appropriate for the job level being examined, and that they are operationally oriented when possible. Refer to Appendix A (Section C.2) and Appendix B (Section C.1.a and Section B of Attachment 2) for additional discussion of and examples to illustrate the concept of operational validity.

Establish a level of difficulty that discriminates between applicants who have and have not mastered the required knowledge, skills, and abilities. See Appendices A (Section C.3) and B (Section C.1.e and Section C of Attachment 2) for further guidance on setting the level of difficulty for individual test questions. The applicants should be able to complete and review the RO examination within 4 hours, and the SRO-only examination within 2 hours. (Refer to ES-402, "Administering Initial Written Examinations," for actual administration time limits.)

In order to maintain examination quality and consistency, between 50 and 60 percent of the questions on the RO examination shall be written at the comprehension/analysis level. The SRO examination, overall, could exceed 60 percent because the K/A categories emphasized on the SRO-only examination are generally consistent with the higher cognitive levels. The cognitive level of any question drawn directly from a bank will be counted at its face value. Refer to Appendix B (Section C.1.d and Section A of Attachment 2) for further guidance regarding the levels of knowledge and sample questions written at each level.

d. The 25 SRO-level questions shall evaluate the additional knowledge and abilities required for the higher license level in accordance with 10 CFR 55.43(b) or the facility licensee's learning objectives. Questions related to 10 CFR 55.41(b) topics may also be appropriate SRO-level questions if they evaluate knowledge and abilities at a level that is unique to the SRO job position. The SRO-only questions are not required to be written at the higher cognitive levels (comprehension/analysis) discussed in the previous item, but shall be consistent with the cognitive level of the approved K/A statement.

e.   All test questions shall be in the multiple choice format described in Appendix B. Each question shall have four possible answer choices and be worth one point.

f.   To avoid compromising the integrity and security of the examination and to enhance consistency, observe the following limits on bank use when preparing the examination:

- Take no more than 75 percent of the questions for the examination (i.e., 56 for the RO and 19 for the SRO-only) directly from the facility licensee's or *any* other written examination question bank without significant modification.

- If the bank contains more than one question that fits a specific K/A statement, randomly select from among the available questions unless there is an appropriate basis for selecting a specific question (e.g., higher cognitive level, better discrimination validity, more operationally oriented, or site-specific priority).

- Write at least 10 new questions (i.e., 8 for the RO examination and 2 for the SRO-only) at the comprehension/analysis level, as described in Appendix B.

- Select the remaining questions for the examination (nominally 11 for the RO and 4 for the SRO-only) from the facility licensee's or *any* other bank, but significantly modify each question by changing at least one pertinent condition in the stem and at least one distractor. Changing the conditions in the stem such that one of the three distractors in the original question becomes the correct answer would also be considered a significant modification. The intent or objective of the question does not necessarily have to be changed. Adding or deleting irrelevant information and making minor changes (e.g., the unit number, component train, or power level when it makes no difference) would not be considered a significant modification to the question.

g.   A technical reference, including the reference's revision or version number (if applicable) and a cross-reference to the facility licensee's examination question bank, if applicable, shall be noted for every question. If the facility licensee has a learning objective applicable to the question, it should also be referenced. However, the absence of a learning objective does not invalidate the question, provided that it has an appropriate K/A and technical reference. Refer to ES-201 for additional instructions regarding documenting the source of questions on facility-written examinations.

To facilitate the review process, examination authors should consider providing a brief explanation of why the answer is correct, and each of the distractors is plausible but incorrect. This *optional* practice increases the efficiency of the examination review process and promotes the detection and correction of problem questions before the examinations are administered.

Reference materials (such as diagrams, sketches, and portions of facility procedures) may be used on a selective basis as attachments to the written examination. Ensure that any reference material used in the examination is easy to read and clearly marked, provides an effective and objective way for the applicant to demonstrate knowledge of the topic or concept, and does not give away the answers to other questions on the examination or improve the applicant's chances of guessing the correct answer by eliminating incorrect distractors.

Form ES-401-5 is a sample worksheet for use in preparing the written examination questions. Facility licensees may use that or a similar form to document the information related to each proposed question that is submitted to the NRC for review and approval.

3.   **Review and Submit the Examination**

a.   Review the entire examination to ensure that it satisfies the criteria on Form ES-401-6, "Written Examination Quality Checklist."

b.   Forward the examination package, including all proposed attachments and the completed quality checklist, to the first reviewer. Section E provides instructions for conducting the quality reviews.

Facility-developed examinations must be reviewed by a supervisor or manager before they are sent to the NRC's regional office in accordance with ES-201. Facility authors shall submit their examinations for management review in time to support their delivery to the NRC's regional office approximately 45 days before the scheduled examination date.

NRC examiners shall submit their examinations to the chief examiner for review at least 1 week before the scheduled pre-review by the facility licensee. (Refer to ES-201.)

4.   **Assemble the Examinations**

a.   Format the examinations using the one-question-per-page layout specified in Appendix B or by placing as many complete questions as possible on each page.

b.   Use a cover sheet in the format shown in Form ES-401-7 (or 8), "Site-Specific RO (or SRO) Written Examination Cover Sheet," as applicable, for all RO and SRO written examinations. Fill out all items in the upper section of the cover sheet, except the name of the applicant, when preparing the examinations.

E.   Quality Reviews

When reviewing questions, reviewers should try to put themselves in the position of the applicants by attempting to answer the questions without using reference material or referring to the answer key. Reviewers should ensure that the conditions and requirements posed in the question are complete and unambiguous, all necessary information is provided, all unnecessary information is deleted, the intended answer clearly follows from what is asked in the question, and all of the distractors are plausible.

1. **Facility Management Review**

   If the examination was prepared by the facility licensee, it shall be independently reviewed by a supervisor or manager before it is submitted to the NRC's regional office for review and approval in accordance with ES-201. The reviewer should evaluate the examination using the criteria on Form ES-401-6 and include the signed form in the examination package submitted to the NRC. Facility licensees are responsible for ensuring that contractor-prepared examinations meet the guidelines specified herein and are encouraged to verify the origins of the questions used to construct the examination.

2. **NRC Examiner Review**

   a. The NRC's regional office staff shall review the examination as soon as possible after receipt so that supervisory approval can be obtained before the final review with the facility licensee, which is normally scheduled about 2 weeks before the examination date. It is especially important for the regional office to promptly review examinations prepared by a facility licensee because of the extra time that may be required if extensive changes are necessary. The chief examiner shall consolidate the comments from all NRC reviewers and submit one set of comments to the author or facility contact. Refer to Section C.3 of ES-201 for additional guidance regarding examination reviews.

   b. If the NRC prepared the examination, the NRC's chief examiner shall ensure that a second examiner independently reviews all examination questions for content, wording, operational validity, and level of difficulty. As a minimum, the independent reviewer shall check the items listed on Form ES-401-6. The facility reviewer blocks in Column "b" are not applicable for NRC-prepared examinations.

   c. If the facility licensee developed the examination, the licensee is primarily responsible for ensuring compliance with the items listed on Form ES-401-6. However, the regional office staff is expected to take reasonable measures, including the selective review of reference materials, individual questions, and past examinations, to verify these items when reviewing the examination; exclusive reliance on the facility author's and reviewer's initials is *not* adequate. Depending upon the expected technical quality of the examination and the time available before the scheduled review with the facility licensee, the regional office staff shall independently review and verify the technical accuracy of a sample of the written examination questions. The regional office staff shall also confirm that the question content for a selected sample of the questions accurately implements the intent of the associated K/A statements from the previously approved examination outline. The sample shall include at least 30 questions[1] with an emphasis on those questions that were drawn directly from the facility licensee's examination bank. If more than 20 percent[1] of the sampled questions clearly do not match the intent of the associated

---

[1] The sample rates apply only to RO and RO/SRO combination exams. If the license class consists entirely of SRO-upgrade applicants who have been granted waivers of the RO examination pursuant to ES-204 or SROs limited to fuel handling, review the entire exam.

K/A statement, the region shall verify the K/A conformance on the remainder of the examination and, as appropriate, discuss its findings with the NRR operator licensing program office and the facility licensee, and assess the number of questions that were repeated from the applicants' audit examination and the last two NRC licensing examinations at the facility.

With regard to assessing the psychometric quality of the proposed examination questions, the regional office shall begin by systematically selecting a sample of questions for detailed review. The sample is based on the nominal bank/modified/new question distribution discussed in Section D.2.f above and the question background information provided by the facility licensee (using Form ES-401-5 or similar method). The sample shall include 10 of the new questions[1] on the examination and 20 additional questions[1] that are randomly selected from among the remaining questions that have not been pre-validated through successful use on an NRC licensing examination administered at the given facility since October 1, 1995. The regional office shall conduct and document the review of the 30 selected questions[1] using Form ES-401-9, "Written Examination Review Worksheet."

When the sample review is complete, the chief examiner shall consult with the responsible supervisor and proceed as directed to evaluate the remainder of the examination.

d.   There are no minimum or maximum limits on the number or scope of changes the regional office may direct the author or facility contact to make to the proposed examinations, provided that they are necessary to make the examinations conform with established acceptance criteria. All unacceptable flaws identified by using Form ES-401-9 (including questions that do not match the intent of the approved K/A, have more than one implausible distractor, or are intended as SRO-only questions but are not at the SRO license level as discussed in Section D.2.d) shall be corrected by rewriting or replacing the questions before the examination is administered. Questions that do not match the intent of the approved K/A statement, but are otherwise good questions, shall, nonetheless, be replaced with questions that match the K/A. Other flaws of a less serious nature (e.g., editorial clarifications or enhancements, single implausible distractors) should still be corrected before the examination is administered, but they will not be categorized as unacceptable for purposes of documentation in the examination report in accordance with Section E.3 of ES-501, "Initial Post-Examination Activities."

e.   Upon supervisory approval, generally at least 14 days before the examinations are scheduled to be given, the chief examiner will review the written examinations with the facility licensee in accordance with ES-201.

When providing feedback to the facility licensee regarding unacceptable questions, the chief examiner shall, at a minimum, *explain* how the Appendix B psychometric quantitative and qualitative attributes are not being met. For example, if the question is determined to have more than one implausible distractor, the attendant explanation shall articulate the reasons the examiner believes each of the faulty distractors is not credible.

Examinations that are written by the NRC shall be clean, properly formatted, and "ready-to-administer" before they are reviewed with the facility licensee. The region shall not rely on the facility licensee to ensure that the quality of the examination is acceptable for administration.

f. After reviewing the examination with the facility licensee, the chief examiner will ensure that any comments and recommendations are resolved and the examination is revised as necessary. If the facility licensee developed the examination, it will generally be expected to make whatever changes the NRC recommends.

g. After the necessary changes have been made and the chief examiner is satisfied with the examination, he or she will sign the quality checklist and forward the examination package to the responsible supervisor for final approval. If the examination was written by the facility licensee, the chief examiner should include a copy of the original submittal with the examination package.

3. **NRC Supervisory Review**

a. The responsible supervisor shall review all questions that are determined to have unacceptable flaws in accordance with Form ES-401-9 before any comments are provided to the facility licensee. The responsible supervisor shall review the entire examination before authorizing the chief examiner to proceed with the facility pre-review in accordance with ES-201. The supervisory review is not intended to be another technical review, but rather a general assessment of examination quality, including a review of the changes being recommended by the chief examiner, and a check to ensure that all applicable administrative requirements have been implemented.

b. Based on the results of the sampling review conducted in accordance with Section E.2.c (above), the responsible supervisor (in coordination with regional management and the NRR operator licensing program office, as appropriate) will continue the examination review as follows:

   • If fewer than 6 of the 30 sampled questions[1] contain unacceptable flaws as determined by using Form ES-401-9, the regional office shall review in detail the remainder of the examination (excluding those questions that were pre-validated by the NRC) using Form ES-401-9, and shall provide comments to the facility licensee for rework and correction. The NRC-validated questions need not be reviewed in detail, but will be evaluated as necessary to complete Form ES-401-6 (including the identification and correction of technical and psychometric flaws that cause the question to have no or multiple correct answers) before reviewing the examination with the facility licensee.

---

[1] The sample rates apply only to RO and RO/SRO combination exams. If the license class consists entirely of SRO-upgrade applicants who have been granted waivers of the RO examination pursuant to ES-204 or SROs limited to fuel handling, then review the entire exam.

The responsible supervisor will review and approve each comment that would require the facility licensee to rework an NRC-validated question.

- If 6 or more of the 30 sampled questions[1] contain unacceptable flaws as determined by using Form ES-401-9, the regional office may return the written examination (with explanatory comments) to the facility licensee for rework and correction without reviewing the remainder of the examination. (Refer to Section C.2.h of ES-201 for additional guidance regarding examination delays.) The facility licensee will be expected to correct the unacceptable flaws in the sampled questions and like-kind flaws that exist in the remainder of the examination. When the facility licensee resubmits the examination, every question (excluding the NRC-validated questions) will be subject to NRC review using Form ES-401-9. The NRC-validated questions will be reviewed as discussed above.

  Alternatively, if the responsible supervisor concludes that the remainder of the examination (excluding the NRC-validated questions) can be reviewed and corrected in time for the scheduled examination date, the regional office should continue the review using Form ES-401-9 and provide comments to the facility licensee for correction.

c. The responsible supervisor should ensure that any significant deficiencies in the original examinations submitted by a facility licensee are evaluated in accordance with ES-201 to determine the appropriate course of action. At a minimum, the supervisor should ensure that they are addressed in the final examination report in accordance with ES-501.

d. Following the facility review, the responsible supervisor should again review the examination to ensure that the concerns expressed by the facility licensee and the NRC have been appropriately addressed. The supervisor shall not sign Form ES-401-6 until he or she is satisfied that the examination is acceptable to be administered.

4. **Facility Peer Review**

As a final check of the examination's technical accuracy, facility management should consider administering the examination (under security agreements) to one or more licensed personnel who were previously uninvolved in developing the examination. In light of examination security concerns, the NRC discourages the use of certain individuals (e.g., the applicants' supervisors or coworkers) to validate the examination. Any comments made and problems identified during the trial administration shall be discussed with the NRC's chief examiner and resolved before the examination is administered to the license applicants. The intent of the review is to identify and correct deficiencies that may affect the validity of the examination.

F.   Attachments/Forms

Attachment 1,   "Example Systematic Sampling Methodology"
Form ES-401-1,   "BWR Examination Outline"
Form ES-401-2,   "PWR Examination Outline"
Form ES-401-3,   "Generic Knowledge and Abilities Outline (Tier 3)"
Form ES-401-4,   "Record of Rejected K/As"
Form ES-401-5,   "Sample Written Examination Question Worksheet"
Form ES-401-6,   "Written Examination Quality Checklist"
Form ES-401-7,   "Site-Specific RO Written Examination Cover Sheet"
Form ES-401-8,   "Site-Specific SRO Written Examination Cover Sheet"
Form ES-401-9,   "Written Examination Review Worksheet"

The following process, which uses the BWR outline (Form ES-401-1) for illustration, *may be used* for each group in Tiers 1 and 2 of the RO examination outline.

1.   Review each group and delete those items [emergency/abnormal plant evolutions (E/APEs) for Tier 1 and systems for Tier 2] that clearly do not apply to the facility for which the examination is being written; be prepared to explain the basis for the deletions to the NRC's chief examiner. Add any operationally-important systems or E/APEs that pertain to the facility but are not included in the generic lists on Form ES-401-1.

2.   Sequentially number the remaining items in the group and sequentially annotate the same number of tokens. If we assume that none of the 20 E/APEs in Tier 1, Group 1 was deleted in Step 1, there should be 20 tokens, numbered from 1 to 20.

   a.   Since the number of items remaining in the group (in this case 20) is the same as the required number of points for the group specified in the right-hand column of the examination outline, each item in the group would be sampled one time.

   b.   If the number of items remaining in the group is smaller than the required number of points for the group (e.g., Tier 2, Group 1 has 23 items but requires 26 points), sample each item once, and determine the rest of the sample by randomly selecting and removing tokens (in this case 3 of the 23) until the required total number of points is reached. Update Form ES-401-1 to note the selected items.

   c.   If the number of items remaining in the group is larger than the required number of points for the group (e.g., Tier 1, Group 2 has 20 items but only requires 7 points), randomly select and remove the required number of tokens and note them on Form ES-401-1.

3.   After selecting the topics to be sampled in each group as described in Step 2, count the number of K/A categories in the group [e.g., 6 for each group in Tier 1 (i.e., K1, K2, K3, A1, A2, and G)] and sequentially annotate the same number of tokens (in this case 6). For each E/APE (and system) selected in Step 2, randomly select and remove a token and note the K/A category on Form ES-401-1. If the E/APE (or system) was sampled more than once in accordance with Step 2.a, randomly select a second K/A category. If the selected K/A category contains no K/A statements having an importance rating above 2.5, systematically select another K/A category, unless the lower importance is justified based on plant-specific priorities. Then replace all tokens in the container and repeat the process for every selected item in each group.

4.   Use a similar method to randomly select from among the K/A statements under each selected K/A category. Describe each K/A topic in the space provided on Form ES-401-1 and enter the importance rating. K/As having importance ratings less than 2.5 can be used if justified based on plant priorities; the facility contact should be prepared to explain the basis to the NRC's chief examiner.

For Tier 3 (plant-wide generics) of the examination outline, randomly select K/As from Section 2 of the NRC's K/A catalog so that each of the four K/A categories (i.e., "Conduct of Operations," "Equipment Control," Radiation Control," and "Emergency Procedures/Plan") has at least two items.

Repeat Steps (1) through (4), above, to select the required number of topics for the SRO-only portion of the exam. With respect to Step (3), select topics from the shaded portions of the Tier 1 and 2 outlines [i.e., the "A2" and "G" K/A categories, which are linked to 10 CFR 55.43, and the fuel handling equipment, which is specifically identified for sampling in 10 CFR 55.43(b)(7)]. For Tier 3, select seven K/As linked to 10 CFR 55.43; sample one of the categories only once.

Facility:												Date of Exam:				
Tier	Group	RO K/A Category Points											SRO-Only Points			
		K1	K2	K3	K4	K5	K6	A1	A2	A3	A4	G*	Total	A2	G*	Total
1. Emergency & Abnormal Plant Evolutions	1				N/A				N/A				20			7
	2												7			3
	Tier Totals												27			10
2. Plant Systems	1												26			5
	2												12			3
	Tier Totals												38			8
3. Generic Knowledge and Abilities Categories		1		2		3		4					10	1 2 3 4		7

Note:
1. Ensure that at least two topics from every applicable K/A category are sampled within each tier of the RO and SRO-only outlines (i.e., except for one category in Tier 3 of the SRO-only outline, the "Tier Totals" in each K/A category shall not be less than two).

2. The point total for each group and tier in the proposed outline must match that specified in the table. The final point total for each group and tier may deviate by ±1 from that specified in the table based on NRC revisions. The final RO exam must total 75 points and the SRO-only exam must total 25 points.

3. Systems/evolutions within each group are identified on the associated outline; systems or evolutions that do not apply at the facility should be deleted and justified; operationally important, site-specific systems/evolutions that are not included on the outline should be added. Refer to Section D.1.b of ES-401 for guidance regarding the elimination of inappropriate K/A statements.

4. Select topics from as many systems and evolutions as possible; sample every system or evolution in the group before selecting a second topic for any system or evolution.

e. Absent a plant-specific priority, only those K/As having an importance rating (IR) of 2.5 or higher shall be selected. Use the RO and SRO ratings for the RO and SRO-only portions, respectively.

6. Select SRO topics for Tiers 1 and 2 from the shaded systems and K/A categories.

7.* The generic (G) K/As in Tiers 1 and 2 shall be selected from Section 2 of the K/A Catalog, but the topics must be relevant to the applicable evolution or system. Refer to Section D.1.b of ES-401 for the applicable K/As.

8. On the following pages, enter the K/A numbers, a brief description of each topic, the topics' importance ratings (IRs) for the applicable license level, and the point totals (#) for each system and category. Enter the group and tier totals for each category in the table above; if fuel handling equipment is sampled in other than Category A2 or G* on the SRO-only exam, enter it on the left side of Column A2 for Tier 2, Group 2 (Note #1 does not apply). Use duplicate pages for RO and SRO-only exams.

9. For Tier 3, select topics from Section 2 of the K/A catalog, and enter the K/A numbers, descriptions, IRs, and point totals (#) on Form ES-401-3. Limit SRO selections to K/As that are linked to 10 CFR 55.43.

ES-401	BWR Examination Outline Emergency and Abnormal Plant Evolutions - Tier 1/Group 1 (RO / SRO)							Form ES-401-1	
E/APE # / Name / Safety Function	K 1	K 2	K 3	A 1	A 2	G	K/A Topic(s)	IR	#
295001 Partial or Complete Loss of Forced Core Flow Circulation / 1 & 4									
295003 Partial or Complete Loss of AC / 6									
295004 Partial or Total Loss of DC Pwr / 6									
295005 Main Turbine Generator Trip / 3									
295006 SCRAM / 1									
295016 Control Room Abandonment / 7									
295018 Partial or Total Loss of CCW / 8									
295019 Partial or Total Loss of Inst. Air / 8									
295021 Loss of Shutdown Cooling / 4									
295023 Refueling Acc / 8									
295024 High Drywell Pressure / 5									
295025 High Reactor Pressure / 3									
295026 Suppression Pool High Water Temp. / 5									
295027 High Containment Temperature / 5									
295028 High Drywell Temperature / 5									
295030 Low Suppression Pool Wtr Lvl / 5									
295031 Reactor Low Water Level / 2									
295037 SCRAM Condition Present and Reactor Power Above APRM Downscale or Unknown / 1									
295038 High Off-site Release Rate / 9									
600000 Plant Fire On Site / 8									
700000 Generator Voltage and Electric Grid Disturbances / 6									
K/A Category Totals:							Group Point Total:		20/7

ES-401							BWR Examination Outline Emergency and Abnormal Plant Evolutions - Tier 1/Group 2 (RO / SRO)	Form ES-401-1	
E/APE # / Name / Safety Function	K 1	K 2	K 3	A 1	A 2	G	K/A Topic(s)	IR	#
295002 Loss of Main Condenser Vac / 3									
295007 High Reactor Pressure / 3									
295008 High Reactor Water Level / 2									
295009 Low Reactor Water Level / 2									
295010 High Drywell Pressure / 5									
295011 High Containment Temp / 5									
295012 High Drywell Temperature / 5									
295013 High Suppression Pool Temp. / 5									
295014 Inadvertent Reactivity Addition / 1									
295015 Incomplete SCRAM / 1									
295017 High Off-site Release Rate / 9									
295020 Inadvertent Cont. Isolation / 5 & 7									
295022 Loss of CRD Pumps / 1									
295029 High Suppression Pool Wtr Lvl / 5									
295032 High Secondary Containment Area Temperature / 5									
295033 High Secondary Containment Area Radiation Levels / 9									
295034 Secondary Containment Ventilation High Radiation / 9									
295035 Secondary Containment High Differential Pressure / 5									
295036 Secondary Containment High Sump/Area Water Level / 5									
500000 High CTMT Hydrogen Conc. / 5									
K/A Category Point Totals:							Group Point Total:		7/3

ES-401												BWR Examination Outline Plant Systems - Tier 2/Group 1 (RO / SRO)	Form ES-401-1	
System # / Name	K 1	K 2	K 3	K 4	K 5	K 6	A 1	A 2	A 3	A 4	G	K/A Topic(s)	IR	#
203000 RHR/LPCI: Injection Mode														
205000 Shutdown Cooling														
206000 HPCI														
207000 Isolation (Emergency) Condenser														
209001 LPCS														
209002 HPCS														
211000 SLC														
212000 RPS														
215003 IRM														
215004 Source Range Monitor														
215005 APRM / LPRM														
217000 RCIC														
218000 ADS														
223002 PCIS/Nuclear Steam Supply Shutoff														
239002 SRVs														
259002 Reactor Water Level Control														
261000 SGTS														
262001 AC Electrical Distribution														
262002 UPS (AC/DC)														
263000 DC Electrical Distribution														
264000 EDGs														
300000 Instrument Air														
400000 Component Cooling Water														
K/A Category Point Totals:												Group Point Total:		26/5

ES-401												BWR Examination Outline Plant Systems - Tier 2/Group 2 (RO / SRO)	Form ES-401-1	
System # / Name	K 1	K 2	K 3	K 4	K 5	K 6	A 1	A 2	A 3	A 4	G	K/A Topic(s)	IR	#
201001 CRD Hydraulic														
201002 RMCS														
201003 Control Rod and Drive Mechanism														
201004 RSCS														
201005 RCIS														
201006 RWM														
202001 Recirculation														
202002 Recirculation Flow Control														
204000 RWCU														
214000 RPIS														
215001 Traversing In-core Probe														
215002 RBM														
216000 Nuclear Boiler Inst.														
219000 RHR/LPCI: Torus/Pool Cooling Mode														
223001 Primary CTMT and Aux.														
226001 RHR/LPCI: CTMT Spray Mode														
230000 RHR/LPCI: Torus/Pool Spray Mode														
233000 Fuel Pool Cooling/Cleanup														
234000 Fuel Handling Equipment														
239001 Main and Reheat Steam														
239003 MSIV Leakage Control														
241000 Reactor/Turbine Pressure Regulator														
245000 Main Turbine Gen. / Aux.														
256000 Reactor Condensate														
259001 Reactor Feedwater														
268000 Radwaste														
271000 Offgas														
272000 Radiation Monitoring														
286000 Fire Protection														
288000 Plant Ventilation														
290001 Secondary CTMT														
290003 Control Room HVAC														
290002 Reactor Vessel Internals														
K/A Category Point Totals:												Group Point Total:		12/3

Facility:		Date of Exam:														
Tier	Group	RO K/A Category Points											SRO-Only Points			
		K1	K2	K3	K4	K5	K6	A1	A2	A3	A4	G*	Total	A2	G*	Total
1. Emergency & Abnormal Plant Evolutions	1				N/A				N/A				18			6
	2												9			4
	Tier Totals												27			10
2. Plant Systems	1												28			5
	2												10			3
	Tier Totals												38			8
3. Generic Knowledge and Abilities Categories		1		2		3		4		10	1	2	3	4		7

Note:

1. Ensure that at least two topics from every applicable K/A category are sampled within each tier of the RO and SRO-only outlines (i.e., except for one category in Tier 3 of the SRO-only outline, the "Tier Totals" in each K/A category shall not be less than two).

2. The point total for each group and tier in the proposed outline must match that specified in the table. The final point total for each group and tier may deviate by ±1 from that specified in the table based on NRC revisions. The final RO exam must total 75 points and the SRO-only exam must total 25 points.

3. Systems/evolutions within each group are identified on the associated outline; systems or evolutions that do not apply at the facility should be deleted and justified; operationally important, site-specific systems/evolutions that are not included on the outline should be added. Refer to Section D.1.b of ES-401 for guidance regarding the elimination of inappropriate K/A statements.

4. Select topics from as many systems and evolutions as possible; sample every system or evolution in the group before selecting a second topic for any system or evolution.

5. Absent a plant-specific priority, only those K/As having an importance rating (IR) of 2.5 or higher shall be selected. Use the RO and SRO ratings for the RO and SRO-only portions, respectively.

6. Select SRO topics for Tiers 1 and 2 from the shaded systems and K/A categories.

7.* The generic (G) K/As in Tiers 1 and 2 shall be selected from Section 2 of the K/A Catalog, but the topics must be relevant to the applicable evolution or system. Refer to Section D.1.b of ES-401 for the applicable K/As.

8. On the following pages, enter the K/A numbers, a brief description of each topic, the topics' importance ratings (IRs) for the applicable license level, and the point totals (#) for each system and category. Enter the group and tier totals for each category in the table above; if fuel handling equipment is sampled in other than Category A2 or G* on the SRO-only exam, enter it on the left side of Column A2 for Tier 2, Group 2 (Note #1 does not apply). Use duplicate pages for RO and SRO-only exams.

9. For Tier 3, select topics from Section 2 of the K/A catalog, and enter the K/A numbers, descriptions, IRs, and point totals (#) on Form ES-401-3. Limit SRO selections to K/As that are linked to 10 CFR 55.43.

ES-401						PWR Examination Outline	Form ES-401-2	
						Emergency and Abnormal Plant Evolutions - Tier 1/Group 1 (RO / SRO)		
E/APE # / Name / Safety Function	K 1	K 2	K 3	A 1	A 2	G	K/A Topic(s)	IR	#
000007 (BW/E02&E10; CE/E02) Reactor Trip - Stabilization - Recovery / 1									
000008 Pressurizer Vapor Space Accident / 3									
000009 Small Break LOCA / 3									
000011 Large Break LOCA / 3									
000015/17 RCP Malfunctions / 4									
000022 Loss of Rx Coolant Makeup / 2									
000025 Loss of RHR System / 4									
000026 Loss of Component Cooling Water / 8									
000027 Pressurizer Pressure Control System Malfunction / 3									
000029 ATWS / 1									
000038 Steam Gen. Tube Rupture / 3									
000040 (BW/E05; CE/E05; W/E12) Steam Line Rupture - Excessive Heat Transfer / 4									
000054 (CE/E06) Loss of Main Feedwater / 4									
000055 Station Blackout / 6									
000056 Loss of Off-site Power / 6									
000057 Loss of Vital AC Inst. Bus / 6									
000058 Loss of DC Power / 6									
000062 Loss of Nuclear Svc Water / 4									
000065 Loss of Instrument Air / 8									
W/E04 LOCA Outside Containment / 3									
W/E11 Loss of Emergency Coolant Recirc. / 4									
BW/E04; W/E05 Inadequate Heat Transfer - Loss of Secondary Heat Sink / 4									
000077 Generator Voltage and Electric Grid Disturbances / 6									
K/A Category Totals:							Group Point Total:		18/6

ES-401	PWR Examination Outline Emergency and Abnormal Plant Evolutions - Tier 1/Group 2 (RO / SRO)							Form ES-401-2		
E/APE # / Name / Safety Function	K 1	K 2	K 3	A 1	A 2	G	K/A Topic(s)	IR	#
000001 Continuous Rod Withdrawal / 1									
000003 Dropped Control Rod / 1									
000005 Inoperable/Stuck Control Rod / 1									
000024 Emergency Boration / 1									
000028 Pressurizer Level Malfunction / 2									
000032 Loss of Source Range NI / 7									
000033 Loss of Intermediate Range NI / 7									
000036 (BW/A08) Fuel Handling Accident / 8									
000037 Steam Generator Tube Leak / 3									
000051 Loss of Condenser Vacuum / 4									
000059 Accidental Liquid RadWaste Rel. / 9									
000060 Accidental Gaseous Radwaste Rel. / 9									
000061 ARM System Alarms / 7									
000067 Plant Fire On-site / 8									
000068 (BW/A06) Control Room Evac. / 8									
000069 (W/E14) Loss of CTMT Integrity / 5									
000074 (W/E06&E07) Inad. Core Cooling / 4									
000076 High Reactor Coolant Activity / 9									
W/EO1 & E02 Rediagnosis & SI Termination / 3									
W/E13 Steam Generator Over-pressure / 4									
W/E15 Containment Flooding / 5									
W/E16 High Containment Radiation / 9									
BW/A01 Plant Runback / 1									
BW/A02&A03 Loss of NNI-X/Y / 7									
BW/A04 Turbine Trip / 4									
BW/A05 Emergency Diesel Actuation / 6									
BW/A07 Flooding / 8									
BW/E03 Inadequate Subcooling Margin / 4									
BW/E08; W/E03 LOCA Cooldown - Depress. / 4									
BW/E09; CE/A13; W/E09&E10 Natural Circ. / 4									
BW/E13&E14 EOP Rules and Enclosures									
CE/A11; W/E08 RCS Overcooling - PTS / 4									
CE/A16 Excess RCS Leakage / 2									
CE/E09 Functional Recovery									
K/A Category Point Totals:							Group Point Total:		9/4

| ES-401 | PWR Examination Outline
Plant Systems - Tier 2/Group 1 (RO / SRO) | Form ES-401-2 |

System # / Name	K 1	K 2	K 3	K 4	K 5	K 6	A 1	A 2	A 3	A 4	G	K/A Topic(s)	IR	#
003 Reactor Coolant Pump														
004 Chemical and Volume Control														
005 Residual Heat Removal														
006 Emergency Core Cooling														
007 Pressurizer Relief/Quench Tank														
008 Component Cooling Water														
010 Pressurizer Pressure Control														
012 Reactor Protection														
013 Engineered Safety Features Actuation														
022 Containment Cooling														
025 Ice Condenser														
026 Containment Spray														
039 Main and Reheat Steam														
059 Main Feedwater														
061 Auxiliary/Emergency Feedwater														
062 AC Electrical Distribution														
063 DC Electrical Distribution														
064 Emergency Diesel Generator														
073 Process Radiation Monitoring														
076 Service Water														
078 Instrument Air														
103 Containment														
K/A Category Point Totals:												Group Point Total:		28/5

ES-401												PWR Examination Outline Plant Systems - Tier 2/Group 2 (RO / SRO)	Form ES-401-2	
System # / Name	K 1	K 2	K 3	K 4	K 5	K 6	A 1	A 2	A 3	A 4	G	K/A Topic(s)	IR	#
001 Control Rod Drive														
002 Reactor Coolant														
011 Pressurizer Level Control														
014 Rod Position Indication														
015 Nuclear Instrumentation														
016 Non-nuclear Instrumentation														
017 In-core Temperature Monitor														
027 Containment Iodine Removal														
028 Hydrogen Recombiner and Purge Control														
029 Containment Purge														
033 Spent Fuel Pool Cooling														
034 Fuel Handling Equipment														
035 Steam Generator														
041 Steam Dump/Turbine Bypass Control														
045 Main Turbine Generator														
055 Condenser Air Removal														
056 Condensate														
068 Liquid Radwaste														
071 Waste Gas Disposal														
072 Area Radiation Monitoring														
075 Circulating Water														
079 Station Air														
086 Fire Protection														
K/A Category Point Totals:												Group Point Total:		10/3

Facility:		Date of Exam:					
Category	K/A #	Topic	RO		SRO-Only		
			IR	#	IR	#	
1. Conduct of Operations	2.1.						
	2.1.						
	2.1.						
	2.1.						
	2.1.						
	2.1.						
	Subtotal						
2. Equipment Control	2.2.						
	2.2.						
	2.2.						
	2.2.						
	2.2.						
	2.2.						
	Subtotal						
3. Radiation Control	2.3.						
	2.3.						
	2.3.						
	2.3.						
	2.3.						
	2.3.						
	Subtotal						
4. Emergency Procedures / Plan	2.4.						
	2.4.						
	2.4.						
	2.4.						
	2.4.						
	2.4.						
	Subtotal						
Tier 3 Point Total				10		7	

Tier / Group	Randomly Selected K/A	Reason for Rejection

Examination Outline Cross-Reference:

		RO	SRO
Level			
Tier #		_____	_____
Group #		_____	_____
K/A #		_____	
Importance Rating		_____	_____

Proposed Question:

Proposed Answer: _____

Explanation (Optional):

Technical Reference(s): _____
(Attach if not previously provided) _____
(including version/revision number) _____

Proposed references to be provided to applicants during examination: _____

Learning Objective: _____ (As available)

Question Source: Bank #   _____
                   Modified Bank # _____ (Note changes or attach parent)
                   New   _____

Question History: Last NRC Exam   _____
(Optional: Questions validated at the facility since 10/95 will generally undergo less rigorous review by the NRC; failure to provide the information will necessitate a detailed review of every question.)

Question Cognitive Level: Memory or Fundamental Knowledge   _____
                            Comprehension or Analysis   _____

10 CFR Part 55 Content: 55.41 _____
                          55.43 _____

Comments:

Facility:	Date of Exam:	Exam Level: RO ☐ SRO ☐

			Initial		
	Item Description		a	b*	c#
1.	Questions and answers are technically accurate and applicable to the facility.				
2.	a. NRC K/As are referenced for all questions. b. Facility learning objectives are referenced as available.				
3.	SRO questions are appropriate in accordance with Section D.2.d of ES-401				
4.	The sampling process was random and systematic (If more than 4 RO or 2 SRO questions were repeated from the last 2 NRC licensing exams, consult the NRR OL program office).				
5.	Question duplication from the license screening/audit exam was controlled as indicated below (check the item that applies) and appears appropriate: __ the audit exam was systematically and randomly developed; or __ the audit exam was completed before the license exam was started; or __ the examinations were developed independently; or __ the licensee certifies that there is no duplication; or __ other (explain)				

6.	Bank use meets limits (no more than 75 percent from the bank, at least 10 percent new, and the rest new or modified); enter the actual RO / SRO-only question distribution(s) at right.	Bank	Modified	New			
		/	/	/			

7.	Between 50 and 60 percent of the questions on the RO exam are written at the comprehension/ analysis level; the SRO exam may exceed 60 percent if the randomly selected K/As support the higher cognitive levels; enter the actual RO / SRO question distribution(s) at right.	Memory	C/A			
		/	/			

8.	References/handouts provided do not give away answers or aid in the elimination of distractors.				
9.	Question content conforms with specific K/A statements in the previously approved examination outline and is appropriate for the tier to which they are assigned; deviations are justified.				
10.	Question psychometric quality and format meet the guidelines in ES Appendix B.				
11.	The exam contains the required number of one-point, multiple choice items; the total is correct and agrees with the value on the cover sheet.				

	Printed Name / Signature	Date
a. Author	_____	_____
b. Facility Reviewer (*)	_____	_____
c. NRC Chief Examiner (#)	_____	_____
d. NRC Regional Supervisor	_____	_____

Note: * The facility reviewer's initials/signature are not applicable for NRC-developed examinations.
        # Independent NRC reviewer initial items in Column "c"; chief examiner concurrence required.

U.S. Nuclear Regulatory Commission

Site-Specific RO Written Examination

Applicant Information

Name:

Date: Facility/Unit:

Region: I ☐   II ☐   III ☐   IV ☐   Reactor Type: W ☐   CE ☐   BW ☐   GE ☐

Start Time: Finish Time:

Instructions

Use the answer sheets provided to document your answers. Staple this cover sheet
on top of the answer sheets. To pass the examination, you must achieve a final grade
of at least 80.00 percent. Examination papers will be collected 6 hours after the examination begins.

Applicant Certification

All work done on this examination is my own. I have neither given nor received aid.

_____
Applicant's Signature

Results

Examination Value	_____ Points
Applicant's Score	_____ Points
Applicant's Grade	_____ Percent

U.S. Nuclear Regulatory Commission

Site-Specific SRO Written Examination

Applicant Information

Name:

| Date: | Facility/Unit: |

| Region: I ☐   II ☐   III ☐   IV ☐ | Reactor Type: W ☐   CE ☐   BW ☐   GE ☐ |

| Start Time: | Finish Time: |

Instructions

Use the answer sheets provided to document your answers. Staple this cover sheet on top of the answer sheets. To pass the examination you must achieve a final grade of at least 80.00 percent overall, with 70.00 percent or better on the SRO-only items if given in conjunction with the RO exam; SRO-only exams given alone require a final grade of 80.00 percent to pass. You have 8 hours to complete the combined examination, and 3 hours if you are only taking the SRO portion.

Applicant Certification

All work done on this examination is my own. I have neither given nor received aid.

_____
Applicant's Signature

Results

RO/SRO-Only/Total Examination Values	_____ / _____ / _____ Points
Applicant's Scores	_____ / _____ / _____ Points
Applicant's Grade	_____ / _____ / _____ Percent

Q#	1. LOK (F/H)	2. LOD (1-5)	3. Psychometric Flaws						4. Job Content Flaws					5. Other		6. B/M/N	7. U/E/S	8. Explanation
			Stem Focus	Cues	T/F	Cred. Dist.	Partial	Job-Link	Minutia	#/ units	Back-ward	Q= K/A	SRO Only					

Instructions

[Refer to Section D of ES-401 and Appendix B for additional information regarding each of the following concepts.]

1. Enter the level of knowledge (LOK) of each question as either (F)undamental or (H)igher cognitive level.

2. Enter the level of difficulty (LOD) of each question using a 1 – 5 (easy – difficult) rating scale (questions in the 2 – 4 range are acceptable).

3. Check the appropriate box if a psychometric flaw is identified:

   • The stem lacks sufficient focus to elicit the correct answer (e.g., unclear intent, more information is needed, or too much needless information).
   • The stem or distractors contain cues (i.e., clues, specific determiners, phrasing, length, etc).
   • The answer choices are a collection of unrelated true/false statements.
   • The distractors are not credible; single implausible distractors should be repaired, more than one is unacceptable.
   • One or more distractors is (are) partially correct (e.g., if the applicant can make unstated assumptions that are not contradicted by stem).

4. Check the appropriate box if a job content error is identified:

   • The question is not linked to the job requirements (i.e. the question has a valid K/A but, as written, is not operational in content).
   • The question requires the recall of knowledge that is too specific for the closed reference test mode (i.e., it is not required to be known from memory).
   • The question contains data with an unrealistic level of accuracy or inconsistent units (e.g., panel meter in percent with question in gallons).
   • The question requires reverse logic or application compared to the job requirements.

5. Check questions that are sampled for conformance with the approved K/A and those that are *designated SRO-only* (K/A and license level mismatches are unacceptable).

6. Enter question source: (B)ank, (M)odified, or (N)ew. Check that (M)odified questions meet criteria of ES-401 Section D.2.f.

7. Based on the reviewer's judgment, is the question as written (U)nsatisfactory (requiring repair or replacement), in need of (E)ditorial enhancement, or (S)atisfactory?

8. At a minimum, explain any "U" ratings (e.g., how the Appendix B psychometric attributes are not being met).

Q#	1. LOK (F/H)	2. LOD (1-5)	3. Psychometric Flaws					4. Job Content Flaws					5. Other		6.	7.	8.
			Stem Focus	Cues	T/F	Cred. Dist.	Partial	Job-Link	Minutia	#/ units	Back-ward	Q= K/A	SRO Only	B/M/N	U/E/S	Explanation

A.   Purpose

This standard specifies the requirements and procedures for administering written examinations for the initial licensing of reactor operator (RO) and senior reactor operator (SRO) applicants at power reactor facilities. As such, this standard includes instructions for proctoring the examinations and conducting post-examination reviews of NRC-developed examinations.

B.   Background

As noted in ES-201, "Initial Operator Licensing Examination Process," facility licensees will generally prepare the written operator licensing examinations, subject to review and approval by the NRC. Generally, examinations that are prepared by the facility licensee will also be administered by the facility licensee in accordance with the instructions contained herein.

C.   Responsibilities

1.   Facility Licensee

   a.   The facility licensee shall safeguard the integrity and security of the examinations in accordance with ES-201.

   b.   The facility licensee shall provide a single room suitable for administering the written examination. To ensure examination integrity, the room shall be large enough so that there is only one applicant per table, with a 1-meter (3-foot) space between tables.

   The examination room and supporting restroom facilities (i.e., the examination area) shall be located to prevent the applicants from having contact with all other facility and contractor personnel during the written examination.

   c.   If desired and compatible with examination security requirements, the facility licensee may arrange for the applicants to have lunch, coffee, or other refreshments during the examination.

   d.   Before the scheduled examination date, the facility licensee should familiarize the applicants with the examination policies and guidelines contained in Appendix E.

   e.   The facility licensee shall provide the necessary number of copies of the approved examinations, answer sheets, and handouts (e.g., equation sheets, selected technical specifications, and steam tables) for each applicant, as directed and approved by the NRC chief examiner. An English dictionary should also be available in the examination room.

The facility licensee may use machine-gradable answer sheets if desired, but this is *not* required.

f.   If the facility licensee developed the examination, it shall also administer the examination to the applicants identified on the "List of Applicants" (Form ES-201-4) as arranged with the NRC chief examiner and in accordance with the specific instructions in Section D.

g.   The facility licensee will send a letter to the NRC regional administrator to formally withdraw the applications of any individuals whose names appear on Form ES-201-4 but who will not be taking the examination.

h.   As discussed in Section E, the facility licensee should provide the NRC's regional office with formal comments for consideration during the grading process (refer to ES-403, "Grading Initial Site-Specific Written Examinations"). The facility licensee may also request an informal meeting with the NRC's chief examiner to discuss the examination questions and resolve facility concerns.

2.   NRC Regional Office

a.   The NRC's regional office may administer the examination, at its discretion, in accordance with the specific instructions in Section D, even if the examination was developed by the facility licensee. However, the regional office will generally arrange for the facility licensee to administer the examination. (Refer to ES-201 for further instructions on examination scheduling.)

If the NRC developed the examination, the regional office may arrange for an NRC examiner or the facility licensee to administer the examination.

b.   If the facility licensee will conduct the examinations while the NRC examiners are on site, the chief examiner should inspect the examination facilities to ensure their adequacy. In addition, the NRC examiners should periodically monitor the exam to ensure that the proctor is appropriately addressing the applicants' questions. If this is not feasible, the regional office should consider having an examiner check the facilities during the preparatory site visit (if one is deemed necessary) or upon arriving at the site for the operating tests.

If the facility licensee will conduct the examinations when no NRC examiners are on site, the chief examiner will ensure that an NRC point of contact is available in the regional office to respond to facility questions while the examinations are being given. If the NRC prepared the examination, an examiner familiar with the examination content must be available to respond to the applicants' questions by telephone.

The written examinations may be administered as soon as they and the license applications (including any applicable waivers) have been approved. The region shall not allow the written examination and operating test dates to diverge by more than 30 days without obtaining concurrence from the NRR operator licensing program office. (Refer to ES-201 for additional guidance regarding examinations that have to be rescheduled to achieve an acceptable product.)

c.   When the applicants have completed the written examination, the chief examiner may conduct an examination review with the facility staff as described in Section E, below.

D.   Examination Administration Instructions

1.   Make Preparations

a.   Arrange for the applicants to be proctored at all times while taking the written examination. Ensure that the proctor clearly understands his or her responsibilities (refer to Section D.2) before the examinations are distributed.

If the NRC will administer the examinations, the chief examiner should consider using the following resources to ensure adequate proctoring:

- NRC secretarial help
- another examiner
- other NRC employees

The examiner may arrange for facility employees to proctor the examination for brief periods if it is necessary for the examiner to go to the restroom.

b.   At least one individual who is familiar with the intent of the questions (i.e., an NRC examiner or facility employee who took part in developing the examination) shall be available to clarify examination questions for the applicants during the examination.

c.   Remove from the examination area, or otherwise remove from the applicants' view, any wall charts, models, or other training materials that might compromise examination integrity.

d.   Only NRC-approved applicants are allowed to take the examination. If applicable, the NRC examiner shall verify each applicant's identity and examination level against Form ES-201-4 before beginning the examination. Any errors or absences shall be resolved with the facility staff, and the form shall be updated as required.

e.   If possible, the RO and SRO applicants shall be seated at alternate tables. The proctor shall construct a chart illustrating the seating arrangement of the applicants during the examination.

f.   If the applicants will record their answers on machine gradable forms
that offer more than four answer choices (e.g., "a" through "e"), use a straight edge
to line out the inapplicable column(s) before distributing the forms.

2.   Start the Examination

a.   Remind the applicants that they may use calculators to complete the examination,
and that only reference materials provided with the examination are allowed
in the examination area (i.e., the examination room and supporting restroom
facilities).

b.   Pass out the examinations, blank answer sheets, and all required handouts
approved by the NRC chief examiner (e.g., steam tables, equation sheets,
and selected technical specifications). Instruct the applicants not to review
the examination until told to do so.

c.   Provide each applicant with a copy of Appendix E, "Policies and Guidelines
for Taking NRC Examinations," and brief the applicants on the rules and guidelines
that will be in effect during the written examination (i.e., review Parts A and B
of Appendix E). If time permits and the operating tests have not yet been
administered, review those policies and guidelines (i.e., Parts C, D, and E
of Appendix E) as well; this will save time later and give the applicants
greater opportunity to resolve any questions they may have.

d.   Instruct the applicants to verify the completeness of their copies by checking
the appropriate cover sheet (Form ES-401-7, ES-401-8, or ES-701-8)
and each page of the examination. RO applicants should have a 75-question exam
and SROs should have a 100-question exam, unless they have obtained
a waiver (per ES-204, "Processing Waivers Requested by Reactor Operator
and Senior Reactor Operator Applicants") to upgrade their RO licenses
with a 25-question SRO-only exam or they are taking the 40-question
SRO examination limited to fuel handling.

e.   Answer any questions that the applicants may have regarding the examination
policies. Start the examination, and record the time.

3.   Monitor the Examination

a.   The proctor shall give full attention to the applicants taking the examination.
The proctor shall not read procedures or other material, grade examinations,
or engage in any other activities in a manner that may divert his or her attention
from the applicants and possibly cause the examination to be compromised.

b.   Personnel responding to questions raised by applicants during the examination
must be extremely careful not to lead the applicants or give away answers
when clarifying questions. If the proctor has any doubt about how to respond
to an applicant's question, it is best to withhold additional guidance and instruct
the applicant to do his or her best with the information that is provided.

Any question changes or clarifications shall be made on a chalk board or white board, if available, and called to the attention of all the applicants. Changes made to questions during the examination should be made in ink on the NRC's master copy and a copy that is retained by the facility staff after the examination is administered. Changes shall be reviewed and approved by the NRC's chief examiner as part of the grading process (refer to ES-403).

All applicant questions regarding specific written examination test items and all statements of clarification shall also be documented (verbatim if possible) for future review by the NRC's chief examiner and for reference in resolving grading conflicts.

c.   The proctor shall periodically advise the applicants of the time that remains to complete the examination. Normally, a chalk board or white board is available and can be used for this purpose.

4.   Complete the Examination

a.   As the applicants complete the examination, ensure that they sign the examination cover sheet and staple it on top of their answer sheets. Collect the examination packages, including the questions and answer sheets, and any reference material provided with the examination. Verify that all applicants have entered their names on both the answer and cover sheets, and record the official start time and the time at which each applicant completed the examination in the space provided on the examination cover sheet.

b.   Retain the cover and answer sheets for grading in accordance with ES-403. The question books may be distributed to the applicants after the last examination has been collected.

c.   Remind the applicants to leave the examination area, as previously defined.

d.   When the allotted time for the examination (3 hours for the 25-question SRO-upgrade exam, 4 hours for the SRO exam limited to fuel handling, 6 hours for the RO exam, and 8 hours for the combined RO/SRO exam) has elapsed, instruct the remaining applicants to stop work, sign their examination cover sheets, and turn in their examinations. The facility licensee may extend the time allowed to complete the examination, but shall first notify the NRC's regional office to ensure that a point of contact remains available to respond to questions. The facility licensee shall inform the NRC when all of the applicants have completed the examination.

e.   Deliver the completed examination packages, the marked-up master examinations, the list of applicant questions and answers, and the seating chart to the NRC's chief examiner or the appropriate facility representative, as applicable, for review and grading in accordance with ES-403.

E. Post-Examination Reviews

1. If the NRC administered the examination, the chief examiner shall ensure that the master copy of the examination reflects all changes made to questions during the administration of the examination. The chief examiner will then provide a copy of the master examination and answer key to the facility staff and answer any questions they may have regarding the NRC's examination review and comment process.

2. If the NRC developed the examination, the chief examiner will also provide the facility licensee with a copy of the examination as edited during the facility prereview. If the facility reviewers believe that the NRC did not adequately resolve the prereview comments, they should address those concerns in a formal comment letter.

3. The NRC's chief examiner will ask the facility prereviewers to confirm that they did not divulge any information about the examination(s) by having them sign the post-examination security statement (Form ES-201-3) after the examinations are completed.

4. The facility licensee should submit formal comments within 5 working days after the examination is administered. However, the facility licensee may expedite the grading process by giving draft comments to the NRC chief examiner before he or she leaves the site. The NRC will consider comments not submitted within the requested time on a case-by-case basis; however, late comments may delay the examination grading process.

    The facility licensee should collect all comments from the license applicants and submit them to the NRC. When submitting applicant comments to the NRC, the facility licensee should identify by docket number which applicant made the comment (which may be useful to the NRC should the applicant request an informal review or a hearing), and include a facility position for each applicant comment. Note that the NRC examination report (refer to ES-501, Section E.3) will not identify examination comments by applicant docket number.

5. The facility licensee should submit all comments in the following format:

    *   List the question, answer, and reference.

    *   State the comment and make a recommendation as to whether the answer should be changed or the question should be deleted. If the facility licensee does not support an applicant's comment, it should briefly explain the reason for its rejection.

    *   Support the comment with a reference, and provide a copy if it was not included in the original reference material submittal. (Note: The NRC will not change the examination without a reference to support the facility's comment.)

6. Formal comments should be signed by an authorized facility representative and addressed to the responsible NRC regional office, with a copy to the NRC's chief examiner.

7. Although the NRC will review all post-examination comments submitted by a facility licensee, the agency is likely to approve only certain kinds of comments. In the interest of efficiency, facility licensees should consider the guidance contained in ES-403 Section D.1, before submitting post-examination comments to the NRC.

A.   Purpose

This standard explains the requirements and procedures for grading site-specific written examinations for the initial licensing of reactor operator (RO) and senior reactor operator (SRO) applicants at power reactor facilities. As such, this standard includes instructions for evaluating and revising the examinations after they are administered, grading the examinations, and conducting the first review of the graded examinations.

B.   Background

As discussed in ES-201, "Initial Operator Licensing Examination Process," facility licensees will generally develop and administer the initial operator licensing written examinations, subject to review and approval by the NRC. Facility licensees will also be expected to grade the written examinations; evaluate the outcome; and submit the examination results to the responsible NRC regional office for review, approval, and licensing action in accordance with ES-501, "Initial Post-Examination Activities."

C.   Responsibilities

1.   Facility Licensee

   a.   If the facility licensee developed and administered the written examinations, the licensee is also expected to perform the following grading activities, as described in Section D:

   •   Review and resolve any questions and comments that arose during and/or after the examination (refer to ES-402, "Administering Initial Written Examinations").

   •   Grade the examinations and review the grading using Form ES-403-1, "Written Examination Grading Quality Checklist."

   •   Evaluate the applicants' performance on the examination.

   Facility management will review the examination grading based on the guidance in ES-501 and will forward the graded examinations and all associated documentation to the NRC's chief examiner so that it is received, when practical, within 5 working days after the examination was administered.

   b.   If the NRC developed the examinations, the facility licensee's responsibility is limited to providing the NRC's chief examiner with comments and recommendations regarding question deletions and answer key changes. Such comments and recommendations should normally be received within

5 working days after the exit meeting; any delay in submitting the comments will likely result in a comparable delay in the final licensing actions. (Refer to ES-402 for additional instructions regarding the post-examination review and comment process.)

2.   NRC Regional Office

a.   If the facility licensee grades the examinations, the regional office shall provide guidance and assistance, as necessary, to ensure that the facility licensee complies with the instructions in Section D.

b.   If the NRC developed the examinations, the regional office should grade the examinations in accordance with Section D after receiving any comments and recommendations from the facility licensee (refer to ES-402). The regional office may take advantage of the facility licensee's machine grading capability if it is available.

c.   After the examinations have been graded, the regional office shall review the grading, process the documentation, and complete the licensing actions in accordance with ES-501.

D.   Grading Instructions

The author of the examination should normally grade the examination; however, the examination may be graded by another equally qualified individual if the author is not available, the number of applicants is unusually large, or the NRC regional office or facility licensee wishes to expedite the grading process. The examinations shall be graded as expeditiously as possible, in accordance with the following instructions:

1.   Evaluate Questions and Comments

a.   Evaluate all questions posed by applicants during the examination, any pen-and-ink changes made on the master examination during its administration, and any post-examination comments or recommendations received from the facility licensee and applicants after the examination was administered. Determine whether any questions should be deleted from the examination, or any answers need to be changed. Do not delete any question or change any answer unless there is a valid reference to support the change. An unreasonable assumption on the part of an applicant does not justify the acceptance of an alternative answer.

If there is some doubt as to whether the NRC's chief examiner will accept a proposed change, the grader is encouraged to discuss the matter with the chief examiner before proceeding with the grading process. This may help to minimize the need for grading corrections during the quality reviews.

For each comment and recommendation, the NRC's chief examiner shall document the reason that the question was changed or the comment was not accepted; this information will be included in the examination report, as discussed in ES-501.

b.   Despite the extensive reviews performed by both the NRC and the facility licensee prior to examination administration (refer to ES-201, Attachment 5), it is possible that a few isolated errors may be discovered only after an examination has been administered. The following types of errors, if identified and adequately justified by the facility licensee, are most likely to result in post-examination changes agreeable to the NRC:

   •   a question with an unclear stem that confused the applicants or did not provide all the necessary information

   •   unintended typographical errors in a question or on the answer key

   •   newly discovered technical information that supports a change in the answer key

   •   a question that is at the wrong license level (RO versus SRO) or not linked to job requirements

Given that both the NRC and the facility licensee agreed that the examination met NUREG-1021 prior to examination administration, the following types of question errors, identified *after* examination administration, are less likely to result in examination changes:

   •   a question which does not exactly match it's referenced K/A statement

   •   a question for which references would be needed to provide the correct answer, even though the facility licensee and the NRC previously agreed that the question should be closed-reference.

   •   a question that contains psychometric errors that do not increase its difficulty or make the question confusing. For example, a question with two implausible distractors or a collection of true false answers would be unsatisfactory during examination pre-review, but neither problem would justify deleting a question after examination administration.

Although the NRC will review all post-examination comments submitted by a facility licensee, in the interest of efficiency, facility licensees should consider the above examples prior to submitting post-examination comments to the NRC. Facility licensees with post-examination comments are encouraged to discuss them with the chief examiner prior to formally submitting any comments in writing.

c.   If it is determined that there are two correct answers, both answers will be accepted as correct. If, however, both answers contain conflicting information, the question will likely be deleted. For example, if part of one answer states that operators are required to insert a manual reactor scram, and part of another answer states that a manual scram is not required, then it is unlikely that both answers will be accepted as correct, and the question will probably be deleted.

If three or more answers could be considered correct or there is no correct answer, the question shall be deleted.

Annotate the recommended changes on the master examination and answer key, and document the reason for every change or deletion.

d. Those applicant questions, facility comments, and recommendations that do not result in answer key changes or question deletions, should be evaluated to determine whether the associated test questions might benefit from editorial changes before they are used on another examination.

e. Before depositing the questions in any examination bank, revise the questions to incorporate all changes, comments, and enhancements, as appropriate.

2. **Grade the Examinations**

a. Copy each applicant's answer sheet, and set the copies aside for later use during the grading review process.

b. On each applicant's original answer sheet, indicate in *red pen or pencil* which questions were answered incorrectly, note their correct answers, and indicate which questions (if any) were deleted. If the answer sheet is more than one page long, it is helpful to note the total number of incorrect answers on each page to aid in tabulating the final grade.

If the examinations are graded by machine, attach a copy of each applicant's profile report to his or her answer sheet, or manually annotate the answer sheet as noted above.

c. If it is necessary to change a grade during the grading process, do so by lining out the original grade in such a way that it remains legible. Briefly explain the reason for the change on the applicant's answer sheet, and initial the change. Under no circumstances will a grader use "white-out" or other methods that obscure the change.

d. After grading all the questions, enter the applicable "Examination Value(s)" (i.e., the original test point total minus the point value of any deleted questions) for the RO, SRO-only, and overall exams in the "Results" section of the applicant's written examination cover sheet (Form ES-401-7 for ROs, ES-401-8 for SROs, or ES-701-8 for SROs limited to fuel handling). Also enter the "Applicant's Score" and "Applicant's Grade" (i.e., the Applicant's Score divided by the Examination Value) on each part of the examination (RO, SRO, and overall) in the spaces provided on the form.

If a facility chooses to share its preliminary grades with the applicants, it should caution them that the outcome may change if the NRC does not accept all of the facility licensee's recommended changes to the examination answer key.

3. **Evaluate and Review the Grading**

    a. Evaluate the applicants' performance on each examination question to identify any indications of a problem with the question or a deficiency in the applicants' training program. A table that summarizes the applicants' answers on each question, or a computerized item analysis (if the examinations were graded by machine) may be used to identify items with which the applicants had problems.

        If it appears that a test question was faulty, determine whether the question should be deleted, the answer key should be changed, and/or the question should be revised before reuse. Then regrade the examinations as necessary.

        If it appears that the training program was deficient, determine the need for remedial training and/or a program upgrade.

    b. After evaluating the examinations, review the grading *in detail* and complete Form ES-403-1, "Examination Grading Quality Checklist."

    c. Forward the examination package (i.e., the master examination and answer key, justification for any examination changes, any item analysis that was performed, the applicant's examination cover and answer sheets (the graded original and one clean copy), and Form ES-403-1) to the designated facility representative (if applicable) or to the NRC's chief examiner for review in accordance with ES-501.

E. Attachments/Forms

Form ES-403-1,   "Written Examination Grading Quality Checklist"

Facility:	Date of Exam:		Exam Level: RO ☐ SRO ☐		
			Initials		
Item Description			a	b	c
1.	Clean answer sheets copied before grading				
2.	Answer key changes and question deletions justified and documented				
3.	Applicants' scores checked for addition errors (reviewers spot check > 25% of examinations)				
4.	Grading for all borderline cases (80 ±2% overall and 70 or 80, as applicable, ±4% on the SRO-only) reviewed in detail				
5.	All other failing examinations checked to ensure that grades are justified				
6.	Performance on missed questions checked for training deficiencies and wording problems; evaluate validity of questions missed by half or more of the applicants				

	Printed Name/Signature	Date
a. Grader	_____	_____
b. Facility Reviewer(*)	_____	_____
c. NRC Chief Examiner (*)	_____	_____
d. NRC Supervisor (*)	_____	_____

(*)   The facility reviewer's signature is not applicable for examinations graded by the NRC; two independent NRC reviews are required.

ES-501
INITIAL POST-EXAMINATION ACTIVITIES

A.   Purpose

This standard describes and coordinates the activities that must be completed after the initial operating tests and written examinations have been administered and graded in accordance with the ES-300 and ES-400 series of the examination standards, respectively. Specifically, this standard includes instructions for assembling and reviewing the examination package, notifying the facility licensee and applicants of the examination results, preparing the examination report, and retaining examination records.

B.   Background

The goal of the NRR operator licensing program office is to complete licensing or denial actions within 30 days after the facility licensee submits the graded examinations or its formal written examination comments to the NRC. The NRC and facility licensee staffs should establish their priorities and schedules to achieve this goal.

C.   Responsibilities

1.   Facility Licensee

a.   If the facility licensee participated in developing, administering, and grading the examination, the licensee shall forward the following examination documentation to the NRC's chief examiner (marked "addressee only") as soon as possible (within 5 working days, when practical) after administering the examinations:

- the graded written examinations (i.e., each applicant's original answer and examination cover sheets) plus a clean copy of each applicant's answer sheet (ES-403, "Grading Initial Site-Specific Written Examinations")

- the master examination(s) and answer key(s), annotated to indicate any changes made while administering and grading the examination(s) (ES-402, "Administering Initial Written Examinations," and ES-403)

- any questions asked by and answers given to the applicants during the written examination (ES-402)

- any substantive comments made by the applicants following the written examination, with an explanation concerning why the comment was accepted or rejected (this item is encouraged but not required) (ES-402)

- the written examination seating chart (ES-402)

- a completed Form ES-403-1, "Written Examination Grading Quality Checklist" (ES-403 and Section D.1)

- the results of any written examination performance analysis that was performed, with recommended substantive changes (ES-403)

- original Form(s) ES-201-3, "Examination Security Agreement," with a pre- and post-examination signature by every individual who had detailed knowledge of any part of the operating tests or written examination before they were administered.

Refer to the referenced examination standards for a more detailed discussion of each documentation requirement.

b. If the facility licensee did not participate in developing, administering, and grading the examination, the licensee should submit comments and recommendations regarding the NRC-developed written examination to the NRC's regional office as soon as possible (within 5 working days, when practical) after the exit meeting. The facility licensee should also include and consider comments made by the license applicants who took the examination. (Refer to ES-402 for more detailed instructions.)

2. NRC Regional Office

a. The NRC's regional office shall ensure that the operating tests and written examinations are graded in accordance with ES-303, "Documenting and Grading Initial Operating Tests," and ES-403, respectively.

b. The NRC's regional office shall ensure that the examination results and licensing recommendations receive the required reviews and approvals in accordance with Section D, that the associated administrative requirements are completed in accordance with Section E, and that the required records are retained in accordance with Section F.

The regional office may use Form ES-501-1, "Post-Examination Check Sheet," to track completion of the administrative items after the examinations are administered.

c. NRC regional management should also review the overall examination results and any generic findings, deficiencies, or issues to determine whether any followup action is required.

If the facility licensee recommends deleting or changing the answers to four or more of the questions on an RO written examination (or two or more on an SRO-only exam) that it developed, the regional office should ask the facility licensee to explain why so many post-examination changes were necessary and what actions will be taken to improve future license examinations. As discussed in Section E.3.a, below, the regional office will also consider post-examination deletions and changes when evaluating the quality of the facility licensee's proposed examination for documentation in the examination report.

If seven or more of the questions on an RO examination (or two or more on an SRO-only exam) are deleted during the grading process, the regional office shall evaluate the remainder of the examination to ensure that it still satisfies the test outline sampling requirements in ES-401, "Preparing Initial Site-Specific Written Examinations." The regional office should consult with the training and assessment specialist in the NRR operator licensing program office if the validity of the examination is in question.

If the content validity of the examination is affected [e.g., several knowledge and ability (K/A) topics are not covered, or the majority of the remaining K/As are associated with a small number of systems) as a result of deleting questions, the NRR operator licensing program office will make a decision concerning whether the examination should be voided.

D.   Examination Reviews and Licensing Action

Except as noted below, the quality reviews generally constitute spot checks, or sampling, to follow up on the work performed by the operating test and written examination graders in accordance with ES-303 and ES-403, respectively. If the quality reviews indicate significant problems, additional detailed review will be necessary.

Reviewers should discuss all grading discrepancies with the grader or previous reviewer before making any changes. In addition, the reviewers shall document any changes by carefully lining out the original entry so that it remains legible, entering the revision with a brief explanation, and initialing the change. Reviewers shall not use "white-out" or other methods that obscure the original entry.

1.   <u>Facility Management</u>

If the facility licensee graded the written examinations, a supervisor or manager shall confirm the quality of the grading and sign the bottom of Form ES-403-1 before sending the examinations to the NRC's regional office.

The NRC will consider the signed form to represent facility management concurrence with the individual and collective examination results, including the justification(s) for any examination change(s).

2. NRC Chief Examiner (or Designee)

The written examination grading shall be independently reviewed by at least two NRC personnel using Form ES-403-1 as a guide. If the examination was graded by the chief examiner, another examiner shall conduct the independent review. If the chief examiner conducts the independent review, he or she may not perform the supervisory review required by Section D.3.

a. If the facility licensee graded the written examinations, the chief examiner shall immediately inventory the examination package to ensure that all required materials have been submitted. The chief examiner shall inform the responsible supervisor of any obvious deficiencies, and shall contact the facility licensee to determine the status of any missing documentation.

b. The chief examiner shall independently analyze *each* examination and answer key change that was made or recommended by the facility licensee or a license applicant to determine whether it is justified. During the analysis, the chief examiner will keep in mind that both the facility licensee and the NRC had previously agreed that the examination met the requirements of NUREG-1021 (refer to ES-201, Attachment 5). Therefore, as discussed in Section D.1 of ES-403, certain kinds of post-examination comments and recommendations are less likely to justify grading or answer key changes.

   The chief examiner shall ensure that the reason for accepting or rejecting each change or recommendation is documented in the examination report. The report shall briefly state the region's basis for accepting or rejecting each facility comment; simply stating concurrence with no explanation is not sufficient. The chief examiner will not accept a change to the examination unless the facility licensee submits a valid reference to support its recommendation.

c. The chief examiner shall review the remaining items on Form ES-403-1. In so doing, the chief examiner should apply his or her judgment when reviewing the examination results and should adjust the level of the review based on the performance of the applicants and the facility licensee (e.g., the number of questions changed or deleted, the average grade, the number of borderline or failing grades, etc.). If the examination was graded by machine or using a template, the chief examiner shall ensure that the template accurately parallels the approved answer key.

   The chief examiner shall independently grade every borderline examination [i.e., those between 78 and 82 percent overall and between 66 and 74 percent on the SRO-only portion (or 76 and 84 percent if the RO portion was waived), as applicable] using the final, approved answer key and the clean applicant answer sheets provided by the facility licensee.

d. The chief examiner should review the written examination results and the facility licensee's performance analysis (if applicable) for indications of the following:

   • deficiencies in the applicants' training program, so that they may be addressed in the examination report

- poor question construction, so that the applicants are not graded unfairly, any significant problems can be addressed in the examination report, and the questions are corrected before reuse

- any indications that the examination was compromised

e. When satisfied with the examination grading, the chief examiner and written examination grader/reviewer (as applicable) shall complete the following actions:

- Sign and date Form ES-403-1 and pass it on to the responsible supervisor for management review (see Section D.2.h).

- Record the written examination results (including RO, SRO, and total points and grades from each applicant's Form ES-401-7, ES-401-8, or ES-701-8) and the names of the NRC examiners who wrote, graded, or reviewed the examinations in the "Written Examination Summary" section of each applicant's Form ES-303-1, "Individual Examination Report."

- Check the written examination "Pass," "Fail," or "Waive" block in the "Examiner Recommendations" section of each applicant's Form ES-303-1 and sign in the space provided. In order to pass the examination, applicants must achieve an overall grade of at least 80.00 percent, with a 70.00 percent or better on the SRO-only items, if applicable. Applicants who only take the SRO portion of the exam (as a retake or with an upgrade waiver of the RO exam) must achieve an 80.00 percent or better to pass. SRO-upgrade applicants who do take the RO portion of the exam and score below 80.00 percent on that part of the exam can still pass overall, but may require remediation (refer to Section E.4.a). SRO-instant applicants who pass the operating test and the written examination overall but fail the SRO portion of the written exam are not automatically eligible for an RO license; however, they may reapply for an RO license, and request an examination waiver, after accepting a final denial of their SRO application (refer to Section D.1.a of ES-204, "Processing Waivers Requested by Reactor Operator and Senior Reactor Operator Applicants").

f. The chief examiner shall also review, *in detail*, the other examiners' operating test documentation to ensure that the test (as given) and its grading meet the requirements in ES-301 and ES-303. In so doing, the chief examiner shall ensure that the other examiners' operating test comments support the pass or fail recommendations and check for consistent documentation and grading among the operators tested on the same simulator crew.

If the documentation is accurate and complete, and the licensing recommendation is appropriate, the chief examiner shall check "Pass" or "Fail" and sign and date the "Final Recommendation" block on Form ES-303-1. By contrast, if the licensing recommendation is not appropriate based on the documentation presented, the chief examiner shall discuss the examination findings with the NRC examiner of record and resolve any disagreement.

If the chief examiner administered the operating test, the responsible regional supervisor shall designate another examiner to independently review the documentation and sign the "Final Recommendation" block on Page 1 of Form ES-303-1.

g. The chief examiner shall record the results of the operating tests and written examinations (including the RO, SRO, and overall grades for each applicant) on Form ES-501-2, "Power Plant Examination Results Summary."

h. The chief examiner shall ensure that the examination documentation is complete and contains all of the items identified in Section F before forwarding the entire package to the responsible supervisor for review and approval in accordance with Section D.3.

    If the written examinations were administered much before the operating tests, the chief examiner should enter that data on the form and forward it with the completed written examination package to the responsible supervisor for review and approval in advance of the operating test results.

3. **NRC Management Review and Licensing Action**

a. The responsible supervisor shall ensure that all examination results and documentation are complete. The supervisor shall evaluate the written examination results, ensure that the required quality reviews were completed, work with the chief examiner and the facility licensee (as necessary) to resolve any grading problems, and then sign and date Form ES-403-1 to document approval of the process.

    Every written examination shall have at least two levels of NRC review. Therefore, the NRC examiner who performed the regional quality review is disqualified from also performing the supervisory review.

b. The responsible supervisor will also independently review the operating test results, check the "Issue License" or "Deny License" block in the "License Recommendation" section of each applicant's Form ES-303-1, and sign and date each form. Under no circumstances will the same individual sign all three levels of recommendation on Form ES-303-1 (i.e., operating test administrator, chief examiner, and NRC supervisor).

    If the responsible supervisor (or licensing official) does not believe that the operating test documentation supports the final recommendation, he or she shall consult the NRC examiner of record and the chief examiner to discuss and resolve any disagreements.

    As discussed in Section C.2 of ES-303, any operating test licensing recommendation that deviates from the nominal grading instructions in Section D.2 of ES-303 (e.g., recommending a simulator test failure based on a single error with serious safety consequences or a passing grade despite multiple errors related to the same rating factor) requires written concurrence from the NRR operator licensing program office before completing the licensing action.

If a recommendation is overturned during the review by regional management (or the NRR operator licensing program office), the responsible supervisor will line out and initial the affected summary evaluations. The supervisor will then enter the new summary evaluation in the appropriate block, explain the change on Form ES-303-2, "Operating Test Comments," and attach that comment form to the applicant's Form ES-303-1.

c. After making the licensing recommendations, the responsible supervisor will have the operator licensing assistant prepare a license, denial, or notification letter for each examined applicant and forward the examination package to the regional licensing official. However, the operator licensing assistant shall not send a denial letter to applicants who withdrew before taking any part of the license examination. Attachments 3 and 4 to this examination standard provide sample RO and SRO (conditional) license letters, as well as a sample denial letter.

Attachment 5 to this examination standard is a sample letter for use in notifying applicants that they passed the examination, but that their licensing action will be delayed. For example, if the NRC granted an applicant a waiver and allowed him or her to take the examination before completing all of the training and experience requirements, the regional office shall normally not issue a license to the applicant until the facility licensee has certified in writing that the applicant has completed all of the waived items. (Refer to ES-202, "Preparing and Reviewing Operator License Applications," and ES-204.) Likewise, if any of the applicants failed the written examination, the regional office shall analyze the question-by-question performance of those applicants who scored 82 percent or lower on the examination overall (or 74/84 percent, as applicable, or lower on the SRO-only items) to ensure that any question deletions or changes will not affect the licensing decision. In addition, if necessary, the regional office shall delay issuing licenses to those applicants until any written examination appeals have been reviewed for impact on the licensing decisions.

Before issuing a license in either instance, the regional office shall ensure that the applicant (1) has been determined to be medically fit within the past 24 months; (2) has not developed any permanent physical or mental condition that would be reportable under Title 10, Section 55.25, of the *Code of Federal Regulations* (10 CFR 55.25); and (3) is up-to-date in the requalification training program. Moreover, the regional office shall advise the facility licensee to properly activate the individual's license in accordance with 10 CFR 55.53(f) if more than 3 months have passed since the examination results were issued. If a licensing action is delayed for any reason, the effective date of the license will be the date on which it is issued; licenses will not be backdated.

d. The final licensing decision is made by the NRC's regional administrator or his or her designee, who must be at or above the branch chief level; short-term designees shall not make licensing decisions. The licensing official will consider all recommendations; make changes as described above; and sign each applicant's license, denial, or notification letter, as applicable.

E.   Examination Followup

1.   <u>Notify Facility Licensee of Results</u>

The NRC's regional office will notify the facility licensee and applicants of the examination results (as described below) only after they are reviewed and approved by the licensing official.

a.   The regional office should normally notify the facility licensee's designated representative of the examination results by telephone, and may confirm the results by mailing a copy of Form ES-501-2 under a separate cover letter. For each applicant who failed or had significant deficiencies that warrant further evaluation and retraining by the facility licensee, the regional office will also send the facility licensee a copy of the applicant's Form ES-303-1 and written examination answer sheet. These form(s) shall *not* be placed in the NRC's Public Document Room or distributed with the final examination report.

If the written examinations were administered much before the operating tests and management has approved the results of those examinations, the regional office may notify the facility licensee of those results rather than waiting until the operating tests are completed.

b.   After the licensing official has signed the license, denial, and notification letters, the regional office shall send each applicant's letter along with the following materials:

* a copy of Forms ES-303-1, ES-303-2, and ES-D-1 (and Form ES-D-2 if the applicant failed the simulator operating test) reflecting the "as run" scenario conditions but *without* any rough examiner notes regarding the applicant's performance (pen-and-ink markups of the original, approved scenarios are acceptable)

* a copy of the applicant's written examination cover and answer sheets (as well as a copy of the master written examination and answer key if the applicant failed the written examination)

c.   The regional office shall send a copy of Form ES-501-2 to the NRR operator licensing program office. If any of the examinations are later regraded in response to an applicant's request for review (refer to ES-502, "Processing Requests for Administrative Reviews and Hearings After Initial License Denial"), the original Form ES-501-2 on file in the regional office shall be corrected by lining out the old grade, entering the new grade, and initialing the change. Whenever a change is made, the regional office shall mail a copy of the revised form to the program office.

d.   The responsible supervisor should consider phoning the facility licensee management counterpart to discuss the examination outcome and lessons learned. Any pertinent feedback on the examination process should be forwarded to the operator licensing program office for consideration.

2.   Return the Facility Reference Material

If the facility licensee desires, the NRC's chief examiner shall ensure that the reference materials provided for NRC examiners to use in preparing for the examinations are returned to the facility licensee as soon as possible. If none of the applicants failed the examination, the materials should be returned as soon as the NRC issues the licenses. If any applicant was denied a license based on an examination failure, the reference materials should be retained pending expiration of the 20-day period during which the applicant may request a regrade. If an applicant requests a regrade in accordance with ES-502, the chief examiner shall determine which reference materials need to be retained and should return all unnecessary materials. All reference materials should be returned to the facility licensee within 30 days following the resolution of any appeals.

3.   Prepare the Examination Report

The NRC's chief examiner shall prepare the final examination report when all portions of the examination have been graded and documented. If the regional office delays some licensing actions in accordance with Section D.3, it should issue and later amend the examination report. The examiner should follow the principles in NRC Manual Chapter 0612, "Power Reactor Inspection Reports," when preparing the report.

a.   The final examination report shall document the following considerations:

- whether the quality of the submitted examination material was within the range of acceptability expected by the NRC. This will be determined as follows:

  - The NRC will evaluate the submitted written examination questions (RO and SRO questions shall be considered separately) using the guidance in Sections E.2-3 of ES-401 to determine the percentage of submitted questions that required replacement or significant modification or that clearly did not conform with the intent of the approved K/A statement. Any questions that were deleted during the grading process, or for which the answer key had to be changed, will also be included in the count of unacceptable questions.

  - The NRC will evaluate the submitted operating test material by combining the scenario events and JPMs (e.g., an operating test composed of 5 administrative JPMs, 10 systems JPMs, and 2 scenarios with 6 events or malfunctions each would total 27 proposed test items for evaluation). For the combined total, the NRC will determine the percentage of submitted test items that required replacement or significant modification to conform with the acceptance criteria in Section D of ES-301.

–   Note: If the review indicated that a specific event in a scenario did not require significant, discriminatory operator actions, it should not be included in the total unless that event was one of the required minimum events for any of the applicants according to Form ES-301-5 or the entire scenario was inadequate. Specific malfunctions that were added to the scenarios to provide complications or distractions for other events should not be judged solely on their individual merits.

–   If 20 percent or fewer of the test items for the submitted operating test, RO written examination, and SRO written examination (assessed separately) required replacement or significant modification, the report will simply state that the facility licensee's submittal was within the range of acceptability expected for a proposed examination. If applicable, an observation shall be included, indicating that the examination changes agreed upon between the NRC and the facility were made according to NUREG-1021.

–   Note: NRC-validated questions, JPMs, and scenario events that required replacement or substantial modification will not be counted unless the facility licensee caused the current unacceptable flaw since the time the NRC previously approved the test item. (For example, the question's reference changed, but the question was not revised accordingly.)

–   If more than 20 percent[1] of the submitted test items (with the operating test and RO/SRO written exams assessed separately) required replacement or significant modification, the report shall include a factual description of the test item changes (observations), including the number and types of test items replaced and/or significantly modified as a result of the joint NRC and facility licensee examination review process. The report shall also note that the overall submittal was outside the acceptable quality range expected by the NRC and that future examination submittals should incorporate any lessons learned from this effort.

–   Negative observations regarding the adequacy of the facility licensee's proposed examination (e.g., stating that the proposed examination was not adequate for administration) shall only be made if the examination was not the facility's first submittal and the NRR operator licensing program office has concurred in the evaluation.

---

[1]   Note that the nominal 20 percent threshold may be raised or lowered, based on the specific circumstances, with NRR operator licensing program office concurrence. For example, no comment may be warranted if the same error was made in a number of questions; conversely, a comment may be warranted based on the egregious nature of the deficiencies even though the 20 percent threshold was not reached.

- any delay in administering the examination and the reason for the delay, and any extensions of the written examination time beyond the nominal time limits specified in ES-402

- the results of the examination, including any significant grading deficiencies if the facility licensee graded the examinations

- an overview of the examination security measures and activities evaluated while preparing and administering the examinations and any examination security issues and incidents or other matters requiring facility attention

  Note that initial examination security issues will generally be documented in the examination report if (1) the potential or actual compromise was discovered while developing the examination and resulted in replacing or modifying any proposed test item(s); (2) the potential or actual compromise was discovered after the examination was administered, but would have resulted in replacing or modifying test items if the NRC had known about it earlier; (3) two or more lesser security issues were discovered, but did not necessitate the replacement of test material; or (4) other security issues were discovered with extenuating circumstances (with concurrence from the NRR operator licensing program office).

- any other issues or findings discussed at the exit meeting

b. The report shall include (or cite the accession number for) the following items, as applicable:

- a copy of the final written examination(s) and answer key(s) with all changes (during and post-examination) incorporated

- a copy of the facility licensee's (and applicants') specific comments and recommended changes regarding the operating tests and written examination that were administered. If applicant comments were submitted, redact the applicant docket number from the examination report. The NRC regional office shall retain a non-redacted version, indicating applicant docket numbers, until any informal administrative reviews or hearings are complete (refer to ES-502).

- the specific NRC explanation for accepting or rejecting each facility recommendation and a specific justification for every additional item deletion or change (refer to Attachment 1 for examples of facility comments and NRC resolutions)

- a simulator fidelity report (as described below, when applicable)

Generic comments submitted by the facility licensee about the examinations or the administration process should also be included in the report, accompanied by regional office responses, as appropriate.

c.   The simulator fidelity report shall document the NRC examiners' evaluation of the performance or fidelity of the simulation facility during the preparation or conduct of the operating tests. Attachment 2 provides a sample report.

All previously undocumented simulator deficiencies encountered while preparing or conducting the operating tests should be described in sufficient detail to allow followup the next time the NRC staff conducts Inspection Procedure (IP) 71111.11, "Licensed Operator Requalification Program," at the facility. The NRC examiners may include in the simulator fidelity report any concerns about physical fidelity (hardware or equipment discrepancies) or functional fidelity (performance of the simulation facility during normal, surveillance, abnormal, or emergency events). Each deficiency should include a description of the operation, event, or transient that was in progress, and how the simulation facility failed to accurately model the expected performance of the reference plant.

d.   The applicants' names and specific grades (i.e., Form ES-501-2) shall *not* be published in the examination report.

e.   The NRC's regional office shall send the final examination report to the facility licensee and ensure that a copy is made available to the public.

4.   Perform Other Activities

a.   If an applicant did not complete the SRO upgrade training program or failed the upgrade examination, regional management should ensure that the RO licensee complies with the requirements of 10 CFR 55.53(e), (f), and (h) and 10 CFR 55.59(a) before resuming active duties as an RO.

Similarly, the regional office should ensure that SRO upgrade applicants who did not participate in RO requalification training while they were enrolled in the upgrade training program comply with 10 CFR 55.59(a). If an applicant missed the annual operating test and/or the comprehensive written requalification examination required by 10 CFR 55.59(a)(2) and then did not take the RO portion of the written licensing examination, the applicant must complete additional training in accordance with 10 CFR 55.59(b) and must make up the missed requalification examination to verify proficiency in the 10 CFR 55.41 topics before resuming licensed duties as an RO **or an SRO** (which requires testing on both 10 CFR 55.41 and 55.43 items). However, the NRC would consider the requirements of 10 CFR 55.59(a)(2)(i) to be satisfied if the applicant repeats the applicable portions (to be determined using a systematic approach) of the license training program and passes a comprehensive audit examination covering the topics required by 10 CFR 55.41.

SRO applicants who passed the written examination overall but scored below 80 percent on either the RO or SRO-only portion will require additional review to determine the nature of their deficiencies and the need for additional training. Pursuant to 10 CFR 55.7, the NRC may, by rule, regulation, or order, impose upon any licensee additional requirements deemed appropriate or necessary to protect public health and to minimize danger to life and property. If the SRO applicant's deficiencies pose such a threat, the NRC may require the facility licensee to provide remedial training and reevaluation and to submit evidence of its completion to the NRC.

b. Once the licensing decisions are complete, the NRC examiners should discard any marked-up documentation or rough notes for those applicants who receive licenses (except as noted below). In accordance with ES-502, NRC examiners should retain all applicable notes and documentation associated with proposed denials until the denials become final; this may include simulator operating test notes regarding crew members who passed the test if the notes contain information relevant to the failing applicant's performance. Examiners are advised that such notes would be subject to disclosure if requested under the Freedom of Information Act.

c. Agency policy requires that all documents that are not classified, proprietary, sensitive or otherwise protected (e.g., under the Privacy Act or Freedom of Information Act) must be made available to the public. Therefore, the NRC's regional office shall ensure that all documents associated with the licensing examination (i.e., those listed in Section F.1, below), excluding those containing the applicants' names or grades, are placed in the NRC's Public Document Room as soon as possible after the examinations have been completed. NRC Manual Chapter 0620, "Inspection Documents and Records," and SECY-04-0191, "Withholding Sensitive Unclassified Information Concerning Nuclear Power Reactors from Public Disclosure," provide additional policies and guidance in this area.

F. NRC Record Retention

1. The NRC's regional office shall ensure, for the most recent initial examination at each facility, that originals (whenever possible) or copies of the following items either are retained in the facility's master examination file or are electronically available via the NRC's Agencywide Documents Access and Management System (ADAMS). The italicized items should be retained or available for the last two examinations at each facility so that examiners can verify compliance with the guidelines for test item repetition.

a. ES-201, Attachment 4, "Corporate Notification Letter"

b. ES-201, Attachment 5, "Examination Approval Letter," with pen-and-ink changes on Form ES-201-4, "List of Applicants," to identify the applicants who were actually examined

c. Form ES-201-1, "Examination Preparation Checklist"

d. the written examination and operating test outline(s), along with Form ES-201-2, "Examination Outline Quality Checklist," and Form ES-401-4, "Record of Rejected K/As" (or the equivalent LSRO forms from ES-701)

e. the proposed NRC- or facility-developed operating tests and written examination (including comments made by the facility licensee or the NRC, as applicable)

f. *the final written examination and answer key* with all changes incorporated (the pen-and-ink corrections made for the applicants while the examination was administered may be changed to typewritten corrections; however, all changes shall be annotated in such a way that they are evident),

Forms ES-401-6, "Written Examination Quality Checklist," and ES-401-9, "Written Examination Review Worksheet" (or the equivalent LSRO forms from ES-701), and any reference handouts (or a list thereof) provided to the applicants

g.   *the as-given scenarios including Forms ES-D-1, "Scenario Outline," and ES-D-2, "Required Operator Actions," for each scenario set administered, as well as the as-given walk-through tests including Forms ES-301-1, "Administrative Topics Outline," and ES-301-2, "Control Room/In-Plant Systems Outline," and the JPMs for each walk-through test* (all record copies should reflect the "as run" test conditions; pen-and-ink markups of the original, approved forms are acceptable)

h.   for each operating test administered, Form ES-301-3, "Operating Test Quality Checklist," Form ES-301-4, "Simulator Scenario Quality Checklist," Form ES-301-5, "Transient and Event Checklist," and Form ES-301-6, "Competencies Checklist" (or the equivalent LSRO forms from ES-701)

i.   Form ES-403-1, "Written Examination Grading Quality Checklist"

j.   Form ES-501-2, "Power Plant Examination Results Summary Sheet"

k.   *the final "Examination Report," with all enclosures*

l.   Form ES-201-3, "Examination Security Agreements"

2.   The NRC's regional office shall place the following items[2] in each applicant's docket file:

a.   Forms ES-303-1, "Individual Examination Report," ES-303-2, "Operating Test Comments" (original copies, all pages, including strip charts and other attachments that support the licensing decision), and ES-D-1, "Scenario Outline," as well as Form(s) ES-D-2, "Required Operator Actions," if the applicant failed the simulator operating test (all record copies should have the required signatures and reflect the "as run" test conditions; pen-and-ink markups of the original, approved forms are acceptable)

b.   all correspondence with the applicant

c.   the applicant's original written examination cover sheet (Form ES-401-7, ES-401-8, or ES-701-8) and answer sheet

---

[2]   These paper documents are official agency records and need not be placed in ADAMS. If they are placed in ADAMS, the regional office shall exercise caution to ensure that they are not accessible to the public because they contain information that is protected under the Privacy Act.

G.   Attachments/Forms

Attachment 1,   "Sample Facility Comments and NRC Resolutions"
Attachment 2,   "Sample Simulator Fidelity Report"
Attachment 3,   "Sample License Letters"
Attachment 4,   "Sample Proposed Denial Letter"
Attachment 5,   "Sample Notification Letter"
Form ES-501-1,   "Post-Examination Check Sheet"
Form ES-501-2,   "Power Plant Examination Results Summary"

Question #28

Comment: The question asks for the required method of securing a diesel generator and ensuring that an auto restart does not recur following auto initiation on receipt of a valid loss-of-coolant accident (LOCA) signal with offsite power still available to its associated emergency bus. The question is recommended for deletion because the system operating procedure directs that the diesel be unloaded, verifying that the 4KV bus auto transfer annunciator is reset, and then secured by placing the handswitch in "pull to lock." Therefore, the key answer (i.e., ensure that the "4KV AUTO TRANSFER INOP" annunciator is *lit* before placing the control switch in PULL TO LOCK) is incorrect.

NRC Resolution: Recommendation accepted. The question is deleted because there is no correct answer. The intended answer specified that the annunciator be confirmed as "lit" when it should have specified "reset" in accordance with System Operating Procedure No. 123, Section 5.1 (Rev. 29).

Question #51

Comment: The question asks for a description of the operation of the residual heat removal (RHR) Loop B outboard injection valve if the level rapidly decreases to 119.5 inches with RHR Loop B operating in the shutdown cooling mode. The question is recommended for deletion because the outboard injection valve reopens automatically when the Group 4 isolation is reset, if a low-pressure coolant injection (LPCI) loop selection is sealed-in. Therefore, the key answer (i.e., the operator must reset the shutdown cooling isolation and manually reopen the RHR Loop B outboard injection valve) is incorrect.

NRC Resolution: Recommendation not accepted. The RHR Loop B outboard injection valve will not auto-open unless the operator manually resets the shutdown cooling isolation signal. Therefore, the use of the phrase "manually reopen" is correct, and the key answer is correct. The facility-provided reference supports that manual action is required to open the injection valve.

Facility Licensee: _____(Facility name)_____

Facility Docket No.: _____(number)_____

Operating Tests Administered on: _____(date)_____

This form is to be used only to report observations. These observations do not constitute audit or inspection findings and, without further verification and review in accordance with IP 71111.11, are not indicative of noncompliance with 10 CFR 55.46. No licensee action is required in response to these observations.

While conducting the simulator portion of the operating tests, examiners observed the following items:

(EXAMPLES)

Item	Description
HPSI Header B pressure (PI-301)	The pressure instrument read mid-scale regardless of actual pressure.
Head bubble	During a scenario that caused a rapid depressurization during natural circulation, the vessel head level indication indicated a void (bubble). The confirming indications (i.e., pressurizer level and pressure) failed to verify or confirm the bubble.
Steam Generator A wide-range level	The meter has been out of service for the last three operating tests (approximately 18 months).

NRC Letterhead

(Date)

LICENSE

(Applicant's name)
(Street address)
(City, State Zip code)

Pursuant to the *Atomic Energy Act of 1954*, as amended; the *Energy Reorganization Act of 1974* (Public Law 93-438), as amended; and subject to the conditions and limitations incorporated herein, the U.S. Nuclear Regulatory Commission hereby licenses you to manipulate all controls of the (Name of facility, facility license number).

Your License No. is OP-(number), and your Docket No. is 55-(number). The effective date is (date). Unless sooner terminated, renewed, or upgraded, this license shall expire 6 years from the effective date.

This license is subject to the provisions of Title 10, Section 55.53, of the *Code of Federal Regulations* (10 CFR 55.53), with the same force and effect as if fully set forth herein.

While performing licensed duties, you shall observe the operating procedures and other conditions specified in the facility license authorizing operation of the facility.

The issuance of this license is based upon examination of your qualifications, including the representations and information contained in your application for this license.

A copy of this license has been made available to the facility licensee.

For the U.S. Nuclear Regulatory Commission,

(Name and title of licensing official)

Docket No. 55-(number)

cc: (Facility representative who signed the applicant's NRC Form 398)

NRC Letterhead

<u>(Date)</u>

LICENSE

<u>(Applicant's name)</u>
<u>(Street address)</u>
<u>(City, State Zip code)</u>

Pursuant to the *Atomic Energy Act of 1954*, as amended; the *Energy Reorganization Act of 1974* (Public Law 93-438), as amended; and subject to the conditions and limitations incorporated herein, the U.S. Nuclear Regulatory Commission hereby licenses you to direct the [licensed] [[fuel handling]] activities of [licensed] operators at, and to manipulate [all] [[fuel handling]] controls of the <u>(Name of facility, facility license number)</u>.

Your License No. is SOP-<u>(number)</u>, and your Docket No. is 55-<u>(number)</u>. The effective date is <u>(date)</u>. Unless sooner terminated, renewed, or upgraded, this license shall expire 6 years from the effective date.

This license is subject to the provisions of Title 10, Section 55.53, of the *Code of Federal Regulations* (10 CFR 55.53), with the same force and effect as if fully set forth herein.

While performing licensed duties, you shall observe the operating procedures and other conditions specified in the facility license authorizing operation of the facility. You shall also comply with the following condition<u>(s)</u>:

•   You shall wear corrective lenses while performing the activities for which you are licensed.

The issuance of this license is based upon examination of your qualifications, including the representations and information contained in your application for this license.

A copy of this license has been made available to the facility licensee.

For the U.S. Nuclear Regulatory Commission,

<u>(Name and title of licensing official)</u>

Docket No. 55-<u>(number)</u>

cc: <u>(Facility representative who signed the applicant's NRC Form 398)</u>

[ ]   Include only for unrestricted senior operators.
[[ ]]   Include only for senior operators limited to fuel handling.

NRC Letterhead

(Date)

(Applicant's name)
(Street address)
(City, State, Zip code)

Dear (Name):

This is to inform you that your grade on the (operating test, written examination, or both) taken on (date(s)), in connection with your application for a (reactor operator, senior reactor operator) license for the (facility name), indicates that you **did not** pass that (test, examination, or both). As a result, the U.S. Nuclear Regulatory Commission (NRC) proposes to deny your application. Enclosed is a copy of the (operating test, written examination, or both) results indicating those areas in which you exhibited deficiencies. (A copy of the master answer key is also provided.)

If you accept the proposed denial and decline to request either an informal NRC staff review or a hearing within 20 days, as discussed below, this proposed denial will become a final denial. You may then reapply for a license in accordance with Title 10, Section 55.35, of the *Code of Federal Regulations* (10 CFR 55.35), subject to the following conditions:

*   a.   Because you passed (a written examination and/or the administrative/systems/ simulator operating test) on (date(s)), you may request a waiver of (that or those) portion(s).

*   b.   Because you did not pass the (SRO portion of or written examination overall or administrative/systems/simulator operating test) administered to you on (date), you will be required to retake that portion.

*   c.   You may reapply for a license 2 months from the date of this letter.

**   a.   Because this is your (second, subsequent) examination failure, you will be required to retake both the written examination and the operating test.

**   b.   You may reapply for a license (6, 24) months from the date of this letter.

***   a.   Because you did not pass either the operating test or the written examination administered to you on (date(s)), you will be required to retake both the operating test and the written examination.

***   b.   You may reapply for a license (2, 6, 24) months from the date of this letter.

If you do not accept the proposed denial, you may, within 20 days of the date of this letter, take either of the following actions:

•   You may request an informal NRC staff review of the grading of your examination. Send your written request to Director, Division of Inspection and Regional Support, Office of Nuclear Reactor Regulation, U.S. Nuclear Regulatory Commission, Washington, DC 20555-0001. If submitting via private courier (e.g., FedEx, UPS), send your request to 11555 Rockville Pike, Rockville, Maryland 20852, instead of using the Washington, DC, address. Your request must identify the portions of your examination that you believe were graded incorrectly or too severely. In addition, you must provide the basis, including supporting documentation (such as procedures, instructions, computer printouts, and chart traces), in as much detail as possible, to support your contention that certain of your responses were graded incorrectly or too severely.

The NRC will review your contentions, reconsider your grading, and inform you of the results. If the proposed denial is sustained, you will have the opportunity to request a hearing pursuant to 10 CFR 2.103(b)(2) at that time.

- You may request a hearing pursuant to 10 CFR 2.103(b)(2). Submit your request, in writing, to the Office of the Secretary, U.S. Nuclear Regulatory Commission, Washington, DC 20555-0001, Attention: Rulemakings and Adjudications Staff, with a copy to the Associate General Counsel for Hearings, Enforcement, and Administration, Office of the General Counsel, at the same address. (Refer to 10 CFR 2.302 for additional filing options and instructions.) If submitting via private courier (e.g., FedEx, UPS), send your request to 11555 Rockville Pike, Rockville, Maryland 20852, instead of using the Washington, DC, address.

Pursuant to 10 CFR 55.35, you may not reapply for a license until your license has been finally denied. Failure on your part to exercise either of the above options within 20 days constitutes a waiver of your opportunity for informal review and your right to demand a hearing. For the purpose of re-application under 10 CFR 55.35, such a waiver renders this letter a notice of final denial of your application, effective as of the date of this letter.

If you have any questions, please contact (name) at (telephone number).

Sincerely,

(Name and title of licensing official)

Docket No. 55-(number)

Enclosures: As stated

cc: (Facility representative who signed the applicant's NRC Form 398)

CERTIFIED MAIL, RETURN RECEIPT REQUESTED

---

*   Use for initial RO or SRO license applicants who passed either the operating test or the written examination but failed the other.

**   Use for second and subsequent retake applicants.

***   Use for applicants who failed both the operating test and the written examination.

NRC Letterhead

(Date)

(Applicant's name)
(Street address)
(City, State Zip code)

Dear (Name):

The purpose of this letter is to forward the results of the site-specific operating test and written examination administered to you during the week of (date) in connection with your application for a (reactor operator, senior reactor operator, limited senior reactor operator) license for the (facility name). Copies of your operating test and written examination answer sheets are enclosed.

However, as explained in paragraph D.3.c of Examination Standard (ES) 501 in NUREG-1021, "Operator Licensing Examination Standards for Power Reactors," Revision 9, we will not issue your license [until your employer certifies in writing that you have acquired all of the training and experience for which you were previously granted a waiver.] [[until we determine that your medical condition and general health are satisfactory for licensing.]] [[[because any written examination with a passing grade of 82 (74 for SRO-only) percent or below is normally held for review until those applicants who failed the examination have had an opportunity to appeal their license denials.

After resolving potential changes from any appeal, the NRC will issue your license if your final grade remains above 80 (70 for SRO-only) percent. Should changes result in your final grade being below 80 (70 for SRO-only) percent, the NRC will send you a proposed denial letter, which will outline your response options.]]]

If you have any questions, please contact (name) at (telephone number).

Sincerely,

(Name and title of licensing official)

Docket No. 55-(number)

Enclosures: As stated

cc: (Facility representative who signed the applicant's NRC Form 398)

[ ]   Use only for applicants who need to complete training or experience prior to licensing.
[[ ]]   Use only for applicants whose medical condition is still under review.
[[[ ]]] Use only for applicants whose final licensing action is pending the resolution of written examination appeals.

Post-Examination Check Sheet	
Facility: **Date of Examination:**	
Task Description	**Date Complete**
1. Facility written exam comments or graded exams received and verified complete	
2. Facility written exam comments reviewed and incorporated and NRC grading completed, if necessary	
3. Operating tests graded by NRC examiners	
4. NRC chief examiner review of operating test and written exam grading completed	
5. Responsible supervisor review completed	
6. Management (licensing official) review completed	
7. License and denial letters mailed	
8. Facility notified of results	
9. Examination report issued (refer to NRC MC 0612)	
10. Reference material returned after final resolution of any appeals	

PRIVACY ACT INFORMATION — FOR OFFICIAL USE ONLY

Power Plant Examination Results Summary					
Facility:			Plant Status:　Hot ☐　　Cold ☐		
Written Examination Date: Prepared by:　Facility ☐　　NRC ☐			Operating Test Date(s): Prepared by:　Facility ☐　　NRC ☐		
NRC Examiners:					
Overall Results					
Applicants:　Total #	# Passed		% Passed	# Failed	% Failed
RO					
SRO					

Individual Results						
Name	Docket # 55-(_____)	Type (1)	Written Grade RO / SRO / TOT	Operating Test(2)		
				W-T	ADM	SIM
			/ /			
			/ /			
			/ /			
			/ /			
			/ /			
			/ /			
			/ /			
			/ /			
			/ /			
			/ /			
			/ /			

NOTES:
(1)　　1=RO; 2=SRO-I; 3=SRO-U; 4=RO-Retake; 5=SRO-I-Retake; 6=SRO-U-Retake; 7=SRO-Fuel
(2)　　P=Passed; F=Failed; W=Waived

PRIVACY ACT INFORMATION — FOR OFFICIAL USE ONLY

PRIVACY ACT INFORMATION — FOR OFFICIAL USE ONLY

Power Plant Examination Results Summary
(Continuation Sheet)

Facility:

Examination Date(s):

Individual Results

Name	Docket # 55-(____)	Type (1)	Written Grade RO / SRO / TOT	Operating Test(2) W-T	ADM	SIM
			/ /			
			/ /			
			/ /			
			/ /			
			/ /			
			/ /			
			/ /			
			/ /			
			/ /			
			/ /			
			/ /			
			/ /			

NOTES:
(1)   1=RO; 2=SRO-I; 3=SRO-U; 4=RO-Retake; 5=SRO-I-Retake; 6=SRO-U-Retake; 7=SRO-Fuel
(2)   P=Passed; F=Failed; W=Waived

PRIVACY ACT INFORMATION — FOR OFFICIAL USE ONLY

ES-502
PROCESSING REQUESTS FOR ADMINISTRATIVE REVIEWS AND HEARINGS
AFTER INITIAL LICENSE DENIAL

A.   Purpose

This standard describes the options and associated responsibilities regarding administrative reviews and hearings related to license application denials and license denials resulting from examination failures. This standard also addresses license re-applications after a denial becomes final.

B.   Background

Operator license applicants who are denied the opportunity to take an NRC licensing examination because they do not meet the eligibility requirements for a license pursuant to Title 10, Part 55, of the *Code of Federal Regulations* (10 CFR Part 55) and those applicants who are denied a license because they failed a written examination or operating test administered pursuant to 10 CFR Part 55 are notified of their denials in writing. The proposed denial letters describe the nature of the deficiencies noted and inform the applicants of their available response options. Applicants may reapply pursuant to the provisions of 10 CFR 55.35. However, the NRC will not accept a re-application as long as a request is pending for either an administrative NRC review or a hearing.

C.   Responsibilities

1.   Applicant

    a.   An applicant who does not appear to meet the experience and training requirements for a license may be asked to provide additional information to the NRC's regional office in accordance with ES-202, "Preparing and Reviewing Operator License Applications." If the NRC still denies the application after the applicant provides the additional information requested by the NRC, the applicant may exercise one of the following options within 20 days after the date on the proposed denial letter from the regional office:

        (1)   Do nothing. The proposed denial letter then becomes the final denial. The applicant may reapply after obtaining the requisite training or experience.

        (2)   Request reconsideration of the application denial. Applicants must submit such requests to the Director, Division of Inspection and Regional Support, Office of Nuclear Reactor Regulation, U.S. Nuclear Regulatory Commission, Washington, DC 20555-0001. If submitting via private courier (e.g., FedEx, UPS), send your request to 11555 Rockville Pike, Rockville, Maryland 20852, instead of using the Washington, DC, address. The applicant's submittal must clearly state the basis for the request.

(3)   Request a hearing pursuant to 10 CFR 2.103(b)(2). Applicants must submit such requests to the Office of the Secretary, U.S. Nuclear Regulatory Commission, Washington, DC 20555-0001, Attention: Rulemakings and Adjudications Staff, with a copy to the Associate General Counsel for Hearings, Enforcement, and Administration, Office of the General Counsel, at the same address. (Refer to 10 CFR 2.302 for additional filing options and instructions.) If submitting via private courier (e.g., FedEx, UPS), send your request to 11555 Rockville Pike, Rockville, Maryland 20852, instead of using the Washington, DC, address.

b.   If an applicant fails the operator licensing written examination or operating test (or both) and receives a proposed license denial letter issued by an NRC regional office in accordance with ES-501, "Initial Post-Examination Activities," the applicant has 20 days from the date on the letter to exercise one of the following three options:

(1)   Do nothing. The proposed denial letter then becomes the final denial. The applicant may reapply, pursuant to 10 CFR 55.35, 2 months after the date on the first denial letter, 6 months after the second denial, and 24 months after each successive denial.

(2)   Request that the NRC administratively regrade the written examination, the operating test, or both, in light of new information to be provided by the applicant. Applicants must submit such requests to the Director, Division of Inspection and Regional Support, Office of Nuclear Reactor Regulation, U.S. Nuclear Regulatory Commission, Washington, DC 20555-0001. If submitting via private courier (e.g., FedEx, UPS), send your request to 11555 Rockville Pike, Rockville, Maryland 20852, instead of using the Washington, DC, address. If the applicant submits such a request, the NRC will not consider a re-application pursuant to 10 CFR 55.35 until a denial is final.

The applicant's request for administrative review must identify the item(s) for which additional review is requested and must include documentation supporting the item(s) in contention. The applicant is responsible for ensuring that the request and the supporting documentation are sent to the NRR operator licensing program office within 20 days after the date on the proposed denial letter.

If the NRC administratively reviews a failure and determines that the applicant did not provide sufficient basis to justify passing grades on all sections of the licensing examination, the NRC will issue a letter to the applicant sustaining the proposed denial. The applicant may then request a hearing pursuant to 10 CFR 2.103(b)(2). In such instances, the applicant must submit a request for a hearing after an administrative review within 20 days after the date on the letter from the NRR operator licensing program office sustaining the proposed denial. In addition, the applicant must submit the hearing request in accordance with Section C.1.b(3), below.

If the applicant does not request a hearing when the NRR operator licensing program office sustains the proposed denial, the proposed denial becomes the final denial. The applicant may then reapply for a license, pursuant to 10 CFR 55.35, 2 months after the date of the first sustained denial letter, 6 months after the second denial, and 24 months after each successive denial.

(3)   Request a hearing as provided by 10 CFR 2.103(b)(2). The applicant must submit the hearing request to the Office of the Secretary, U.S. Nuclear Regulatory Commission, Washington, DC 20555-0001, Attention: Rulemakings and Adjudications Staff, with a copy to the Associate General Counsel for Hearings, Enforcement, and Administration, Office of the General Counsel, at the same address. (Refer to 10 CFR 2.302 for additional filing options and instructions.) If submitting via private courier (e.g., FedEx, UPS), send your request to 11555 Rockville Pike, Rockville, Maryland 20852, instead of using the Washington, DC, address. If the applicant requests a hearing, the NRC will not consider a re-application pursuant to 10 CFR 55.35 until the denial is final.

2.   **Facility Licensee**

a.   The NRC may ask the facility licensee to provide reference materials, technical support, and (if the facility licensee prepared the examination) a confirmation of the validity of the test items, as necessary for the NRC staff to evaluate and resolve any concerns raised by a license applicant who asked the NRC to reconsider a proposed denial of an application or license.

b.   If the facility licensee prepared the examination, it should ensure that any written examination questions that are determined to be invalid (e.g., those that have no or multiple correct answers) are retrieved from any examination bank into which they have been deposited and corrected or discarded.

3.   **NRC**

a.   The NRC will conduct administrative reviews of Part 55 license application denials based on eligibility as described in Section D.1, below.

b.   The NRC will conduct administrative reviews of Part 55 license denials based on examination failures as described in Section D.2, below.

c.   The NRC will conduct Part 55 operator licensing hearings in accordance with 10 CFR Part 2, "Rules of Practice for Domestic Licensing Proceedings and Issuance of Orders."

D. Administrative Review Procedures

1. Application Denial

If an applicant requests an administrative review in accordance with Section C.1.a, the NRR operator licensing program office will generally complete its review of the applicant's eligibility within 60 days of receiving the request. Upon completing its review, the NRR operator licensing program office will notify the applicant in writing as to whether he or she will be allowed to take the license examination. If the review leads the NRR operator licensing program office to sustain the original denial, the applicant may request a hearing pursuant to 10 CFR 2.103(b)(2).

2. Examination Results

If an applicant requests an administrative review in accordance with Section C.1.b, the NRR operator licensing program office will generally complete its review, as follows, within 75 days after receiving the request.

a. The NRR operator licensing program office will determine whether to (1) review the appeal internally,; (2) have the regional office review the appeal, or (3) convene a three-person board to review the applicant's documented contentions. The appeal board will normally be composed of a branch chief and two examiners or subject matter experts; it may also include a representative from the affected region, but no one who was involved with the applicant's licensing examination.

For written examinations, the review will generally focus only on those questions that the applicant is contesting. The review shall evaluate the original grading of the applicant's (or applicants') examination(s), the reference material supplied by the facility licensee, and the contentions and supporting documentation provided by the applicant(s). If there are multiple appeals, all question deletions and answer key changes will be applied equally to each appellant's examination, without regard to who submitted the complaint. Moreover, in those rare instances when a generic finding results in an answer key change (e.g., failure to provide a print or other reference necessary to answer a question), the corrective action may be applied, as appropriate, to adjust the grading of other questions that were not contested.

For operating tests, the review shall evaluate the examiner's comments, the examination report, the test that was administered, and the contentions and supporting documentation provided by the applicant or facility licensee (e.g., plant system descriptions, operating procedures, logs, chart recorder traces, and process computer printouts).

b. Based on the findings and recommendations from the review, the NRR operator licensing program office will decide whether to sustain or overturn the applicant's license examination failure. The NRR operator licensing program office will then notify the applicant in writing of the results of the review.

c. When the NRR operator licensing program office has concurred in the results of the review, the NRC's regional office will (1) issue a license if the proposed denial was overturned, (2) review the examination results of the other applicants to determine whether any of the licensing decisions are affected, (3) update the master examination file to reflect any test item deletions or answer key changes, and (4) consider the need to correspond with the facility licensee regarding the quality of the examination, as discussed in Section C.2.c of ES-501.

ES-601
CONDUCTING NRC REQUALIFICATION EXAMINATIONS

A.   Purpose

Title 10, Section 55.59(a), of the *Code of Federal Regulations* [10 CFR 55.59(a)] requires licensed operators and senior operators to complete a requalification program developed by the facility licensee and to pass a comprehensive requalification written examination and an annual operating test. In lieu of accepting the facility licensee's certification that the operator has passed the required examinations and tests administered within the facility licensee's Commission-approved program, the NRC may administer a comprehensive requalification written examination and an annual operating test.

This standard provides *general* guidance and requirements for conducting NRC requalification examinations. In addition this standard provides guidance and procedures for evaluating the facility licensee's licensed operator requalification training program to ensure it is effectively maintaining the competency of the licensed operators. Specific guidance and requirements for conducting the comprehensive requalification written examinations and the annual operating tests (including both the plant walk-through and dynamic simulator sections) are provided in ES-602 through ES-604. These standards are not a substitute for the operator licensing regulations and are subject to revision and other changes to the internal operator licensing program policy.

B.   Background

Section 306 of the *Nuclear Waste Policy Act of 1982* (NWPA) authorized and directed the NRC to promulgate regulations, or other appropriate guidance, for training and qualifying nuclear power plant operators. Those regulations were to include requirements governing the administration of requalification examinations and operating tests at nuclear power plant simulators. The NRC's requalification evaluation program consists primarily of periodic, onsite requalification inspections, supplemented with NRC examinations at facilities where the NRC believes that ineffective training is causing operators to commit errors. The NRC's Office of the General Counsel has concluded that this program satisfies the statutory requirements in Section 306 of the NWPA. The oversight program will require the NRC to actively oversee each facility licensees' requalification training programs, and the Commission's regulations will continue to contain legally binding requirements that apply to the conduct of operator requalification examinations by facility licensees.

When determining the scope of a facility's requalification inspection and examination activities, regional managers will consider overall facility performance, the results of the NRC's inspection programs (e.g., requalification, emergency operating procedure, and resident), the results of routine initial and requalification examinations, and other factors. Generally, the facility will only need to meet the inspection requirements of Inspection Procedure (IP) 71111.11, "Licensed Operator Requalification Program"; however, when necessary, the NRC can initiate augmented activities in accordance with program office guidance to ensure safe plant operation. Those activities could include a training program inspection in accordance with IP 41500, "Training and Qualification Effectiveness," operational evaluations of on-shift crews, or NRC examinations conducted in accordance with this series of examination standards.

The NRC will conduct requalification examinations only when it has lost confidence in the facility licensee's ability to conduct examinations, or when the staff believes that the inspection process will not provide the needed insight. Regional management should consider conducting requalification examinations or operational evaluations when any of the following conditions exist:

- requalification inspection results indicate an ineffective operator requalification program
- operator errors are a major contributor to operational problems
- allegations have been raised regarding significant training program deficiencies

The decision to conduct NRC examinations should be implemented through the normal resource planning system, because an inspection activity will be replaced with examinations that are more resource-intensive. Using the existing inspection planning process will ensure that the regional office and the NRC's Office of Nuclear Reactor Regulation (NRR) will consider the need to conduct examinations, as well as the alternative expanded inspection tools, when allocating the required resources. Operational evaluations should be considered as a reactive effort based on immediate safety concerns.

C.   Scope

The NRC-conducted requalification examinations will measure the effectiveness of a facility licensee's requalification program by evaluating the licensee's ability to adequately prepare written examination questions, job performance measures (JPMs), and simulator scenarios, as well as its ability to properly evaluate its operators' performance. The examination procedures are based on a systems approach to training (SAT) program, as defined in 10 CFR 55.4. To the extent possible, these procedures rely on existing requalification program standards for developing and implementing the NRC's examinations. The SAT approach allows the NRC to conduct requalification examinations that are fundamentally consistent with existing facility-developed programs. As such, this approach reduces the impact on the facilities and improves the reliability of the NRC's assessment of requalification training programs.

The NRC-conducted requalification examination will normally be composed of three parts, including a two-section open-reference written examination, a walk-through evaluation, and a dynamic simulator evaluation. The three examination parts are further described in ES-602, "Requalification Written Examinations," ES-603, "Requalification Walk-Through Examinations," and ES-604, "Dynamic Simulator Requalification Examinations," respectively. The NRC will consider preferentially using the facility licensee's requalification examination structure or methodology if it is different from that described herein, provided that it complies with 10 CFR 55.59 and is free of significant flaws; the NRC's regional office shall consult with the NRR operator licensing program office to determine the appropriate examination procedure.

To the extent practical, the examination will be based on the facility licensee's requalification program and learning objectives. The facility licensee is expected to use the plant-specific job and task analyses (JTAs) as the basis for developing the examination materials and substantiating the importance rating factors for each task. The facility licensee may also refer to the NRC's "Knowledge and Abilities Catalog[s] for Nuclear Power Plant Operators: Pressurized- [or Boiling-] Water Reactors" (NUREG-1122 or 1123, respectively), for additional guidance on identifying job-specific importance rating factors. The use of a JTA will result in more technically sound and operationally oriented examinations.

An examination team, composed of NRC examiners and facility representatives, will develop, review, and conduct each requalification examination. Parallel evaluation of operator performance by NRC examiners and facility evaluators will enhance the NRC's ability to assess both individual and program performance.

The administrative guidelines and procedures for conducting an NRC requalification examination are outlined in Attachment 1, "Examination Timetable."

D.   Examination Preparations

1.   Communication

a.   When the NRC determines that it is necessary to conduct a requalification examination, the regional office will notify the facility licensee to be evaluated at least 90 but preferably 120 days before the examination start date, using the corporate notification letter shown in Attachment 2. If possible, the NRC will schedule the site visits to coincide with the facility's requalification training cycle. Depending on the number of operators and crews at the facility, it may be necessary to conduct the examinations over a period of 2 or more weeks to attain the required sample size. The requalification training cycle, referenced herein and throughout NUREG-1021, is that continuous period of time (not to exceed 24 months) within which the facility licensee conducts its operator requalification training program.

If the purpose of the examination is to retest operators who previously failed an NRC-conducted requalification examination, the regional office should modify the corporate notification letter, as appropriate.

b.   The facility licensee is expected to respond to the corporate notification letter at least 60 days before the evaluation by submitting the materials and information requested in the letter.

The facility licensee may request that the NRC's chief examiner or another NRC representative meet with appropriate facility licensee managers and the operators to be examined. Such a meeting should be scheduled during the examination preparation week as discussed in Section D.5.

c.   At least 30 days before the examination, the NRC will confirm with the facility licensee which operators have been selected to participate in the evaluation.

2. Selection of Operators

a. The NRC expects facility licensees to train and examine their operators in the same crew configurations with which they normally operate the plant. Generally, the NRC expects the crew to include no more than five operators, but the agency will consider larger crews on a case-by-case basis.

At times, to ensure an adequate sample size, the examination team may configure crews that do not routinely work together to perform shift duties. Mixed crews of shift and non-shift operators should not be configured unless the facility licensee routinely evaluates mixed crews in its requalification training program, or the facility licensee's normal crew size is so large that it is necessary to separate a normal crew for examination purposes.

b. All crew members for requalification dynamic simulator examinations must be currently licensed on the facility and up-to-date in the facility licensee's requalification program.

c. The selections will be made to minimize perturbation of the facility licensee's schedules and plant operations. Operating crew(s) in training will be given first priority during the examination week(s). If the NRC is reevaluating the facility's program after an unsatisfactory evaluation, the selection process should favor operators who either failed their previous NRC-conducted examinations or were not previously examined.

d. During retake examinations, the dynamic simulator crew evaluation may include operators who have passed an NRC requalification examination. However, these operators will not be required to take the written or walk-through portions of that examination. The operators' performance on the simulator examination will be evaluated in accordance with the guidance of ES-604.

e. A shift technical advisor (STA) may be added to the crew if the facility normally uses an STA during requalification training. In such instances, the NRC expects the STA's duties and responsibilities to be the same as those assigned during requalification training and plant operations.

f. The NRC will review the list of crews and operators submitted by the facility licensee, and will recommend any necessary changes.

3. **Reference Material**

    a. The NRC expects the facility licensee to supply the reference materials requested in the corporate notification letter (see Enclosure 1 to Attachment 2). The NRC will evaluate the facility's reference materials for adequacy before the scheduled preparation week, using the "Evaluation Checklist for Facility Reference Material," Form ES-601-2.

    b. The NRC reserves the right to prepare the requalification examinations using the facility's background reference materials if the facility licensee's test items are inadequate for examination preparation. If the NRC prepares the examination, the staff may require reference materials comparable to those listed in ES-201, Attachment 3, "Reference Material Guidelines for Initial Operator Licensing Examinations."

    c. The NRC expects the facility licensee to provide a sample plan that meets the guidelines of Attachment 3, "Examination Sample Plan," for the NRC's use in developing the examination.

4. **Examination Team Selection**

    a. The NRC will contribute no fewer than two examiners to the examination team. The regional office should consider assigning additional examiners if the operating crews for the dynamic simulator examinations contain five or more operators. To promote consistency in requalification program administration, regional office management should try to assign an examiner who participated in a prior requalification inspection or examination at the facility to be part of the NRC's examination team.

        In most cases, the NRR operator licensing program office will send a representative to observe the examination process or an examiner to participate as an additional member of the examination team. The program office will work with the responsible regional supervisor to make the necessary arrangements.

    b. The facility licensee is expected to provide an employee to work with the NRC as part of the requalification examination team. The employee must be drawn from the operations staff, and must be an active senior reactor operator (SRO) as defined in 10 CFR 55.53(e) or (f). The NRC encourages the facility licensee to designate another employee from the training staff to be a member of the examination team. This employee should also be a licensed SRO, but may be a certified instructor. If the facility licensee desires, and the chief examiner agrees, the facility licensee may also include additional employees from the operations or training staffs who have qualifications comparable to the facility's other examination team members.

The function of these examination team members is to provide facility-specific technical assistance to the NRC in developing and reviewing the written examination items, plant walk-through topics, and dynamic simulator scenarios. If necessary, the facility representatives may participate as facility evaluators in conducting the operating test or written examination. However, the facility representatives should only be used as evaluators if they routinely perform that function during the administration of the facility licensee's requalification program.

5.   **Examination Development**

The facility licensee may develop proposed written examinations and operating tests and forward them to the NRC as part of its reference material submittal (see Attachment 2). In accordance with 10 CFR 55.59(a)(2)(ii), the facility licensee must ensure that the operating tests require the operators to demonstrate an understanding of and ability to perform the actions necessary to accomplish a comprehensive sample of the items specified in 10 CFR 55.45(a)(2) – (13), inclusive, to the extent applicable to the facility.

Approximately 2 weeks before the scheduled examinations, the NRC examiners will visit the facility to make final preparations for the examination. The written, walk-through, and dynamic simulator examinations will be developed in accordance with ES-602, ES-603, and ES-604, respectively. The examination should distinguish between reactor operator (RO) and SRO knowledge and abilities to the extent that the facility training materials allow the examiners to make these distinctions. The NRC examiners will rely upon the facility licensee's examination team members for site-specific technical assistance in developing, reviewing, and validating the written examination static scenarios and items, plant walk-through topics (JPMs), and dynamic simulator scenarios.

The chief examiner and the responsible regional supervisor will determine the required length of time on site and the required number of examiners. This determination will be based on the experience of the examiners, the quality of the facility licensee's testing material, and the level of effort required to develop new test items.

If requested by the facility licensee, the chief examiner will brief the operators and managers about the requalification examination process. The examiner will use this time to explain the examination and grading processes and to respond to any questions that the operators may ask about the NRC's examination procedures. If the schedule does not allow them to meet during the preparation week, they may meet at any mutually agreeable time.

6. Examination Security

To ensure examination security, each facility representative who acquires knowledge of the content of the NRC's requalification examination before it is administered will be subject to the security restrictions described below from the time he or she first acquires the specific knowledge until the examination exit meeting.

To the maximum extent possible, only the examination team members and a simulator operator should be given specific knowledge about the content of the examination. The facility evaluators should be given the package of simulator scenarios and JPMs the week before the examination to allow them to prepare for their evaluation, including coordinating the use of the simulator to perform JPMs and scenarios. If the facility licensee submits a proposed examination, those who participate in developing the examination become subject to the security restrictions when their involvement begins. Also, if facility representatives other than the examination team members are used to time validate the written examination, they too become subject to the security restrictions as soon as they are exposed to the examination questions.

Facility representatives who acquire specific knowledge of the NRC's examinations will sign Form ES-601-1, "Examination Security Agreement," or a reasonable facsimile before their examination involvement begins and again after the examination process is complete (i.e., following the exit meeting).

E. Operator and Program Evaluation Procedures

1. Examination Administration

a. For each selected operator, conduct a requalification examination using ES-602, ES-603, and ES-604 for the written, walk-through, and simulator portions of the requalification examination, respectively. Document operator performance on Form ES-601-5.

b. The number of persons present during an operating test should be limited to ensure the integrity of the test and to minimize distractions to the operators. Under *no* circumstances will another operator be allowed to witness an operating test. Facility licensees are not to use operating tests as training vehicles for future requalification examinations.

Other examiners may observe an operating test as part of their training or to audit the performance of the examiner administering the operating test. The chief examiner may permit others (such as resident inspectors, regional personnel, researchers, or NRC supervisors) to observe an operating test if the applicant does not object to the observers' presence. Deviations from this policy must be approved, in advance, by the NRR operator licensing program office.

Other non-NRC personnel [e.g., representatives from the Institute of Nuclear Power Operations (INPO) or the Nuclear Energy Institute (NEI)] may observe the operating tests with prior approval from the NRR operator licensing program office. The chief examiner will control the observers' activities in accordance with guidance provided by the program office.

2.   **Examination Grading**

a.   The facility licensee is expected to grade the written examinations and operating tests in parallel with the NRC's examiners.

b.   The facility evaluators are expected to provide preliminary pass/fail results for the simulator and walk-through portions of the examination by the end of each day, and the final results before the exit briefing or at the end of each examination week for multi-week examinations.

c.   The NRC will notify the facility licensee immediately if any operator's performance on the examination is sufficiently poor to require immediate removal from licensed activity. The NRC will also notify the facility licensee of the results of the examination in accordance with ES-605, "License Maintenance, License Renewal Applications, and Requests for Administrative Reviews and Hearings."

d.   The facility licensee will provide the NRC with the final results of the written examinations and an overall summary of the examination results within 2 weeks after the exit meeting.

3.   **Evaluation of Requalification Programs**

A requalification program evaluation requires a minimum sample size of 12 operators. The sample size is determined by counting the number of operators taking the dynamic simulator examination. This total includes those operators who only participate in the simulator examination for the purpose of meeting crew composition requirements, but excludes those operators who are being reexamined after failing a previous NRC-conducted examination.

In instances where fewer than one-half of the operators taking the dynamic simulator examination complete the entire examination, the regional supervisor will determine whether a valid program evaluation can be made. In these instances, the regional supervisor will contact the NRR operator licensing program office.

a.   A satisfactory requalification program meets *each* of the following criteria:

(1)   At least 75 percent of the operators must pass all portions of the examination in which they participate. The pass rate is determined by dividing the number of operators who pass all portions of the examination in which they participate by the total number of operators in the sample.

   In the event of a crew failure, only those operators who receive a satisfactory evaluation in the individual followup evaluation will be counted when calculating the operator pass rate.

When calculating the pass rates, fractions should be rounded up to the next highest whole number. For example, if 15 operators are evaluated, 75 percent passing would be 11.25 operators; thus, 11 of 15 passing would *not* meet the 75 percent requirement, but 12 would.

(2)   At least two-thirds (66 percent) of the crews must pass the simulator examination.

For requalification examinations with more than three crews participating, three-of-four, or four-of-five crews must pass to satisfy this requirement.

b.   The NRC will consider the following areas in the overall program evaluation, and may use the related findings to identify facility weaknesses that will be documented in the examination report:

(1)   The facility evaluators do not concur with the NRC examiners on all unsatisfactory crew evaluations.

(2)   More than one facility evaluator is determined to be unsatisfactory. Section D of Appendix C provides guidance that examiners should use to assess evaluator competence.

(3)   The facility licensee failed to train and evaluate an operator in all positions permitted by the individual's license. (For instance, the facility is required to train and evaluate an SRO in the RO position, as well as in directing operators.) An SRO will not be required to perform RO activities during the simulator portion of the operating test; however, his or her performance will be evaluated if the facility normally places the SRO in a shift RO position during the simulator examination. Otherwise, RO skills will be evaluated during the performance of JPMs.

(4)   The facility licensee has insufficient administrative controls to preclude an RO or SRO with an inactive license from performing licensed duties. Operators must meet the requirements of 10 CFR 55.53 to restore an inactive license to active status.

(5)   The facility licensee has insufficient quality control of its examination bank. The NRC will evaluate the facility's performance in this area if post-examination changes to facility-developed test items result in significant modifications or deletions of more than 10 percent of the questions on the written examination.

(6)   The number of test items duplicated from any past examination or combination of examinations administered during the current requalification training cycle [as described in 10 CFR 55.59(a)(1)], or the number of operating test items repeated on successive days of an examination period, is such that the discrimination validity and integrity of the examination could be affected. When test items are repeated, they should be selected in a distributed manner and approximately equally over all previous examinations to reduce predictability (if a large number of items were taken from the most recent examination).

(7)   The facility licensee's failure decisions are not as conservative as the NRC's. To ensure that the rationale for the evaluation is fully understood, the NRC will review with the facility managers any case where the facility licensee passed an operator whom the NRC failed. In addition, the NRC will assess whether the facility licensee's evaluations are conducted in accordance with documented facility guidance and whether facility managers periodically assess their evaluation process.

The NRC also expects the facility program to explicitly link an operator's examination failure with *unsafe* performance. In this way, all facility failures and NRC failures will agree. In certain instances, the facility licensee's program may have operator performance standards that are *not* explicitly linked to unsafe performance, and thus do not meet the threshold stated in these standards for the operator to fail the examination. In such instances, the facility licensee is expected to differentiate failures in which the operator performed at an unsafe level from those in which the operator failed for reasons other than safety (i.e., not meeting higher facility-established performance standards). In these instances, operators identified as failing for safety reasons would also be considered NRC failures.

4.   Evaluation of Operator Performance

To pass the NRC-conducted requalification examination, the operator must pass a written examination and an operating test consisting of a walk-through examination and a dynamic simulator examination. These examinations are developed and administered in accordance with ES-602, ES-603, and ES-604, respectively, unless the NRR operator licensing program office authorizes the regional office to use the facility licensee's alternative examination methodology. To pass the operating test, the operator must also be a member of a crew that passes the dynamic simulator examination.

F.   Unsatisfactory Operator or Program Evaluation

1.   Actions Following an Unsatisfactory Operator Evaluation

In all cases, a facility licensee's administrative procedures should ensure that an operator who fails a requalification examination is removed from licensed duties, given remedial training, and reexamined before being allowed to return to licensed duties. This also applies to an SRO who performs only RO-level duties at the facility when the failure is caused solely by activities involving SRO responsibility. ES-605 contains the procedure for notifying the operator about his or her performance on the requalification examination, as well as guidance about the actions to be taken for an operator to return to licensed duty.

The NRC has deleted the regulation [10 CFR 55.57(b)(2)(iv)] that required an operator to pass an agency-administered requalification examination as a prerequisite for license renewal. Nonetheless, it would be inappropriate to renew the license of any operator who failed to pass any NRC-conducted requalification examination, without some level of agency involvement in the retesting process. The amount of NRC involvement may include conducting the retest in accordance with the appropriate examination standard(s); inspecting the facility licensee in accordance with IP 71111.11, "Licensed Operator Requalification Program," as it retests the operator; or reviewing the reexamination prepared by the facility licensee. The regional office, in consultation with the NRR operator licensing program office, will determine the appropriate level of involvement on a case-by-case basis depending on the quality of the facility licensee's program. As long as the operator submits a timely renewal application, the term of the license will continue until the renewal requirements are satisfied or the operator fails three NRC-conducted examinations as discussed in ES-605.

If an operator who failed a requalification examination is not prepared for a reexamination after 6 months of remedial training, the regional office will request the following information from the facility licensee:

- confirmation that the facility licensee still has a need for the individual's license

- the expected completion date of the operator's remedial training
  and when the facility licensee will be ready to administer its retake examination

- assurance that the operator will not be returned to licensed duties until
  he or she successfully retakes the examination (or portion thereof) administered
  by the facility licensee with a satisfactory requalification program
  or in accordance with the provisions of the confirmatory action letter (CAL)
  if the facility licensee has an unsatisfactory program and the NRC
  has not determined it to be "provisionally satisfactory."

The NRC will inform the facility licensee that a comprehensive requalification examination may be necessary if the operator is not ready to take a retest within 1 year after failing the examination.

2. Actions Following an Unsatisfactory Requalification Program Evaluation

    The NRC will take the following actions for all requalification programs that the agency evaluates as unsatisfactory:

    a. The NRC expects the facility licensee to identify program deficiencies
        and corrective actions to improve operator performance. The NRC will use
        a CAL to establish a formal dialogue and to document the facility licensee's
        corrective action commitments.

        An operator who fails the requalification examination, as determined by the NRC,
        will be subject to an NRC-administered reexamination before resuming licensed duties.

        An operator whose performance does not meet facility standards,
        as determined by the facility licensee, is expected to be remediated and reevaluated
        by the facility licensee in accordance with the provisions of the facility licensee's

requalification program. The NRC will review and/or monitor the reexamination to ensure the adequacy of the facility licensee's requalification program.

b. The NRC will schedule a meeting with senior facility managers to review the examination results, as well as the identified deficiencies and their root causes, the proposed corrective actions and the schedule for their implementation, and the need for followup inspections and examinations. (Refer to Section F.3 for additional guidance on conducting augmented inspections.)

    The regional administrator will evaluate the examination and inspection results and make a decision regarding the continued operation of the facility and possible enforcement action against the facility licensee. At a minimum, the regional administrator should consider the following factors when making this determination:

    •   the results and corrective actions from previous program evaluations

    •   the significance of generic performance deficiencies identified during the program evaluation

    •   recent facility events that relate to licensed operator performance

    •   recommendations by the NRC staff (including the results of any operational evaluations and inspections)

c. If the unsatisfactory program evaluation is caused by operator performance deficiencies, an operational evaluation is required. The operational evaluation is intended to help the regional administrator determine whether the facility's remaining operating crews are suitably qualified to continue to operate the facility. In this case, the facility licensee identifies the individual operators and shift crews it proposes to use to continue plant operations. The regional office may choose not to evaluate those operators who passed their most recent NRC-conducted initial or requalification examination within the past 12 months. However, the regional office will evaluate all other operators in those areas noted as operational deficiencies during the requalification examination, regardless of whether they have already passed or not yet taken the facility-administered requalification examination. The regional office will conduct the operational evaluations in accordance with the guidance in ES-603 and ES-604, as applicable.

    If the facility licensee proposes to use a shift crew that is significantly different from its normal configuration, even though all of the operators may have recently passed an NRC-conducted examination, the regional office may perform an operational evaluation of this crew.

    The regional office should schedule the operational evaluation as soon as possible after determining that the facility licensee's requalification program is unsatisfactory. The evaluation should not be delayed to accommodate the facility's operating schedule, the completion of programmatic corrective actions, or the completion of remedial training for operators who failed the requalification examination. The operational evaluation may identify further program deficiencies that may need to be reflected in the CAL discussed in Section F.2(a) or may warrant additional inspection by the NRC. Additional operator weaknesses that require remediation may also be identified.

d.   The NRC will review the corrective actions the facility is to perform, the expected followup actions by the NRC, and the schedule for each.

As part of the followup activities, the NRC may conduct additional operational evaluations, requalification retake examinations, and augmented inspections, as necessary. Before these activities, the NRC will verify that the facility licensee has completed the applicable corrective actions, and will obtain a certification of crew readiness from the facility managers. Regional managers should consider using a new chief examiner and having examiners from other regional offices participate on those operational evaluations and requalification retake examinations that have restart approval implications.

e.   The regional administrator will incorporate into the decision concerning followup activities any extraordinary circumstances surrounding the examination that may have a bearing on the validity of the examination results.

f.   Once the NRC determines that a requalification program is unsatisfactory, the program will remain unsatisfactory until the facility licensee completes all identified corrective actions agreed upon by the NRC for restoring the program to satisfactory status, and the NRC completes all related followup activities.

For purposes of allowing facility examiners to perform reexamination functions, however, a facility may attain a status of "provisionally satisfactory" provided that the facility has completed to the NRC's satisfaction all short- and intermediate-term corrective actions agreed on with the NRC.

Once the NRC determines that the facility licensee has satisfactorily implemented these corrective actions, the regional administrator or designee will determine whether to permit the facility to reexamine all operators who failed the NRC-conducted requalification examination for the purpose of returning the operators to licensed duties. Any operator who fails the NRC-conducted examination still needs to pass a future NRC-administered (i.e., conducted, inspected, or approved, as appropriate) requalification examination to renew the license. Long-term corrective actions are expected to be completed before the NRC's next requalification program evaluation.

To attain a satisfactory rating following an unsatisfactory evaluation, the subsequent requalification program evaluation, with a sample size of at least 12 operators, must satisfy the passing criteria in Section E.3.

The regional administrator or designee may specify additional actions, as appropriate. The specific sequence of actions is not critical; however, this sequence of events corresponds to a typical regional response to an unsatisfactory program evaluation. The regional administrator or designee should defer determining whether a plant shutdown is required until he or she reviews all factors listed in Section F.2(b) above.

3.   **Augmented Inspection Guidelines**

If the NRC determines that an augmented requalification program inspection is required, regional management shall define its scope and depth based upon the nature of the deficiencies.

The regional office should consider the following activities in addition to those specified in Section F.2.

a.   The regional office may conduct augmented inspection coverage of all shifts. The inspection procedures for shift coverage should be used as appropriate. Inspection activities should devote particular attention to the following areas:

- operator performance and attitude
- operator overtime
- management oversight
- shift staffing

b.   The regional office may develop a long-term training program inspection plan based on Inspection Procedure (IP) 41500, "Training and Qualification Effectiveness." Such an inspection plan may include the following activities:

- ongoing status reviews of requalification training effectiveness, with an emphasis on known program deficiencies and implementation of short-term corrective actions

- an inspection to determine the root cause(s) for the unsatisfactory requalification program evaluation, and to verify that the facility licensee's proposed corrective action plan should preclude or minimize the probability of recurrence

- an inspection to evaluate the adequacy of the facility licensee's corrective actions, and to determine the effectiveness of the facility licensee's SAT-based requalification program

c.   The regional office may convene an enforcement panel to determine whether action is warranted on the basis of the requirements of 10 CFR 50.54(i-1). The potential exists that a requalification program rated unsatisfactory on two successive NRC evaluations does not meet the minimum requirements of 10 CFR 55.59(c) as required by 10 CFR 50.54(i-1). The basis for any proposed enforcement action will be the inadequate corrective action or requalification program element deficiencies [identified by the inspections related to Section F.3(b)], which led to the successive requalification examination failures.

G.   Requalification Program Evaluation Report

After the regional administrator or designee approves the requalification examination results, the regional office will prepare a final requalification program evaluation report. A copy of the written examination need only be included in the program evaluation report if the report addresses written examination problems. The regional office will issue the report within 30 days following receipt of the facility licensee's final results or the examination exit meeting, whichever is later, and will place a complete copy of the report in the facility's requalification file.

The chief examiner is responsible for completing Forms ES-601-3 and ES-601-4, the "Power Plant Requalification Results Summary (and Continuation) Sheet(s)." The examiner will enter each operator's scores in the appropriate columns. Under the "Simulator" column, the examiner will enter the results of the operator's individual followup evaluation. If the operator did not receive an individual followup evaluation, the examiner will enter a passing score. If an operator was a member of a crew that failed the dynamic simulator examination, but the operator passed or did not receive an individual followup evaluation, the examiner will enter a pass in the "Simulator" column for that operator. Crew failures will be summarized in the overall results at the top of Form ES-601-3.

The regional office will send a copy of the summary (and continuation) sheet(s) to the headquarters' operator licensing assistant (OLA). The NRR operator licensing program office uses the results summary to verify the data in the operator licensing tracking system (OLTS) so that statistics can be maintained on operator performance. As it contains information that is subject to the Privacy Act, the regional office will not include the results summary in the examination report.

If a small number of operators are given retake examinations, the regional office may issue an addendum to the original requalification evaluation report instead of issuing a new report. If the reexaminations are conducted concurrently with initial examinations or inspected during a requalification program evaluation in accordance with IP 71111.11, the results may be reported as part of the initial examination or inspection report.

H.   Individual Requalification Examination Report

After the regional office completes the requalification evaluation, it will keep a copy of each operator's NRC-conducted written, walk-through, and simulator examination results and return the original documents to the facility licensee. The facility licensee is required by 10 CFR 55.59 to maintain records of these examination results along with a copy of the written examination until the operator's license is renewed or 2 years after the license expires.

The NRC's chief examiner will ensure that Form ES-601-5, "Individual Requalification Examination Report," is completed for each operator who takes an NRC-conducted requalification examination. The report will include the following information for each individual:

*   written examination grade
*   crew evaluation from the dynamic simulator examination
*   the individual followup results (P or F) from the dynamic simulator examination
*   the number (and percentage) of JPMs performed correctly, if conducted

The regional office will send a copy of this report to the facility's training manager and the operator with a letter notifying the operator of the examination results in accordance with ES-605. The regional office will also file a copy in the operator's docket file.

I.   Operator License Renewal Policy

Operators are not required to take an NRC-conducted requalification examination in order to renew their licenses. However, if an operator takes, but fails to pass, an NRC-conducted examination, the NRC will not renew the license until the operator passes a retake examination conducted by the NRC, passes a retake examination administered by the facility licensee and inspected by the NRC in accordance with IP 71111.11, or passes an examination approved by the NRC. The regional office, in consultation with the NRR operator licensing program office, will determine the appropriate level of involvement on a case-by-case basis depending on the quality of the facility licensee's requalification program.

ES-605 contains the specific procedures to follow for an operator who fails one or more NRC-administered requalification examinations, as well as the procedure for processing license renewal applications.

J. Records Retention

1. Facility Requalification Examination File

The NRC's regional office shall ensure that the original (whenever possible) or a copy of the following items are either retained in the facility's master examination file or electronically available via the NRC's Agencywide Documents Access and Management System (ADAMS):

a. Examination standard attachments and forms:

- Form ES-403-1, "Written Examination Grading Quality Checklist"
- ES-601, Attachment 2, "Corporate Notification Letter"
- Form ES-601-1, "Examination Security Agreement"
- Form ES-601-3, "Power Plant Requalification Results Summary Sheet"
- Form ES-601-4, "Power Plant Requalification Results Continuation Sheet" (if applicable)
- Form ES-604-2, "Simulator Crew Evaluation"

b. a master list of all JPMs administered and the operators to which they were administered

c. a master list of all scenarios conducted and operators to which they were administered (facility-generated forms or Form ES-D-1, "Scenario Outline," may be used to meet this requirement)

d. a copy of the written examination and answer key

e. a copy of the requalification examination report

The regional office may require that additional documents be retained in the facility's requalification examination file. (Note that paper copies of examinations that were administered before the implementation of ADAMS may be discarded after confirming that the examination report is available in the NRC's Public Document Room.)

2. **Operator Docket Files**

The regional office will retain the following records in each operator's docket file until the license is renewed or 2 years after the license expires or is terminated.

a.   Form ES-601-5, "Individual Requalification Examination Report"
b.   ES-605, Attachment 1, 2, or 3, "Results Notification Letter"
c.   a copy of all failed portions of the NRC-graded examination

3. **Other Files**

The regional office will retain reference materials used to develop each examination until the NRC has resolved with the facility licensee all failures associated with the examination and has sent a notification letter to each operator.

K.   Requalification Stress Feedback

The level of stress perceived by operators and facility personnel can affect their overall performance on the requalification examination. Therefore, the NRR operator licensing program office is interested in monitoring the stress of operators and facility personnel participating in the requalification examination. Regional examiners and other personnel who are involved with an NRC requalification examination should assume the following responsibilities:

•   Monitor the level of stress in operators and facility representatives, and be alert for examination techniques that may be causing examination stress.

•   Recommend to the program office any changes to NUREG-1021 that would further alleviate operator stress. Recommendations should be documented and forwarded to headquarters using "Report on Interaction" (ROI) forms.

L.   Attachments/Forms

Attachment 1,   "Examination Timetable"
Attachment 2,   "Sample Corporate Notification Letter"
Attachment 3,   "Examination Sample Plan"
Form ES-601-1,   "Examination Security Agreement"
Form ES-601-2,   "Evaluation Checklist for Facility Reference Material"
Form ES-601-3,   "Power Plant Requalification Results Summary Sheet"
Form ES-601-4,   "Power Plant Requalification Results Continuation Sheet"
Form ES-601-5,   "Individual Requalification Examination Report"

| Date* | Activity |

-120/90   The NRC notifies the facility licensee.

-60   The facility licensee sends the NRC the materials requested for developing the examination (including written examination questions, simulator scenario banks, and JPMs).

The facility licensee proposes composition of the crews to be evaluated and identifies facility examination team members.

The facility licensee may ask the NRC's chief examiner to review the examination process with operators and managers.

-45   The facility licensee submits its proposed requalification written examination and operating test.

-30   The NRC concurs on the operating crews to be evaluated.

-14   The NRC examiners visit the facility to review the JPMs to be administered, observe the static and dynamic simulator examinations, and validate the test items, as needed. The chief examiner and the regional branch chief determine the length of time on site and the number of examiners required, on the basis of the examiners' experience and the quality of the facility licensee's testing materials.

The facility licensee designates a simulator operator.

If requested, the chief examiner briefs the operators and managers about the requalification examination process.

-7   The facility examination team members finalize the examinations based on preparation week activities. Evaluators review reference material to prepare for the JPMs and simulator scenarios.

0   The NRC administers the examinations to selected crews and operators. The facility licensee notifies the NRC of its final results for crews and individuals at the end of each examination week.

+7   The NRC finalizes the examination results.

+14   The facility licensee transmits the written examination grades and a final summary to the NRC.

+30#   The NRC issues operator results and the final requalification examination report.

---

*   Number of days before or after the examination, except as noted
\#   Number of days after receipt of facility results

NRC Letterhead

(Date)

(Name, Title)
(Name of facility)
(Street address)
(City, State Zip code)

SUBJECT: REQUALIFICATION PROGRAM EVALUATION

Dear (Name):

In a telephone conversation on (date), (Name, title) and (Name, title) arranged to evaluate the requalification program and licensed personnel at the (facility name). The evaluation is scheduled for the week of (date). NRC examiners and evaluators from your facility will conduct requalification examinations, and the NRC will evaluate your requalification program in accordance with Sections ES-601 through ES-604 of NUREG-1021, "Operator Licensing Examination Standards for Power Reactors," Revision 9, Supplement 1. You are encouraged to ensure that your training staff and proposed examinees are familiar with these standards.

For the NRC to adequately prepare for this evaluation, the facility licensee will need to furnish the NRC with the approved items listed in Enclosure 1, "Reference Material Guidelines." You are also requested to submit, at your option, a proposed examination for use during the examination week. However, if you do submit a proposed examination, the personnel participating in its development will become subject to the security restrictions described in this letter.

Please review the guidance promulgated in Revision 9 of NUREG-1021 concerning the content and scope of simulator examination scenarios. The scenario examination bank should cover the entire spectrum of emergency operating procedures (EOPs), including alternative decision paths within the EOPs, and it should incorporate a range of failures with various degrees of severity for the same type of event. Each scenario should contain simultaneous events that require the senior reactor operators (SROs) to prioritize their actions and to assign particular tasks to other crew members. Each scenario should also require the SROs to decide when to make the transition between EOPs and which actions to take within EOPs.

You are requested to designate at least one employee to be a member of a joint NRC-facility examination team. That employee is expected to be an active SRO [as defined by Title 10, Section 55.53(e) or (f) of the *Code of Federal Regulations* (10 CFR 55.53(e) or (f))] from the (facility name) operations department. You are encouraged to designate a second employee from the training staff to be a member of the examination team. This employee should also be a licensed SRO, but may be a certified instructor. If desired and agreed to by the chief examiner, you may designate one additional employee from the training staff who has appropriate qualifications to be a member of the examination team. In addition to these individuals, you will need to designate a simulator operator for scenario preview and validation during the onsite examination preparation week. In some cases, you may also need to designate a simulator operator during the test item review period. All of these individuals will be subject to the examination security agreement.

The NRC restricts any facility licensee representatives under the security agreement from knowingly communicating (by any means) the content or scope of the examination to unauthorized persons, and/or participating in any facility licensee programs (such as instruction, examination, or tutoring) in which an identified requalification examinee will be present. These restrictions apply from the day that the facility licensee representative signs the examination security agreement indicating that the representative understands that he or she has specialized knowledge of the examination. The chief examiner will determine when a facility licensee representative has received specialized knowledge concerning the examination and will execute an examination security agreement. In most cases, the examination team members will not be required to enter into an examination security agreement more than 60 days before the examination week. The simulator operator will normally become subject to the security restrictions during the examination preparation and validation week; however, this may occur as much as 45 days before the examination week.

Sixty days before the examination administration date, please provide the NRC's regional office with a proposed list of operators, including crew composition, for the examination. The list should include at least 12 operators, comprising three or more crews, and the current mailing address for each proposed operator, if different from that listed on the most recent Form 398 submitted to the NRC. Your training staff should send this information directly to the NRC's chief examiner, ensuring that each operator's address is sent in a manner to ensure privacy.

The facility licensee may request that the NRC chief examiner or another NRC representative meet with the licensee managers and the operators to be examined during the examination preparation week, normally 2 weeks before the examination. However, if the schedule does not allow them to meet during the preparation week, they may meet at any mutually agreeable time. The NRC examiner will explain the examination and grading processes and will respond to any questions that the operators may have about the NRC's examination procedures. If such a meeting is desired, your training staff should schedule it with the NRC's chief examiner.

The facility licensee staff is responsible for providing adequate space and accommodations to properly develop and conduct the examinations. Enclosure 2, "Administration of Requalification Examinations," describes our requirements for developing and conducting the examinations. Also, a facility operations management representative above a shift supervisor level should observe the simulator examination process at the site.

This letter contains information collections that are subject to the *Paperwork Reduction Act of 1995* (44 U.S.C. 3501 et seq.). These information collections were approved by the Office of Management and Budget, under approval number 3150-0018, which expires on June 30, 2009.

The public reporting burden for this collection of information is estimated to average 25 hours per response, including the time for reviewing instructions, gathering and maintaining the necessary data, and completing and reviewing the collection of information. Send comments on any aspect of this collection of information, including suggestions for reducing the burden, to the Information and Records Management Branch (T-6 F33), U.S. Nuclear Regulatory Commission, Washington, DC 20555-0001, or by electronic mail to bjs1@nrc.gov; and to the Desk Officer, Office of Information and Regulatory Affairs, NEOB-10202, (3150-0018), Office of Management and Budget, Washington, DC 20503.

The NRC may neither conduct nor sponsor, and a person is not required to respond to, a collection of information unless it displays a currently valid OMB control number.

In accordance with 10 CFR 2.390 of the NRC's "Rules of Practice," a copy of this letter and its enclosures will be available electronically for public inspection in the NRC's Public Document Room or from the Publicly Available Records System (PARS) component of the NRC's Agencywide Documents Access and Management System (ADAMS). ADAMS is accessible from the Electronic Reading Room page of the NRC's public Web site at http://www.nrc.gov/reading-rm/adams.html.

Thank you for your cooperation in this matter. (Name) has been advised of the NRC guidelines and policies addressed in this letter. If you have any questions on the evaluation process, please contact (Name, regional section chief) at (telephone number).

Sincerely,

(Appropriate Regional Title)

Docket No.: 50-(Number)

Enclosures:
1. Reference Material Guidelines
2. Administration of Requalification Examinations

DISTRIBUTION:
    Public
    NRC Document Control System
    Regional Office Distribution

Reference Material Guidelines

1. Sixty days before the examination date, the facility licensee should provide test items to the NRC to support all aspects of the requalification examination.

2. The facility licensee is expected to submit the following reference materials for all NRC-conducted requalification examinations:

    •   an examination sample plan that meets the requirements of Attachment 3 to ES-601

    •   the facility's examination banks (written, simulator, and JPM) and associated reference materials (including, at a minimum, technical specifications, abnormal and emergency operating procedures, and emergency plan procedures utilized in requalification training)

    •   additional reference materials requested by the NRC's chief examiner

3. The facility licensee's examination banks are expected to contain the following information:

    •   a minimum of 700 test items equally divided for use in the two sections of the written examination and covering all safety-related elements of the facility's job task analysis (JTA). The facility licensee is expected to maintain a dynamic bank by reviewing, revising, or generating at least 150 questions a year. New questions should cover equipment and system modifications, as well as recent industry and licensee events and procedural changes.

    •   JPMs that meet the criteria in ES-603 for evaluating each reactor operator (RO) and senior reactor operator (SRO) safety-related task identified in the facility's JTA. The JPM bank should expand at a rate of at least 10 JPMs per year until this goal is reached. It is estimated that 125 to 150 JPMs will be the final result.

    •   a bank of at least 30 simulator scenarios reflecting all abnormal and emergency situations to which an operator is expected to respond or control. At least five scenarios per year should be generated until all aspects of the emergency operating procedures are covered with sufficient variation in the type and scope of initiating events and level of degradation. Emphasis should be placed on scenarios that include applicable industry events.

Enclosure 2

Administration of Requalification Examinations

1. The NRC must evaluate at least 12 operators to perform a program evaluation. The guidelines on crew composition in the simulator are described in Section D.2 of ES-601 and ES-604.

2. The simulator and simulator operators need to be available for examination development. The chief examiner and the facility representatives will agree on the dates and amount of time needed to develop the examinations.

3. The chief examiner will review the reference materials used in the simulator. The NRC will not authorize for use during the simulator test any reference material that is not normally used for plant operation in the control room.

4. The facility licensee will provide a single room for completing Section B of the written examination. The examination room and supporting restroom facilities will be located to prevent the examinees from having contact with any other facility or contractor personnel during the examination.

5. The chief examiner will inspect the examination room to see that it meets the minimum standard that will ensure examination integrity. The minimum spacing standard consists of one examinee per table and a 1-meter (3-foot) space between tables. No wall charts, models, or other training materials are allowed in the examination room.

6. The facility licensee is expected to provide a copy of each reference document for each examinee for Section B of the written examination. The material should include documents that are normally available to the operators in the control room (such as the technical specifications, operating and abnormal procedures, administrative procedures, and emergency plans). The chief examiner will review the reference materials before the examination begins.

7. The NRC's requalification examination will attempt to distinguish between RO and SRO knowledge and abilities to the extent that the facility training materials allow the developers to make these distinctions.

8. Prudent scheduling of examination week activities is important to help alleviate undue stress on the operators. The facility training staff and the NRC chief examiner should attempt to formulate a schedule that will minimize delays while conducting the examination. The following suggestions will help to structure the examination activities to achieve this objective:

    •   Consider allowing operators to stay at home until their scheduled examination times.

- Segregate the group of operators who are completing their examination, instead of the group of operators who are scheduled to start their examination.

- Following simulator scenarios, the facility evaluators and NRC examiners should quickly determine whether followup questioning is required so that the crew members may be released to talk among themselves about the scenario.

- Ensure that time validation of JPMs, particularly those performed in the simulator, is accurate. Establish a reasonable schedule to prevent operators from waiting for simulator availability to complete their JPMs.

9. The NRC does not require the facility licensee to videotape dynamic simulator examinations. If the facility licensee requests to videotape the examination, any use of the tape must be completed before the NRC leaves the site at the end of the examination. If a disagreement over the grading of an operator's examination still exists at the end of the examination week, the facility licensee may retain the tape for the purpose of submitting it to support a request for regrading by the NRC. During the regrading, the NRC will review only the portion of the videotape under contention. After all requalification examination grades are finalized, including the review of any regrading requests, the facility licensee is expected to erase all video tapes made during the examination.

A. Introduction

An examination sample plan provides a systematic approach to selecting and developing test items to determine whether a student has mastered the knowledge, skills, and abilities (KSAs) to be covered in a particular training program. The sample plan should provide an explicit, documented link between the learning objectives associated with the training program and the test items used to perform the evaluation and to verify the relevance to the job task analysis (JTA) associated with the operator's position.

ES-401, "Preparing Initial Site-Specific Written Examinations," gives explicit guidance for developing a sample plan for initial examinations using NUREG-1122 or -1123, the NRC's "Knowledge and Abilities Catalog[s] for Nuclear Power Plant Operators: Pressurized-[or Boiling-] Water Reactors." A similar methodology may be applied to any training program. With respect to a requalification program, the scope of topics is necessarily limited because the amount of material that is covered during a requalification program is less than that covered in an initial licensing training program. However, the NRC permits and encourages reserving 10 to 20 percent of test items for topics that have high importance ratings and contain K/As that operators should retain because of their safety significance, but were not necessarily covered during the requalification cycle.

B. Requalification Test Outline

The facility licensee is expected to develop a test outline for all NRC-administered requalification examinations. At least 80 percent of the test outline must reflect the training curriculum of the most recent requalification cycle in a manner consistent with the distribution of emphasis in the curriculum.

The curriculum of the requalification training cycle for which the examination is being developed should identify the following information:

1.   requalification lecture/classroom topics indicating the percent of the cycle devoted to each

2.   concentration of training exercises using the simulation facility, including the types of scenarios trained (e.g., accident, abnormal, normal) and the number of times each scenario was run

3.   special focus of the training such as plant modifications, licensee event reports (LERs), and major changes to operating practices or policy

4.   practical training such as operation of individual systems or components for requalification training purposes, using either the simulation facility, mockups, or actual systems and components

The format of the sample plan is a matter of training department preference as long as the plan results in a thorough and accurate assessment of the facility's training program and its intended objectives. The sample plan is expected to contain the following information for use in developing or selecting the test items to be used in the requalification examination:

1. identification of the subjects to be evaluated (system, component, procedure, or other training subject)

2. the preferred testing media for evaluating each subject (written, simulator, or walk-through examination); more than one testing method may be used to evaluate a subject

3. the learning objectives intended to be evaluated

4. a list of references used to develop the test items

5. the specific K/A topic or facility JTA KSAs that are closely linked to the learning objectives for each subject and the importance factors for each (the facility licensee may use a site-specific K/A if it exists)

6. all test items used in the examination should have a K/A value of 3 or greater; the facility licensee may propose the use of test items with NRC K/A values less than 3 with appropriate justification

7. the percentage or number of points of the examination that should be devoted to the topic area (e.g., 3 points for technical specification interpretation, or 5 percent on reactor coolant pumps)

8. whether the subject is identified as safety-related in the facility's JTA

9. whether the subject was covered in the cycle for which the examination is being developed

10. the identification code or number for previously developed test items that evaluate the subject

11. recent safety-related issues and events (e.g., relevant LERs)

1. Pre-Examination

I acknowledge that I have acquired specialized knowledge about the NRC requalification examinations scheduled for the week(s) of _____ as of the date of my signature. I agree that I will not knowingly divulge any information about these examinations to any persons who have not been authorized by the NRC's chief examiner. I understand that I am not to instruct, evaluate, or provide performance feedback to those operators scheduled to be administered these examinations from this date until completion of examination administration, except as specifically noted below and authorized by the NRC (e.g., acting as a simulator booth operator or communicator is acceptable if the individual does not select the training content or provide direct or indirect feedback). Furthermore, I am aware of the physical security measures and requirements (as documented in the facility licensee's procedures) and understand that violation of the conditions of this agreement may result in cancellation of the examinations and/or an enforcement action against me or the facility licensee. I will immediately report to facility management or the NRC's chief examiner any indications or suggestions that examination security may have been compromised.

2. Post-Examination

To the best of my knowledge, I did not divulge to any unauthorized persons any information concerning the NRC requalification examinations administered during the week(s) of _____. From the date that I entered into this security agreement until the completion of examination administration, I did not instruct, evaluate, or provide performance feedback to those operators who were administered these examinations, except as specifically noted below and authorized by the NRC.

PRINTED NAME	JOB TITLE / RESPONSIBILITY	SIGNATURE (1)	DATE	SIGNATURE (2)	DATE
1.					
2.					
3.					
4.					
5.					
6.					
7.					
8.					
9.					
10.					
11.					
12.					
13.					
14.					
15.					

NOTES:

The following checklist represents the minimum content of facility-generated reference material. Items marked "optional" should be checked if requested from the facility licensee by the chief examiner. The chief examiner or designee may use this checklist to make a quick, general evaluation of the completeness and adequacy of the facility licensee's references. The chief examiner may resolve any specific questions about the references with the facility staff as necessary.

I. Quantity

	Reference Material	Required Minimum	Actual Submitted
A.	Open-reference written examination items	350 per section; bank is to be dynamic, With at least 150 revised, reviewed, or newly generated questions per year	
B.	Simulator scenarios	25; + 5 per year following the initial requalification exam until at least 30 scenarios covering all aspects of the EOPs are developed	
C.	Job performance measures	95; + 10 per year following the initial requalification exam until the JTA is fully covered	
D.	Technical specifications	1 copy	
E.	Applicable plant procedures	1 set (optional)	
F.	Emergency plan	1 copy	
G.	Applicable administrative procedures	1 copy (optional)	
H.	Sample plan	1 copy	
I.	Requalification cycle training reference material (lesson plans, handouts, etc.)	1 set (optional)	
J.	Appropriate sections of JTA or facility-specific K/A Catalog	1 set (optional)	

Reviewed by: _____   Date: _____

II.   Usability

A.   The reference material is legible.   yes   no

B.   The reference material is properly arranged and labeled for its function.   yes   no

C.   The reference material indicates a SAT program.   yes   no

D.   Reference material is available to verify that test items are appropriate,
      job relevant, and technically accurate.   yes   no

E.   Reference material is available to adequately support the examination topics.   yes   no

Comments

Reviewed by: _____   Date: _____

III.   Quality

Exam Section   Required Standards   Comments

A.   Sample Plan   Subjects covered in requalification cycle
                       are identified.

                       The test outline incorporates:
                       • time spent on topic
                       • relative importance
                       • frequency of performance
                       • job level (RO or SRO)

                       The test outline identifies
                       K/As (or facility equivalent)
                       of sufficient importance.

                       Plant-specific priorities are identified
                       (LERs, procedure changes,
                       system modifications, risk-dominant
                       accident scenarios, risk-important
                       systems and operator actions[1] identified
                       in the facility licensee's PRA/IPE, etc.).

                       Appropriate testing methods
                       are indicated for each K/A
                       (i.e., JPM, written exam, and/or simulator).

                       Applicable learning objectives
                       are associated with K/As.

                       Methodology exists to tie
                       test items to a learning objective
                       and a K/A.

                       Sample plan includes important
                       important topics not covered in
                       the requalification cycle.

                       Test areas appropriate to ROs
                       and SROs only are identified.

Reviewed by: _____   Date: _____

---

[1]   Chapter 13 of NUREG-1560, "Individual Plant Examination Program: Perspectives on Reactor Safety
       and Plant Performance," identifies a number of important human actions that may be appropriate for evaluation.

III.   Quality (cont.)

Exam Section   Required Standards   Comments

B.   Written   At least 10 percent of all
test items shall be reviewed
using Form ES-602-1.

Test items are important to safety.

Test items are clearly written.

Test items are appropriate
to license level.

The criteria for open-reference
examinations are met.

Test items are associated
with K/As of 3 or greater
and are adequate discriminators.

A learning objective
and applicable reference
are identified for each test item.

The facility has identified
SRO-level questions
for both sections of the exam.

If the above criteria are not adequately met, the NRC will conduct further review
of the examination bank utilizing ES-602, Attachment 1, "Guidelines for the Development
and Review of Open-Reference Examinations," and Form ES-602-1, "NRC Checklist
for Open-Reference Test Items."

Reviewed by: _____   Date: _____

III.   Quality (cont.)

Exam Section	Required Standards	Comments

C.   Walk-Through

At least 10 percent of the JPM bank
were reviewed using Form ES-C-2.

Test outline identifies
applicable plant systems:
*   systems covered in
    requalification cycle
*   new or recently modified systems
*   systems in recent facility LERs
    or vendor notices
*   PRA-identified risk-dominant systems
*   systems in NRC generic
    communications

Tasks/abilities for identified systems:
*   are applicable to the facility
*   are at the AO/RO/SRO level
*   have a K/A value of 3 or greater
*   include JPMs pertinent only to SROs

Some JPMs are performed
under low-power or shutdown
operating conditions.

Some JPMs require the operator
to implement alternative paths
within the facility licensee's
procedures.

Facility JPMs contain
the information found on Form ES-C-1.

Reviewed by: _____   Date: _____

III.   Quality (cont.)

Exam Section	Required Standards	Comments

D.   Simulator

At least 10 percent of the scenarios reviewed using Form ES-604-1.

Scenarios are an appropriate measure of the material covered in the sample plan.

Scenarios are based on:
- lessons covered in the requal cycle
- recent industry events
- LERs
- emergency and abnormal procedures
- design and procedural changes

Scenarios exercise crew's ability to use facility procedures in accident prevention and mitigation.

Scenario events have a K/A of 3 or greater.

Some scenarios are based on low-power[2] operations.

Some scenarios are based on the dominant accident sequences (DAS) for the facility as determined by PRA/IPE.

Scenario identifies critical tasks that meet the criteria of Appendix D.

Proposed examination scenarios that were used for training during the most recent training cycle have been reviewed by the NRC and replaced or modified, if appropriate, to ensure the validity of the examination and to minimize the potential for examination compromise.

Reviewed by: _____   Date: _____

---

[2]   NUREG-1449, "NRC Staff Evaluation of Shutdown and Low-Power Operation," defines "low power" to include the range from criticality to 5 percent power.

Privacy Information — For Official Use Only

	Overall Results	Total	Passed	Failed
Facility:		#	# / %	# / %
Exam Date:				
NRC Examiners:	Reactor Operator:			
	Senior Operator:			
	Total:			
	Operating Crews:			

Operator	Docket 55-(___)	Grader	JPMs %	Written (A & B)	Written	Results (P or F)		W/T
						Simulator		
						Crew	Indiv	
		NRC		%				
		FAC		%				
		NRC		%				
		FAC		%				
		NRC		%				
		FAC		%				
		NRC		%				
		FAC		%				
		NRC		%				
		FAC		%				
		NRC		%				
		FAC		%				
		NRC		%				
		FAC		%				

Privacy Information — For Official Use Only

Operator	Docket 55-(___)	Grader	JPMs %	Written (A & B)	Written	Results (P or F) Simulator Crew	Results (P or F) Simulator Indiv	W/T
		NRC		%				
		FAC		%				
		NRC		%				
		FAC		%				
		NRC		%				
		FAC		%				
		NRC		%				
		FAC		%				
		NRC		%				
		FAC		%				
		NRC		%				
		FAC		%				
		NRC		%				
		FAC		%				
		NRC		%				
		FAC		%				
		NRC		%				
		FAC		%				
		NRC		%				
		FAC		%				

U.S. Nuclear Regulatory Commission
Individual Requalification Examination Report
(Privacy Information — For Official Use Only)

Facility:	Operator's Name:	
Docket No: 55-	License No:	Expiration Date:
Exam Type: RO / SRO	Retake: 1st / 2nd / No	Date of Last Exam:

Written Examination Results

Date(s) of Exam:	NRC Examiner (Print):	Facility Evaluator (Print):
	NRC Grading	**Facility Grading**
Section A (Points)	of	of
Section B (Points)	of	of
Overall Score (%)	%	%

Simulator Examination Results

Date(s) of Exam:	NRC Examiner(s) (Print):	Facility Evaluator(s) (Print):
Crew Evaluation	Pass / Fail	Pass / Fail
Individual Followup	Pass / Fail / NA	Pass / Fail / NA

Walk-Through Examination Results

Date(s) of Exam:	NRC Examiner(s) (Print):	Facility Evaluator(s) (Print):
No. of Successful JPMs	of 5	of 5
Exam Results (%)	%	%

NRC Examiner Recommendations

Category	Results	Signature/Date
Written	Pass / Fail	
Simulator	Pass / Fail	
Walk-Through	Pass / Fail	

NRC Supervisor Review

Date:	Pass / Fail	Signature:

ES-602
REQUALIFICATION WRITTEN EXAMINATIONS

A.   Purpose

The NRC staff uses this standard to conduct written requalification examinations in accordance with Title 10, Section 55.59(a)(2)(iii), of the *Code of Federal Regulations* [10 CFR 55.59(a)(2)(iii)]. NRC examiners are to follow this standard, in conjunction with ES-601, "Conducting NRC Requalification Examinations," to prepare and administer all NRC-conducted written requalification examinations.

B.   Scope

The written examination is useful for evaluating the knowledge, skills, and abilities (KSAs) of licensed operators that are difficult to infer from behavior alone, but can be readily tested through written responses to questions that value interpretation and allow the examinee to use references. Additionally, an individual's knowledge of factual information and his or her ability to perform "paper-and-pencil" tasks are best evaluated through a written examination.

The written examination consists of two sections for which the examinee may refer to references (i.e., "open-reference examinations"). Section A, "Plant and Control Systems," is administered using a static simulator. Section B, "Administrative Controls/Procedural Limits," may be administered in a classroom. Each section should be designed to last a minimum of 1 hour, including time for the operator to review his or her work. Combined, the two sections of the written examination should be designed to last 3 hours. The facility licensee will determine the exact number of questions and time allowed to complete each section, on the basis of the requalification sample plan and the license level of the operators taking the examination [reactor operator (RO) or senior reactor operator (SRO)].

Although the examination is designed so that examinees may use references, an examinee should not expect to have time to complete the examination by consulting references to determine each answer. A good mix of test items will contain some questions that evaluate the operators' abilities to determine a correct response without delving into reference material and others that require the use of reference material to select the correct response. By combining test items that require references with those that do not, the written examination can test a broader sample of operator knowledge within a given period.

On both sections of the written examination, certain questions will test the knowledge and abilities (K/As) of an RO, while others will test those of an SRO. In developing the examination, the examiner should consult the facility's job task analysis (JTA) and the NRC's "Knowledge and Abilities Catalog[s] for Nuclear Power Plant Operators: Pressurized- [and Boiling-] Water Reactors" (NUREG-1122 or 1123), to help identify the most suitable topics for an RO or SRO, respectively. Additionally, 10 CFR 55.41 and 55.43 provide further guidance on item selection for RO and SRO written examinations, respectively.

1. **Section A, "Plant and Control Systems" (Static Simulator)**

This section of the written examination is designed for using the simulator as a reference tool to visually provide realistic information and to give the operators an environment that is as close as possible to their normal control room. While administering this section, the simulator will be "frozen" in the middle of an evolution, transient, or accident. In developing the test items for this section, allow the use of references and relate them to plant systems and components, control room indications, instrumentation and controls, and technical specification (TS) limiting conditions for operation (LCOs).

Section A is designed to evaluate the operators' knowledge of plant systems, integrated plant operations, and instrumentation and controls. In addition, it evaluates the operators' abilities to recognize TS LCOs and to determine the effects of postulated events. The NRC encourages facility licensees to include questions that test the operators' abilities to use their facility curves and charts.

While administering Section A, the examination team will use one "frozen" simulator condition or setup. The condition places the simulator in a "snapshot" of the plant following a major transient that resulted in an engineered safety feature (ESF) initiation or a steady-state situation at power. Some equipment should be frozen in an abnormal or failed condition to provide adequate material for test items.

2. **Section B, "Administrative Controls/Procedural Limits"**

Section B of the written examination is designed to evaluate the operators' abilities to analyze a given set of conditions and determine the proper procedural and administrative guidance to use. This section may include theory-related questions that are appropriate to sample the topics listed in 10 CFR 55.41 and 10 CFR 55.43, as long as they are operational in nature or test unique facility characteristics.

Section B is designed to test the operators' knowledge and use of plant procedures and administrative controls, while allowing the use of references. The NRC uses administrative, operating, normal, abnormal, and emergency procedures, the TS, and the emergency plan as sources of test items for this section of the examination. The test items focus on how direction, guidance, and information found in these procedures are used or interpreted, rather than focusing on finding the procedure in which the necessary information is located. Additionally, the test items for Section B of the SRO examination examine the operators' understanding of the reasons and bases for procedural requirements. The use of graphs, charts, tables, and drawings is appropriate. The simulator may be made available to the examinees to make the examination more operationally oriented.

C.   Examination Development

1.   Facility Examination Team Members' Responsibilities

a.   The facility is expected to provide a bank of test items that are developed using the guidance in Attachment 1 and Appendix B. The number of test items should meet the submittal guidelines of ES-601, Attachment 2, Enclosure 1, "Requalification Examination Reference Material Requirements." Form ES-601-2, "Evaluation Checklist for Facility Reference Material," provides information that facility personnel may use to evaluate reference material sets before submitting them to the NRC.

The facility licensee shall maintain its examination question bank up-to-date by reviewing, modifying, or creating at least 150 questions each year to expand the bank and reflect procedure or system changes, new lesson plans, and recent licensee and industry events.

If the facility question bank contains at least 700 items that meet the format guidance of Attachment 1, the facility may release the bank to its operators for review.

b.   The following items should be provided for each test question:

- applicable K/A reference and values (RO/SRO)
- reference JTA (if applicable)
- estimated time to answer
- appropriate learning objectives
- applicable reference [(e.g., lesson plan, emergency operating procedure (EOP)]

c.   The facility is expected to provide a sample plan that meets the guidelines of ES-601, Attachment 3, "Examination Sample Plan," and may submit a proposed examination that conforms to the facility's sample plan. The proposed examination should contain a total of 30 to 40 test items, depending on the time validation (maximum of 3 hours) of the individual questions selected. Sections A and B should each contain 15 to 20 questions, and each section must be designed to last a minimum of 1 hour, with the total examination designed to last 3 hours.

The facility licensee will determine the number of questions in each section, based on the requalification sample plan and the license level of the operators taking the examination (RO or SRO), and subject to the quantitative constraints of the previous paragraph. Plant systems questions that do not directly relate to the static scenario can be included in Section A to meet the facility's sample plan and the requirements of 10 CFR Part 55. In addition, up to 20 percent of the test items may be from topics outside the sample plan, as long as the information stated in Section C.1.b. of this standard is provided.

If the facility licensee submits a proposed examination, those individuals involved in its development become subject to the security restrictions of ES-601 once examination development commences. These restrictions remain in effect until the NRC examination is given. If, after developing a proposed examination, the facility decides not to submit it for use in the NRC-conducted examination, the developers are released from the security restrictions of ES-601.

d.   After the NRC has reviewed the facility's examination bank and commented on the test items selected for the examination, the facility team members are expected to prepare the examination for final NRC review and approval. The examination may be finalized before or during the preparation week.

e.   The facility team representative will evaluate each test item that the NRC revised, in order to assess the following criteria:

- appropriateness
- time required to answer, given the operational context
- technical accuracy
- clarity
- K/A and objective references

Following this evaluation, the facility examination team representatives and the chief examiner need to agree on the final form of the examination. They also need to complete a time validation of the proposed examination. A variety of methodologies have proven effective in accomplishing this task; Attachment 1 provides further information.

Any individual involved in time validating the examination is required to sign the security agreement, Form ES-601-1. The examination team may add or delete items from the examination based on the results of this time validation, and their experiences. If any test items are added, it is not necessary to time validate the entire examination again, as long as a subject matter expert (SME) has reviewed the added questions, indicating the approximate time that an operator should take to answer each question.

f.   The facility licensee is expected to provide a sufficient number of copies of each reference so that each examinee can use the references during the examination and, immediately upon completion of the examination, compile the examinations and reproduce sufficient copies for their own use and that of the NRC.

g.   To help relieve the burden of providing a complete set of references to each operator, the examination may be assembled so that a different sequence of questions appears on each operator's examination. Alternatively, handouts of relevant information (e.g., plant curves, blank forms, etc.) may be provided with the test.

2.   **NRC Examination Team Members' Responsibilities**

a.   The NRC will begin its evaluation of the sample plan, the bank of test items, and the proposed examination as soon as possible after receiving the facility's materials. The NRC will promptly evaluate the materials to allow sufficient time for the NRC or the facility to develop the test items and for the facility to revise them to meet NRC standards, if required. The NRC examiners should review the proposed test items using Form ES-602-1, "NRC Checklist for Open-Reference

Test Items," to ensure appropriateness, clarity, and importance to safety, as described in Attachment 1.

If the facility licensee intends to administer both sections of the examination during a single 3-hour period as noted in Section D.1.c, the examination team members must review the examination as a whole to ensure that the items in either section do not compromise those in the other.

b.   A minimum of 80 percent of the test items will be chosen in accordance with the sample plan. The remaining 20 percent may be substituted by the examination team, using facility examination bank questions or new questions that the exam team develops. Should it be necessary to develop additional items to satisfy the sample plan, the NRC will ask the facility to do so.

c.   If, after reviewing at least 75 percent of the bank, insufficient test items exist to develop an NRC examination that meets the sample plan, the NRC staff will declare the bank of test items inadequate. In that event, the regional managers may either cancel the scheduled examination or administer an examination using NRC-developed test items without consideration for the 20-percent substitution constraints.

d.   If the sample plan does not include topics from outside the requalification cycle, the examination team should consider incorporating 10 to 20 percent non-requalification-cycle-specific test items.

e.   If a test item does not have a clear tie to the JTA, the examiner will discuss the applicability of the test item with the facility representatives.

D.   Examination Administration and Evaluations

1.   Written Examination Conduct

a.   An NRC examiner or knowledgeable facility representative who has signed the security agreement will proctor each section of the examination. As a minimum, an NRC examiner will observe the examination briefing as the operators begin the examination to ensure that all administrative aspects of the examination are adhered to. If an NRC examiner does not continuously proctor the examination, an examiner will periodically visit the examination room to ensure that the proctor appropriately addresses questions on the content or administration of the examination that may have arisen.

b.   Section A is administered on the facility's simulator or an approved simulation facility.

c.   Section B may be administered in the simulator or in a classroom setting as the facility staff and the chief examiner deem appropriate. If both sections of the examination are administered in the simulator during a single 3-hour period, operators may return to a section of the examination that they already completed or retain both sections of the examination until the time has expired.

d.   For Section A of the examination, the facility licensee is responsible for giving the group of examinees at least one copy of all controlled reference materials available in the control room. Examination reference materials will *not* include material that is intended for training use only. The licensee controls all reference materials in accordance with its 10 CFR Part 50, Appendix B, procedure revision control program. The materials should be authorized for use in operating the power plant, agreed upon by the facility and the chief examiner, and in effect at the time of the examination validation (i.e., the preparation week).

e.   During the administration of Section B, each examinee will have available for use the following materials (complete, current issue):

   • technical specifications
   • plant procedures (EOP/AOP/NOP, etc.)
   • emergency plan (as available in the control room)
   • administrative procedures applicable to operations
   • other controlled plant reference materials that are normally available in the control room (e.g., curves and data book, forms, plant drawings, flow charts, etc.) and authorized for use in operating the plant

Note that "non-controlled" reference materials, such as the Emergency Operating Procedure Owner's Group Basis Documents will *not* be provided unless they are authorized to be used by the control room operators during plant operations.

2.   Examination Administration Procedures

The written examinations will begin only after the chief examiner has verified the adequacy of the examination facilities and made arrangements for continuous proctoring of the examination as discussed in Section D.1.a of this examination standard. An NRC examiner may act as proctor during this examination. However, the chief examiner is responsible for ensuring that the actions of D.2.b – D.2.i (below) are complete.

Each section of the written examination will be administered as follows:

a.   An NRC examiner will verify each examinee's identity and examination level against the list provided by the facility licensee. If possible, the ROs and SROs should be seated at alternating tables. Any errors or no shows will be resolved with the facility staff, and the list will be update as required.

b.   The proctor will remind the operators that they may use calculators to complete the examination, and that no reference material other than that provided is allowed in the examination area. The proctor will define the examination area for the examinees.

c.   The proctor will pass out the examinations and answer sheets and instruct the examinees not to turn over the examination until told to do so. The examinees will be informed that pads of scrap paper are available from the proctor upon request.

d. The proctor will brief the examinees regarding the rules and guidelines in effect during the written examination using Parts A and B of Appendix E, "Policies and Guidelines for Taking NRC Examinations." The proctor should inform the examinees that they may refer to the instructions directly beneath their examination cover sheet. The proctor will read the indicated policies *verbatim*.

e. The proctor will ask the examinees to verify the completeness of their copies by checking each page of the examination. The proctor should also have the examinees check to ensure that an equation sheet has been included in their examinations, if required.

f. After answering any questions that the examinees may have about the examination policies, the proctor will start the examination and record the time.

g. The proctor will periodically advise the examinees of the time that remains to complete the examination. Normally, a chalk board or white board is available for this purpose.

h. As the examinees complete the examination, the proctor will ensure that they sign the examination cover sheet and staple it on top of their answer sheets. The proctor will collect the examination packages, including the questions and answer sheets, any references used with the examination, and all scrap and unused paper. The NRC examiner will keep the cover and answer sheets, dispose of the scrap paper, and give the packages of questions to the facility licensee for subsequent use.

i. The proctor will remind the examinees to leave the examination area, as defined by the examination team.

3. <u>Written Examination Evaluations</u>

Using the examination and answer key, the facility and NRC will independently grade each section of the written examination and will complete the grading of all written examinations within 10 working days of the examination administration date. NRC examiners will record the grades on the written examination cover sheet (Form ES-602-2) and complete Form ES-403-1, "Written Examination Grading Quality Checklist."

An individual's grade will be obtained by summing the points credited to the examinee on both sections of the examination, and dividing by the total points available (i.e., compensatory grading methodology.)

To pass the written portion of the examination, operators must achieve an overall score of 80 percent on the written examination.

4.   **Test Item Evaluation**

If a number of test items require significant modification during the grading of the examination (e.g., more than 10 percent of the items are deleted or the answer is changed from the original key) the NRC will determine the root cause and reflect it in the examination report. As discussed in ES-601, if significant deficiencies exist in the facility's quality control of their examination bank, the NRC will consider them as part of the program evaluation.

If technical flaws that have some degree of safety significance are found in procedures while analyzing the answers to the written examination, the facility may institute an immediate procedural change and inform all operators of the change.

E.   Attachments/Forms

Attachment 1,	"Guidelines for Developing and Reviewing Open-Reference Examinations"
Form ES-602-1,	"NRC Checklist for Open-Reference Test Items"
Form ES-602-2,	"Written Requalification Examination Cover Sheet"

A.   Introduction

The following guidelines are intended for use by those who are involved in developing or reviewing test items for the written portion of the NRC's requalification examination. As described in ES-601, "Conducting NRC Requalification Examinations," the written examination consists of two sections. Section A utilizes a static simulator to provide the context for questions on plant and control systems, while Section B focuses on plant procedures and administrative controls. Examinees may use references, including simulator displays, for both sections. Open-reference written examinations are used for two reasons:

1.   Examination Validity

By permitting the use of references that are available to the control room operators, the conditions and requirements of the written examination more closely approximate those of the actual job. The information provided to the operators in the test items should closely parallel the information typically available to them, while the responses elicited by the questions should be related to the decisions, solutions, and actions required for effective job performance. In other words, consulting references more closely correlates *job demands* and *test demands* — a cornerstone of examination validity.

2.   Level of Knowledge

Use of references enhances examination validity by elevating the level of knowledge of the test items. As described later in these guidelines, operator access to references precludes the use of questions that test for the mere recall of facts and specifics. Instead, open-reference test items require test takers to demonstrate that they can find, apply, analyze, evaluate, or otherwise *use* knowledge to handle the problems and issues they may encounter on the job.

B.   Open-Reference Guidelines

Most principles for effective test item construction apply equally to all types of written questions, regardless of format. Therefore, those who develop and review open-reference test items should consult Appendices A and B of this NUREG-series report in addition to the following guidelines.

1.   Selection of Test Topics

Use the following criteria to select test item topics for the NRC's requalification examination:

a.   *Requalification Training Program Curriculum*

Base the test topics on the curriculum of the most recent operator requalification program training cycle. However, the NRC may substitute up to 20 percent of the examination topics selected by the facility with subjects not emphasized during the requalification cycle. These test items should emphasize knowledge that is of high importance in terms of safety significance.

b.   *Performance Basis*

Like the requalification program itself, draw the test topics from a job-task analysis (JTA) for an RO and an SRO. The facility licensee should validate each test item by demonstrating a link between each item and the following JTA products:

- important operator tasks, as identified by the JTA

- important K/As (rated 3.0 or higher), as identified in NUREG-1122/1123 or a facility-specific K/A catalog

- facility learning objectives identified as important to safety

c.   *Adequacy of Test Coverage*

The facility's proposed sample plan (or test outline) should be checked to ensure that it provides balanced, comprehensive coverage of the requalification training cycle topics. The distribution of proposed facility test items on the examination may be revised if the topics under- or over-represent the material covered in the requalification program. Recent safety-related issues and events [e.g., those in relevant licensee event reports (LERs)] should be addressed in the sample plan. ES-601, Attachment 3, "Examination Sample Plan," provides further information on sample plan development.

2.   **General Guidelines for Sections A and B**

Use the following guidelines to construct and review test items for both parts of the written examination. These guidelines are intended to supplement, rather than replace, the good practices stated in Appendices A and B.

a.   *Operational Orientation*

As previously discussed, examination validity is enhanced to the extent that the demands of the test match the demands of the job. Therefore, in addition to being derived from important K/As and testing objectives, the context and stipulations of test items should mirror the situations encountered in the work setting. The following example illustrates effective and ineffective ways to design test items from K/As and learning objectives:

K/A:	*Knowledge of the design attributes of the turbine-driven auxiliary feedwater pump (TDAFWP) differential pressure controller*
Task:	*Operate the TDAFWP controls during all modes of plant operation.*
Learning Objective:	*The student will be able to operate the TDAFWP differential pressure controller without error during a loss-of-feedwater event.*
Enabling Objective:	*After completing this lesson, the student will be able to explain the operation of the TDAFWP differential pressure controller.*
Poor Test Item:	*State the parameters used by the TDAFWP differential pressure controller.*
Better Test Item:	*Before isolating the "C" steam generator (per EPP11), an operator noted that the transducer-fed auxiliary feed flow indicators for the "C" steam generator were reading greater than the flow indicators to the "A" and "B" steam generators. What is the reason for this flow deviation?*

Notice that the "better" test item requires the operator to demonstrate mastery of the knowledge by applying it to an actual job situation. In developing items, ask yourself "Why is the K/A important to satisfactory job performance?" and "In what situation will the operator need this K/A?" The answers to these questions can provide a basis and context for a test item.

b.   *Level of Knowledge*

The operational orientation required of test items on the open-reference examination, as well as the operators' access to controlled documents, *precludes* the use of questions that test for mere recall or memorization. Rather than requiring operators to simply recognize or recall facts and specifics, open-reference test items have the operators *demonstrate* understanding by *using* the knowledge to address real-life situations and problems. A test item at the higher level of knowledge requires operators to determine or identify the appropriate fact, rule, or principle and then correctly apply it to a novel situation. Appendix B describes each level of knowledge. Together with Table 1 (at the end of this attachment) Appendix B also provides sample questions that illustrate the various levels of knowledge.

c.   *Realistic Context*

To additionally ensure examination validity, make the situation or problem posed in the open-reference test item as similar as possible to the actual situations that operators encounter on the job. Situations described in the questions should be realistic, and should also be free of common "context" problems, including "backward logic" and "window dressing."

Backward-logic questions provide operators with information they normally have to produce, while asking them for information they normally receive, as illustrated by the following example:

K/A:	*Ability to calculate shutdown margins*
Backward Logic Item:	*If the shutdown margin is 5.5 percent, how long has the unit been shut down?*
Better Item:	*The unit has been shut down for x hours. Which of the following is the shutdown margin?*

Questions with window dressing have additional, unnecessary information, typically in an attempt to make a memory level item more operationally oriented, as in the following example:

> *The plant has tripped from the effects of a tornado crossing the site boundary. You, as Shift Supervisor, direct the phone talker to complete the 15-minute notification. He informs you that the normal notification network is inoperable. Which of the following do you direct him to use for completing the 15-minute notification?*

Better Item: *If the normal notification network is inoperable, which of the following methods do you use to complete the 15-minute notification after the plant has tripped?*

Another common problem when constructing a question with a realistic context is that "real world" situations often have more than one correct solution or response. Carefully check the question and references to ensure that each test item has only one correct answer.

d.   *Question Novelty*

One of the most effective ways to ensure that an operator has a high level
of knowledge is to present novel situations and require the operator to realize
both what information is relevant and how to apply it. If a test question
does *not* contain unique or varied circumstances different from those presented
in training, the item will be reduced to eliciting simple recall.

When candidates are able to memorize test items and answers (in their static state)
to respond to test items, we cannot really determine if they can truly solve the problems
or if they have merely memorized the answers. Once a test item and its answer
have been seen and rehearsed, then the item ceases to be a viable discriminator
of safe operator performance. It is no longer challenging or testing problem-solving
ability; rather, it is simply testing recall. Therefore, test items must be dynamic,
replacing or substituting items of like kind and difficulty to preserve integrity
in the test discrimination process.

Because an infinite number of combinations of plant or equipment parameters
and malfunctions may exist at any given time, a true test will compensate
for this variation and will become dynamic so that the test can adapt to the infinite number
of combinations and still test the same kinds of responses, but to different situations.

Review the training material to ensure that questions do not include overly
familiar conditions. Keep in mind, however, that all conditions and situations
should be reasonable, realistic, and safety-related.

e.   *Relationship of Open-Reference Examinations and Direct Lookup Questions*

Direct lookup questions are associated with open-reference examinations.
The key phrase here, "direct lookup," conveys the meaning that little mental activity
is involved other than simply copying an answer that is readily available in a reference
(i.e., simple recall of where to find the information). Merely omitting from the item stem
any mention of where to find the answer does not make it an acceptable
open-reference question.

Do not use direct lookup questions for two reasons. First, these items only test memory,
in that the information is readily available; this is an inefficient and less valid means
of testing candidate knowledge. Second, other than demonstrating that
a candidate knows *where* to find information, this type of question does *not* test
the understanding or analysis of the information that can be applied on the job.
Consequently, this type of question will not discriminate the safe operator
from the unsafe operator.

The other option is an "open-reference" question. Use an open-reference examination to test candidate knowledge for the following purposes:

- Which reference to use and where to find it?
- How to apply the information in the reference to the problem?

For an open-reference question, the kind and amount of information required to solve the problem would exceed that which could normally be committed to memory. In other words, the NRC does not expect candidates to remember the information needed to solve the problem.

In regular closed-reference questions, we expect the candidate to know and understand how systems operate to answer a question with the information provided in the stem of the question. For a closed-reference question, the candidate would not need to consult a reference. In other words, the NRC expects the candidate to solve the problem by knowing and understanding how the systems work, given various conditions set in the problem.

Whether an examination is open- or closed-reference, we should, to the extent possible, test problem-solving or decision-making because, at this more complex level of thought, we more closely approximate the job and achieve a valid assessment.

Memory types: Understanding how memory operates relates to understanding why an open-reference question is preferable. Obviously, all that we know or do involves memory. Operationally, however, we can look at memory as falling into two categories:

- simple memory
- complex memory

Simple memory can be viewed as recall or recognition of simple bits and chunks of information. Simple memory may still be involved when the volume of information increases (i.e., the amount of information is large, but the process is basically simple memorization of *more* bits of information). Visualize the type of memory required to memorize 5 letters of the alphabet versus 26 letters, or the recitation of a short poem (or procedure) versus a long one, and so forth. This memorization process does not involve analysis, integration of facts, or problem solving.

Rather, the process requires repetition, practice, and rehearsal. The difference lies in the amount of information to be recalled, not the level of mental processing.

By contrast, complex memory, as the term suggests, involves a higher level of cognitive processing. The bits and chunks of information must now be combined or integrated to create something new, solve a problem, predict a response, or make a decision. Therefore, both the amount of information and what is to be done with it makes the cognitive mental processes complex. Naturally, too, some questions will involve greater complexity than others, but the mental processes will be the same — integrating bits of information, combining and sorting them, and distinguishing the relevant from the irrelevant to arrive at an answer to the question. This is the essence of an analysis/synthesis process.

As previously stated, the NRC should evaluate candidates at this complex level, because this level of thought processing most closely approximates that needed on the job. The complex, problem-solving level subsumes knowledge of the bits and chunks of information frequently tested at the simple memory level. Therefore, by testing at the complex level, we are also implicitly testing at the simple memory level. As a prerequisite to solving the problem, the candidate recalls and integrates these bits and chunks of information. Therefore, testing at the analysis level and is more efficient than testing at the simple memory level.

A Final Note: Undue emphasis is placed on the term "immediately" in the definition of a direct lookup. Speed of knowing where to locate a reference is irrelevant to direct lookup. The NRC expects candidates who have been trained to quickly locate the appropriate reference. The speed issue is relevant to whether the stem of the question contains unnecessary cues to the candidate about where to find the reference. If the intent of the open-reference question is to assess whether the candidate knows where to find the information, a cue regarding the location of information should not be in the stem. Part of the value of an open-reference exam is to test the candidate's evaluative knowledge of *where* to look. If the stem provides unnecessary cues to the reference, a candidate can immediately go to the reference and a value of the open-reference test is lost.

Speed in answering the question proper is a function of the level of difficulty and the thought processes/steps required to answer the question. Obviously, if the question is a direct lookup, by definition, it assesses only simple memory and will be quickly and easily answered. This type of question should never be asked.

References should be considered "tools" that operators use to solve problems. The correct use of these "tools" is what should be tested during the open-reference examination, not just the recall of facts and specifics. As previously stated, "direct lookup" questions should not be included in the examination; rather, questions should be structured to determine whether operators can identify, locate, or select correct reference information to produce organized responses and satisfactory solutions to job-related problems and issues. An example of a lookup question, which should generally be avoided, follows:

Based on the "Alarm Response Procedure" 1ZZ-040-3, what is the setpoint of the high-high containment pressure alarm (PK25) on VB3?

a.   *10 psig*
b.   *15 psig*
c.   *20 psig*
d.   *25 psig*

This question should be rejected because a candidate can easily find the setpoint in the alarm response procedure (ARP). Some may argue that knowing how to look up this data in the ARP makes the item valid; however, no higher-order cognitive skills requiring analysis or synthesis of information were required to determine the correct response. Avoid similar questions on precautions or prerequisites that are listed in procedures. A better question using reference material would be as follows:

Using the current plant conditions (assume ECCS and CS flow rates REMAIN CONSTANT), how much time is available before switchover to containment recirculation?

a.   *3.6 hours*
b.   *4.2 hours*
c.   *4.8 hours*
d.   *5.2 hours*

This is a "lookup" question in a sense, but it certainly requires gathering data from the control boards (e.g., ECCS flow, CS flow, and RWST level) and then identifying the correct emergency procedure and locating and selecting the correct graph to determine how much time is left before a specific level is reached in the RWST. It requires use of both the simulator and the plant procedures as references.

Another appropriate question using facility references is as follows:

Following a LOCA, automatic actions have occurred as follows:

- *The reactor has tripped and is shut down.*
- *AFW has actuated and steam generator pressure is being controlled at 1,005 psig, using steam dumps to the condenser.*
- *Containment pressure has risen to 15 psig, and no additional automatic actions have occurred.*

Which of the following Functional Recovery Procedures should be implemented IMMEDIATELY?

a.   *FR-C1*
b.   *FR-Z1*
c.   *FR-P1*
d.   *FR-I1*

This question requires identifying which systems should have actuated based on the ESFAS setpoints and which critical safety functions are compromised. The operator should refer to the functional recovery procedures to verify which critical safety functions have been compromised. Knowing where to look and what to look for are required to answer this question in a reasonable time.

The item could also be used in the simulator section by requiring the operator to look at the control board in the "frozen" simulator to determine the plant conditions and deduce what critical safety functions were not met. Naturally, the more integration and evaluation required, the more time must be given to answer the question.

Another question that makes effective use of reference material is as follows:

> *While operating at 100-percent power, VCT and pressurizer alarms and indications show decreasing pressurizer level with two charging pumps operating. Also, the blowdown and main steam radiation monitors have alarmed. While following the appropriate abnormal and emergency procedures, you, as the Shift Supervisor must evaluate the existing conditions. What emergency classification would you declare on the basis of this information?*
>
> a.   *Notification of Unusual Event*
> b.   *Alert*
> c.   *Site Area Emergency*
> d.   *General Emergency*

This question requires the operator to consult references to classify an event. It also requires analyzing the situation, finding the correct part of the EPIPs, and selecting the appropriate classification.

f.   *Difficulty Level Versus Discriminatory Value*

Test developers sometimes believe, erroneously, that open-reference questions should be more difficult to compensate for the operators' access to reference material. Frequently, this increased difficulty is in the form of requiring knowledge of more obscure or otherwise unnecessary information. Both open- and closed-reference questions should have the same standard of difficulty; that is, difficulty should be based on the job demands and responsibilities of operators. A question should be constructed so that it effectively discriminates a competent operator from one who is not. A high K/A value should not be confused with the difficulty or discriminating ability of a question.

g. *Time Limits*

Operators take considerably longer to answer open-reference questions than closed. (Weaker operators especially have been found to spend an inordinate amount of time consulting references rather than writing responses). Provide the operators ample time to complete the examination, although not so much time that less-than-competent operators have the opportunity to locate answers without prior familiarization with the topic. Use the following guidelines to determine the appropriate length of the examination:

(1) A competent operator should complete the combination of Sections A and B in 3 hours. Give the operators an appropriate amount of time to review Sections A and B based on the number of questions assigned to each section. For example, if Section A has 15 questions and is validated for 45 minutes, allow operators 15 minutes for review. Likewise, if Section B has 20 questions is validated for 90 minutes, allow 30 minutes for review. The time allocated to review Sections A and B must be included in the 3-hour time limit.

(2) Questions should be developed so that Sections A and B each have approximately 15 to 20 points, for a total test value of 30 to 40 points. The examination sample plan should be used to determine the exact number of questions to be asked in each section. As noted in Appendix B, multiple-choice questions are preferred, but other formats are acceptable. No question will be worth more than 2 points.

(3) In an open-reference examination, every answer need not require the operator to use a reference. The individual developing the questions should make a reasonable estimate of the time required to answer each question and identify any references needed to obtain a response.

Whether and to what extent references are needed affect what constitutes a reasonable amount of time to develop a response. For example, if the static scenario is set up for an abnormal plant transient that requires relatively rapid operator analysis or response, the time allowed to respond to the question should be similar to that required to react to the transient. The NRC does not expect an operator to answer a question as quickly as he or she would react in the plant, but does expect that the operator would consult few references.

Conversely, questions involving scenarios for which an operator would have time to consult many references would allow similar time to develop a response to the question.

(4)   Each proposed examination is expected to be time validated. The best method would result in the examination being taken in near-test conditions by a representative cross-section of plant operators. Then, by taking the average of the time it took each individual to answer each question, a reasonable time may be established for the test. However, if a large deviation occurs among test takers on particular questions, they should be asked why they took either an excessive or relatively short amount of time to answer the question (compared to that anticipated). Responses may lead to eliminating certain operators' times from the averaging process and, thereby, eliminating anomalies associated with individuals (rather than the test items themselves). However, logistics dictate that sometimes only one or a few individual(s) can participate in validating the time to complete the test. In any case, the results need to be carefully evaluated for any unanticipated deviations from the amount of time anticipated to complete each item.

Facility managers responsible for validating the examination are expected to validate the time for each question similarly. When performing time validation of the examination, these expectations should be made clear to the facility representatives validating the examination so that a reasonable estimate can be obtained.

h.   *Correct Mode of Measurement*

No matter how high their importance ratings or operational relevance, certain operator knowledge, skills, and abilities are not amenable to written testing, as in the following example:

Arrange the major steps in the proper sequence to start, parallel, and load DG-2:

_____*Use governor control to increase DG-2 KW.*
_____*Raise DG speed to 900 RPM.*
_____*Match voltage with bus 1A2 voltage.*
_____*Close breaker 1AD2.*

Despite its operational orientation, the underlying skill addressed in this test item would be better assessed by having the operator simulate or perform the steps during either the simulator or walk-through portions of the operating examination.

3.   **Specific Guidelines for Section A, "Plant and Control Systems"**

The following guidelines are specific to the Plant and Control Systems section of the written examination as performed on a static simulator. These guidelines are divided into two sections, namely "Question Development" and "Simulator Setup."

a.   *Question Development*

To ensure that the operators' knowledge of systems and integrated plant response is adequately evaluated, Section A of the written examination should incorporate the behavior of systems and controls in normal, abnormal, and emergency plant conditions. To the extent possible, questions should require the operators to refer to control room indications in formulating their responses, as in the following example:

Which one of the following describes the location of the steam break?

a.   *inside containment, upstream of the steam line flow transmitters*
b.   *inside containment, downstream of the steam line flow transmitters*
c.   *outside containment, between "C" MSIV and "C" main steam line check valve*
d.   *outside containment, between "C" MSIV and "C" main steam line containment penetration*

The scenario used should put the plant at some point in a major plant transient (e.g., LOCA, SGTR, loss of all AC) with several passive or active failures incorporated. However, the number of malfunctions or failures included in the scenario should be limited. No more than four minor failures should be used (e.g., failure of a safety-related pump to start, failed pressurizer pressure indication, nuclear instrumentation failure). Four will provide sufficient effects to test a wide range of objectives. Such a scenario would provide sufficient visual cues to develop a good percentage (at least 50 percent) of questions directly related to the existing plant conditions.

Questions may be used that do not relate to the transient but use the simulator as a frame of reference only, provided the operators are aware of this lack of relationship to the transient.

Carefully ensure that multiple questions stemming from one event do not give each other away. The operator should be able to understand and correctly answer each question, based only on the information given in the question, rather than on the answer to a previous question.

Use of plant diagnostic questions for which the examinee attempts to determine what transient has occurred are generally not suitable, given the purpose of this section of the examination. Having the operator attempt to identify what took place may limit the number of questions you may ask about the transient. Indicate what symptoms or events have occurred, which procedure has been implemented, and the point in the procedure that was reached at the time the simulator was "frozen."

The operator's response should either determine the root cause of the actual system or component failure, or (by using "what if" questions) propose a future event and ask for the expected response.

b.   Simulator Setup

Before the test, advance the simulator recorders to provide clean readings and check the recorders for proper operation. Check all indications (e.g., bulbs, meters, manual loader indications) to ensure they are in proper working order.

When the simulator has been frozen, secure the chart recorder drive power, if necessary.

Before administering the test, verify the simulator indications to ensure that they are what is expected to arrive at the correct answer.

Freeze any "first-out" annunciators that would normally blink to announce first-out conditions and provide them to the operators.

If a transient is stabilized by use of plant procedures, note the step at which the simulator is frozen and record this information on the simulator operations summary sheet. As necessary, give the examinees the progress of the procedure step in effect.

4.   Ideas for Open Reference Formats

Table 2 provides a list of sample formats to assist individuals who are developing performance-based, open-reference questions.

Table 3 provides additional guidance on the process for developing open-reference questions.

Examples of Different Types of Questions   Table 1

1.   Memory level questions are not to be used on open reference examinations.

2.   Comprehension level questions would require the operator to demonstrate
      an understanding of a concept without necessarily relating it to other material,
      or fully comprehending it in depth.

      A spurious safety injection (SI) signal resulted in HHSI flow to the loop cold legs
      when the plant was in Mode 4. After completing corrective actions
      for the inadvertent SI initiation, you must —

      a.   *stroke test the cold leg motor-operated stop valves within 24 hours*
      b.   *test the cold leg injection check valves for leakage within 48 hours*
      c.   *stroke test the cold leg motor-operated stop valves before entering Mode 3*
      d.   *test the cold leg injection valves for leakage before entering Mode 2*

3.   Analysis, synthesis, and application level questions require higher-order cognitive
      thought processes.

      a.   Application level questions may require the operator to apply the knowledge
            to various concrete situations.

            Given the following conditions:

            •   *both main feed pump turbines tripped*
            •   *AFW automatically started*
            •   *AFW valves reset to control steam generator water level*
            •   *AFW suction pressure decreases to 7 psig*

            Which ONE of the following describes AFW pump response
            for the given conditions?

            a.   *the pump suction will automatically shift to nuclear service water*
            b.   *the pump suction will automatically shift to UST*
            c.   *the pump will trip when suction pressure decreases to 5 psig*
            d.   *the pump will trip after a 6-second delay*

2   **Table 1**

b.   Analysis questions require the operator to mentally integrate a number
of conditions, analyze their interrelationships, sort through and discriminate
among distractors, and finally choose the correct answer.

*Which answer below correctly indicates the posting required
for a room using the results of the following radiological survey?*

SURVEY RESULTS:

AIRBORNE ACTIVITY: 6.34 E-9 uci/cc (Co-60)
FLOOR SMEAR: Beta-610 dpm/cm ^{2}; Alpha-4 dpm/cm^2
EQUIPMENT SMEAR: Beta-1800 dpm/cm ^{2}: Alpha-16 dpm/cm^2
GENERAL RADIATION LEVEL: 110 mr/hr

a.   *Radiation Area, Airborne Area and Full Anti-Cs*
b.   *High Radiation Area, Airborne Area and Full Anti-Cs*
c.   *High Radiation Area, Full Anti-Cs*
d.   *Locked High Radiation Area, Airborne Area, Double Anti-Cs*

c.   Problem-solving questions require putting together elements to demonstrate
an understanding of the underlying knowledge.

*The plant is operating at 100-percent power when a LOCA occurs.
The reactor trips automatically, but fast transfer fails and buses 1A1
and 1A2 become de-energized. PPLS and CPHS initiate
and all equipment operates as designed.*

Which ONE of the following is the expected system response?

a.   *OPLS initiates load shed and starts both emergency diesel generators*
b.   *OPLS does NOT actuate; the emergency diesel generators
start and re-energize buses 1A1 and 1A2*
c.   *OPLS does NOT actuate; the emergency diesel generators
do NOT start, and safeguards motors are started by the sequencers*
d.   *OPLS does NOT actuate; the emergency diesel generators
run at idle speed, and safeguards motors are started
by the sequencers*

Example Formats for Open-Reference Questions   Table 2

1.   Given Plant/System/Component Condition/Problem

- diagnose cause of the problem
- identify location of the problem
- predict the effect on the plant/system/component
- identify the precipitating events/actions
- classify/indicate if conditions meet the specified criteria
- indicate and utilize proper procedures/references
- identify appropriate recuperative actions

2.   Given Plant Conditions and Operator Actions/Procedural Steps Implemented

- indicate purpose/reason behind taking these actions
- determine whether the correct actions were taken given available cues
- indicate what further actions are required to achieve a desired effect

3.   Given a Proposed or Hypothetical Course of Action/Recommendation

- determine its appropriateness or acceptability
- predict the expected plant/system/component response
- predict the effect on other systems/components

4.   Given Data Regarding Plant Conditions or Parameters

- compute or determine the status or change in other parameters
- utilize charts, curves, graphs, etc., to perform calculations or estimations

Developing Open-Reference Test Items   Table 3

The decision steps and mental model for developing analysis-level open-reference questions are as follows:

1. Determine the purpose of the test. Do you want to test *knowledge where* and *knowledge what/how*?

2. Determine the information needed to respond to the question. Is the volume and kind of information such that you would *not* normally expect candidates to recall the information from memory to answer the question?

3. If the answer is *yes* to both Questions 1 and 2, develop an open-reference question.

4. Construct the question as two tiers:

Tier Purpose	Process	Criteria	Outcome
1. Knowledge where	Evaluate reference sources	Avoid clues in stem	Locate reference sources
1. Knowledge what/how	Integrate multiple variables/events	Information volume and detail high (not in memory)	Identify correct answer

Question Stem

bits, chunks of stem information
(conditions, set points, components, etc.)

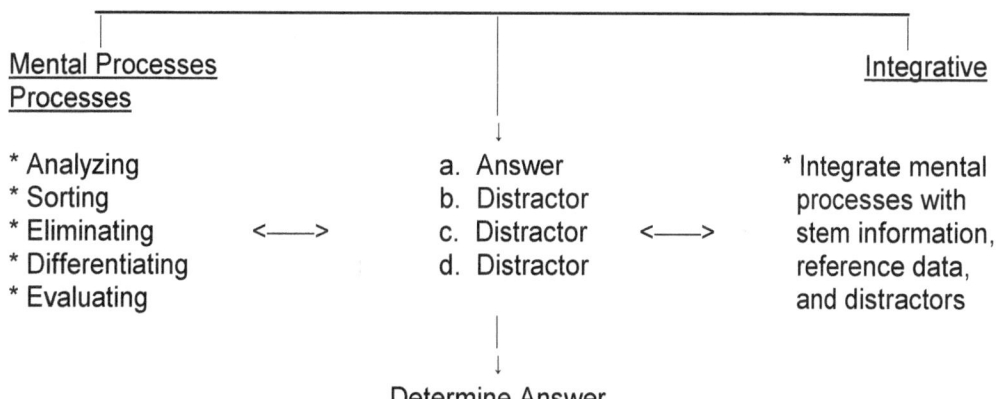

ES-602, Page 25 of 27

Test Item Level

_____ 1. Does each test item have a documented link to important operator tasks, K/As, and/or facility learning objectives?

_____ 2. Is each question operationally oriented (i.e., is there a correlation between job demands and test demands)?

_____ 3. Is the question at least at the comprehension-level of knowledge?

_____ 4. Is the context of the questions realistic and free of window dressing and backward logic?

_____ 5. Does the item require an appropriate use of references (i.e., use of analysis skills or synthesis of information either to discern what procedures are applicable or to consult the procedures to obtain the answer)?

_____ 6. Is the question a "direct lookup" question, or does one question on the examination compromise another? A "direct lookup" question is defined as a question that only requires the examinee to recall where to find the answer.

_____ 7. Does the question possess a high K/A importance factor (3.0 or greater) for the job position?

_____ 8. Does the question discriminate a competent operator from one who is not?

_____ 9. Is the question appropriate for the written examination and the selected format (e.g., short answer or multiple choice)?

_____ 10. Do questions in Section A take advantage of the simulator control room setting?

_____ 11. Does any question have the potential of being a "double-jeopardy" question?

_____ 12. Is the question clear, precise, and easy to read and understand?

_____ 13. Is there only one correct answer to each question?

_____ 14. Does the question pose situations and problems other than those presented during training?

_____ 15. Does the question have a reasonable estimated response time?

U. S. Nuclear Regulatory Commission
Written Requalification Examination

Operator Information

Name:	
Date:	Region: I / II / III / IV
Facility/Unit:	Reactor Type: W / CE / BW / GE
Start Time:	Stop Time:

Instructions

Use the answer sheets provided to document your answers. Staple this cover sheet on top of the answer sheets. Points for each question are indicated in parentheses after the question. The passing grade is 80.00 percent. You have a total of 3 hours to complete both sections of the examination.

Operator Certification

All work done on this examination is my own. I have neither given nor received aid.

_____
Operator's Signature

Results

Results	
Test Value (Points)	Section A: _____ Section B: _____ TOTAL: _____
Operator's Score (Points)	Section A: _____ Section B: _____ TOTAL: _____
Operator's Grade (Combined)	_____ Percent

ES-603
REQUALIFICATION WALK-THROUGH EXAMINATIONS

A.   Purpose

NRC examiners, working with facility evaluators, follow this standard to administer walk-through requalification examinations as authorized by Title 10, Section 55.59(a)(2)(iii), of the *Code of Federal Regulations* (10 CFR 55.59(a)(2)(iii)). The walk-through examination is an effective tool for evaluating the ability of a licensed operator to manipulate system components and controls, interpret references, use administrative procedures, and demonstrate knowledge of component locations.

B.   Scope

This standard provides specific guidance and requirements for NRC examiners to use in preparing, reviewing, and administering walk-through requalification examinations, in which each operator performs five job performance measures (JPMs). Each operator's walk-through examination is designed to test the operator on plant systems that are important to the safe operation of the reactor. NRC examiners and facility evaluators jointly approve the JPMs for each examination. Each JPM consists of several steps, one or more of which is designated as "critical." An operator must properly complete each critical step to pass the JPM.

The examination team will agree on five JPMs so that at least two are conducted in the simulator (or the control room) and at least two are conducted in the plant. To the maximum extent practical, control room JPMs will be conducted using the simulator. When operators perform JPMs in the control room or the plant, they will be cautioned not to manipulate the reactor controls. To successfully complete these JPMs, operators will demonstrate to the examiners the steps or actions they would take to complete the task.

C.   JPM Development

1.   Facility Exam Team Members' Responsibilities

a.   The NRC staff expects the facility licensee to identify the plant systems that are critical to protecting the public health and safety. The systems that are selected for the examination should meet the following criteria:

- systems covered during the current requalification cycle (the facility's sample plan should identify the systems and appropriate learning objectives; see Attachment 3 to ES-601, "Conducting NRC Requalification Examinations")

- new or recently modified systems

- systems that are the subject of recent facility licensee event reports (LERs) or vendor notices
- risk-important systems, components, and operator actions[1] for plant or vendor generic systems, as identified through probabilistic risk assessment (PRA)

- systems that are the subject of NRC information notices

- systems that are important to safety during low-power or shutdown operations

b. For those systems that are identified as being important to safety, the facility representatives are expected to review the job task analysis (JTA), learning objectives, and the "Knowledge and Abilities Catalog for Nuclear Power Plant Operators: Pressurized- [or Boiling-] Water Reactors" (NUREG-1122 or 1123, respectively). The facility representative should highlight for use as JPMs the tasks, abilities, and learning objectives that fulfill the following criteria:

- apply to the facility

- are at the appropriate level for the operator being examined [i.e., the reactor operator (RO) is responsible for auxiliary operator (AO)/RO tasks, and the senior reactor operator (SRO) is responsible for AO/RO/SRO tasks]

- have a knowledge and ability (K/A) rating of 3.0 or higher (tasks and abilities selected for use may have ratings below 3.0 if proper facility justification exists for such ratings)

c. Many tasks that are important to safety are unique to a specific plant and are not referenced in NUREG-1122 or 1123. The NRC staff expects each facility to maintain a list of these plant-specific tasks and develop JPMs that test the operators' knowledge and ability in these areas. The facility is responsible for ensuring that the tasks are appropriate to the applicable license level and have a safety importance rating of at least 3.0, before submitting them to the NRC for review.

If a facility-specific K/A is used in lieu of those specified in NUREG-1122 or 1123, the importance ratings must be based on protecting public health and safety.

d. JPMs should meet the guidelines provided in Appendix C and Form ES-C-2, "Job Performance Measure Quality Checklist." The JPMs should indicate which steps are "critical" to successful completion of the task. Critical steps

---

[1] Chapter 13 of NUREG-1560, "Individual Plant Examination Program: Perspectives on Reactor Safety and Plant Performance," identifies a number of important human actions that may be appropriate for the operating test. In determining important operator actions, do not overlook actions that are relied upon or result in specific events being driven to low risk contribution. This will help identify those human actions, assumed to be very reliable, that might otherwise not show up on a list of risk-dominant actions.

are those that when not performed correctly, in the proper sequence, and/or at the proper time, will prevent the system from functioning properly or preclude successful completion of the task. Form ES-C-1, "Job Performance Measure Worksheet," or an equivalent facility form should be used to construct and format the JPMs.

In accordance with 10 CFR 55.59(a)(2)(ii), requalification operating tests require operators and senior operators to demonstrate an understanding of and ability to perform necessary actions. Therefore, JPMs selected for the walk-through examination shall not test solely for simple recall or memorization. Although it was written in a style to address written examinations, refer to ES-602 Attachment 1, "Guidelines for Developing and Reviewing Open-Reference Examinations," when preparing JPMs as well. Although an operating test does not require every JPM to be alternate path or demonstrate detailed system understanding, simple one-step JPMs or JPMs that only require directly looking up the correct answer are not appropriate. JPMs that incorporate the testing of immediate actions steps from memory are acceptable. However, JPMs should not solely test immediate action steps, and should include testing additional steps or items that are not from memory.

The majority of the JPMs selected for the walk-through examination will cover topics from the most recent requalification training cycle. In addition, the facility is expected to create at least 10 new JPMs each year until they have a JPM bank that is representative of Sections C.1.a and C.1.b of this examination standard. The NRC anticipates that a facility's bank will comprise approximately 125–150 JPMs; however, the exact number will depend on the facility's JTA. New JPMs should generally be based on recent requalification training, industry events, facility changes, and tasks for safety-significant systems.

e.   The NRC staff expects each facility to develop "time-critical" JPMs to evaluate time-critical tasks identified in the facility's JTA for each licensed position. To facilitate the selection of time-critical JPMs for the requalification examination, the facility licensee is expected to uniquely identify these JPMs. To successfully complete a time-critical JPM, the operator must perform the "time-critical" steps within a pre-specified time period, in addition to successfully performing all of the critical steps that are not time-critical. The time period identified in the time-critical JPM should be based on a regulatory requirement or a facility commitment to the NRC.

f.   The NRC staff also expects each facility to develop "alternate-path JPMs" and include them in the JPM bank. To facilitate the selection of alternate-path JPMs for the requalification examination, the NRC staff expects the facility licensee to uniquely identify these JPMs. Appendix C provides guidance for use in developing these JPMs.

2.   **NRC Examination Team Members' Responsibilities**

a.   The NRC's examination team will review and approve the JPMs selected by the facility. The majority of the selected JPMs should be based on the systems covered during the most recent requalification cycle. However, the facility should also select JPMs in systems that are important to safety, regardless of when they were reviewed in requalification training. NRC examiners will review the JPMs submitted by the facility to ensure that 20 percent of the selected JPMs were not covered in the most recent training cycle, as this examination is intended to sample skills and abilities that operators should always be able to display. In general, examiners should select systems in Groups I and II of the appropriate written examination model in ES-401, "Preparing Initial Site-Specific Written Examinations," with Group I comprising at least 50 percent of the selected systems.

b.   The NRC staff will discuss with the facility representatives the selected JPMs that are not identified in NUREG-1122 or 1123 to ensure that the system or task meets the site-specific importance criteria. Any modifications to the selection of JPMs will also be discussed with the facility representatives. The NRC may substitute up to 20 percent of the facility-proposed JPMs with NRC-developed JPMs. The NRC will give facility representatives sufficient time to review any substituted JPMs.

c.   The chief examiner has the authority to decide the content of each examination set. NRC examiners should review the proposed JPMs using the criteria in Appendix C and Form ES-C-2, "Job Performance Measure Quality Checklist."

d.   The chief examiner will ensure that a sufficient number of different JPMs are scheduled during the examination week to avoid compromising the examination.

e.   The chief examiner will ensure that the time validation of each JPM is reasonable and will verify that each JPM is identified as "time-critical" or "not time-critical."

D.   Exam Administration

1.   Conducting JPM Walk-Through Examinations

a.   The facility evaluator is responsible for conducting the walk-through examination while the NRC examiner observes. The NRC examiner and the facility evaluator may ask the operator questions to clarify his or her performance of the JPM after the JPM is completed. In most instances, the NRC examiner will ask the facility evaluator to question the operator about the appropriateness of an action or a response that does not follow the actions specified in the JPM.

b.   The facility evaluator will brief the operator, using Parts A, C, and D of Appendix E. If desired, the evaluator may brief the operators as a group before starting the walk-through examination.

c.   Operators should not be informed of the expected completion time before commencing the JPM. Informing operators of the expected completion time may increase tension as operators approach the time limit. However, the evaluator may inform operators that a JPM is time-critical, if it is normal practice to do so at the given facility.

d.   Time should be allotted during the operating test for evaluating each operator's performance of five JPMs.

Each walk-through examination should last approximately 2 hours. This time includes the validated times associated with each planned JPM, plus any administrative tasks required to conduct the examination.

Administrative tasks may include the following examples:

- transit time to and from the plant site

- time spent complying with facility security and radiological administrative requirements (unless this is part of the JPM being performed)

- transit time within the plant after a JPM is completed to get to a location where the initiating cue for the next JPM is to be given

Note: The JPM sample size will be constrained to the requirements of this examination standard for NRC-conducted examinations. The facility may perform additional evaluation of its operators outside the time frame designated for the NRC examination. However, any additional evaluation by the facility will not be factored into the final requalification evaluation of the operator by the NRC. The criteria for determining requalification program status remain the same.

e.   JPMs that directly relate to the operators' job functions are preferable, particularly for SROs. For example, if an SRO will not perform an emergency action level (EAL) classification during the dynamic simulator or written examinations, the examination team may choose to have the operator perform one JPM that involves classifying an emergency.

f.   The NRC examiner will ensure that the facility evaluator conducts an appropriate examination. Appendix C provides examples of good evaluation techniques to look for during the walk-through examination. If the NRC examiner observes improper evaluation techniques that may render the examination invalid, the NRC examiner will stop the walk-through and counsel the facility evaluator. If the facility evaluator continues to display poor evaluation techniques, the NRC examiner will stop the examination and request that another facility evaluator continue the examination. If necessary, the NRC examiner may conduct the walk-through with the original facility evaluator observing and co-evaluating.

g.   If an evaluator believes that followup questioning is required and is not sure how to phrase the question, he or she should consult the NRC examiner. This will avoid inadvertent prompting of the operator and enhance communication between the facility evaluator and the NRC examiner.

h.   The examiner will document the operator's performance using the applicable portions of a JPM worksheet, Form ES-C-1, or the facility equivalent for each JPM.

Document any questions asked to clarify the operator's performance.
Also, fill out the JPM summary matrix (Form ES-603-1) to maintain operators'
scores during the examination, document which JPM each operator performed,
and fulfill the requirements of ES-601, Section J.1.b.

i.   After completing an operator's JPM set, the NRC and facility evaluators
      shall discuss and resolve any outstanding issues that may result in
      the operator failing the walk-through examination or any individual JPM.
      A discussion of what was observed will often correct a difference of opinion.
      Unresolved differences should be brought to the attention of the chief examiner.

2.   **Grading the Examination**

      a.   To pass the walk-through examination, each operator must successfully complete
            at least four of the five JPMs. To successfully complete a JPM, the operator
            must complete all critical steps and satisfy the completion criteria specified
            in the given JPM.

      b.   An operator is expected to complete each JPM within the validated time period.
            For a JPM that is not time-critical, an operator may exceed the validated time
            if the facility evaluator and the NRC examiner agree that the operator
            is making acceptable progress toward completing the JPM.

            For time-critical JPMs, the facility representatives should identify a period
            that they consider to be the absolute maximum time in which they would
            expect an operator to perform the given task (e.g., locally opening reactor trip
            breakers on an anticipated transient without scram or locally starting
            an auxiliary feedwater pump on a loss of all feedwater). An operator
            who fails to meet the time criteria will receive an unsatisfactory evaluation
            for the given JPM.

E.   Attachments/Forms

Form ES-603-1,   "JPM Summary Matrix"

Operators' Names >>>																		
JPM Number/Brief Description																		
1.																		
2.																		
3.																		
4.																		
5.																		
6.																		
7.																		
8.																		
9.																		
10.																		
11.																		
12.																		
13.																		
14.																		
15.																		

ES-604
DYNAMIC SIMULATOR REQUALIFICATION EXAMINATIONS

A.   Purpose

NRC examiners use this standard in preparing and administering dynamic simulator requalification operating tests in accordance with the provisions of Title 10, Section 55.59(a)(2)(iii), of the *Code of Federal Regulations* [10 CFR 55.59(a)(2)(iii)].

By simulating actual plant operation, the dynamic simulator test provides a comprehensive evaluation of the integrated plant knowledge and skills required of operating crews. It is effective in evaluating a crew's communication skills and team behavior and in identifying any areas in which the licensed operators should be retrained to improve their knowledge and abilities (K/As) in accordance with the provisions of the requalification program developed by the facility licensee.

B.   Scope

The dynamic simulator test consists of two scenarios. Each scenario is constructed to last approximately 45 to 60 minutes. The actual time needed to complete the scenarios will depend upon the specific events within the scenarios, but should allow the crew the time necessary to perform the actions required to respond to each event. To successfully complete this portion of the operating test the crew must demonstrate the ability to operate effectively as a team while completing a series of critical tasks (CTs) that measure the crew's ability to safely operate the plant during normal, abnormal, and emergency situations.

The NRC examiners evaluate the performance of each crew, using standard competency rating scales. Each competency is rated according to the crew's ability to satisfactorily complete the tasks that have been designated as "critical" within that crew's scenario set. Critical means "necessary to place and maintain the reactor in a safe operational or shutdown condition." Each valid CT must meet the criteria specified in Section D of Appendix D. If the crew fails to correctly perform a CT, that failure would indicate a significant deficiency in the knowledge, skill, or ability of that crew to demonstrate team behavior and will be evaluated using the behavioral anchors on Form ES-604-2, "Simulator Crew Evaluation Form."

Facility evaluators will evaluate the performance of the operators during the dynamic simulator test. Because the primary purpose of the dynamic simulator test is to evaluate *crews*, each *individual* is not required to perform a specific number of CTs and may not necessarily receive an individual evaluation by an NRC examiner. However, NRC examiners will follow up on significant individual performance deficiencies on CTs observed during the simulator test in a manner and setting that is compatible with the deficiency. A significant performance deficiency is the omission of or the inability to complete a CT, or the demonstration of a significant lack of knowledge or ability while performing a CT. This followup evaluation will be graded as a component of the individual's operating test. To meet the requirements of 10 CFR 55.59(a)(2), it is the facility licensee's responsibility to conduct its annual operator performance evaluations on the dynamic simulator in accordance with the requirements of its requalification program. The facility licensee may use the NRC-conducted operating test to meet this requirement if the conditions of 10 CFR 55.59(a)(2)(ii) are satisfied [i.e., every individual operating test includes a comprehensive sample of the items specified in 10 CFR55.45(a)].

If an operator demonstrates significant performance deficiencies linked to the execution of CTs during the dynamic simulator portion of the operating test, the facility and NRC examination team members should discuss those deficiencies at the end of the dynamic simulator test.

If the operating crew performs satisfactorily and NRC examiners observe no significant individual performance deficiencies linked to CTs, the individual would pass the dynamic simulator test. In the case of operators who demonstrate significant deficiencies while performing CTs, the facility evaluators and NRC examiners will decide whether the operator would pass or fail by asking the operator followup questions about his or her performance to determine the extent of the knowledge or ability deficiency demonstrated. The number and scope of followup questions to be asked will be agreed to by the NRC examiners and facility evaluators and will be based on the individual's demonstrated knowledge or ability weakness identified during the performance of CTs. The followup questions and individual's answers will be documented and used, along with the individual's performance, as the basis for a pass or fail decision. Section E.2 of this standard describes the method for evaluating and documenting individual performance.

In the rare event that the only way to evaluate the scope and depth of the individual's performance deficiency is by conducting another scenario to gain additional information, the examination team (NRC and facility) will determine the content, critical tasks, operator actions, and crew position rotation necessary to complete the evaluation of the individual's performance. Conducting another scenario is time consuming and may adversely affect the examination process. If an individual operator exhibits only minor deficiencies in performance and satisfactorily completes the testing requirements of 10 CFR 55.59(a), remedial retraining and reevaluation will be conducted in accordance with the facility licensee's requalification program.

C.   Examination Development

Developing the NRC's dynamic simulator requalification examination is a combined effort between the facility representatives and the NRC examiners on the examination team. The responsibilities of the examination team members are outlined below.

1.   Facility Team Member Responsibilities

    a.   The facility licensee develops the dynamic simulator scenarios with identified CTs that meet the guidance specified in Appendix D and Form ES-604-1, "Simulator Scenario Review Checklist." The facility licensee will submit each proposed dynamic simulator test to the chief examiner 45 days before the scheduled examination.

    b.   The facility licensee is expected to provide a qualified simulator operator to assist in developing and administering the simulator examinations. The simulator operator must be available to support the examination team during the examination preparation week, normally 2 weeks before the examination.

The simulator operator will be expected to sign a security agreement at the time that the chief examiner determines that he or she has access to specialized knowledge of any part of the examination.

c.   The scenarios should be based on the training that was conducted during the requalification cycle, recent industry events, licensee event reports (LERs), emergency and abnormal procedures, and design and procedural changes. The scenarios should demonstrate the crew's ability to use facility procedures to prevent and mitigate accidents. Some scenarios should be based on the dominant accident sequences (DAS) for the facility or actual events that have occurred at that or a similar facility. DAS are those sequences that contribute significantly to the frequency of core damage as determined by the facility licensee's probabilistic risk assessment (PRA) or individual plant examination (IPE). The PRA/IPE should also be used to identify risk-important operator actions.[1] In identifying those actions, do not overlook actions that are relied upon or result in specific events being driven to low risk contribution. This will help identify those human actions that are assumed to be very reliable and might not otherwise show up in a list of risk-dominant actions.

d.   The facility representatives on the examination team will be given the opportunity to review any modifications the NRC made to the scenarios. The representatives may recommend changes to events that are critical to plant safety, but must substantiate the reasons for those changes. The examination team has to agree on the validity and content of each scenario before the examination.

e.   The NRC encourages each utility to have its management discuss with the NRC any problems with examination complexity. Utility managers engaged in the examination review will be required to sign a security agreement. Responsibility rests with the utility to resolve any issues *before* administering the examination. This review is to ensure that the final scenarios are (1) consistent with the facility's requalification requirements for operators licensed at the facility, (2) within the capability of the simulation facility, and (3) within the scope of the facility's procedures.

This utility's senior manager or representative should communicate any significant concerns about scenario validity to the chief examiner. If adequate resolution is not reached, the concerns should be brought to the attention of the NRC's regional managers and then, if necessary, to managers in the NRC's Office of Nuclear Reactor Regulation (NRR).

---

[1]   Chapter 13 of NUREG-1560, "Individual Plant Examination Program: Perspectives on Reactor Safety and Plant Performance," identifies a number of important human actions that may be appropriate for evaluation on the dynamic simulator operating test.

2. **NRC Team Member Responsibilities**

   a. At least 2 weeks before the preparation week, the chief examiner or a designee will complete a draft of Form ES-604-1, "Simulator Scenario Review Checklist," for each scenario that the facility proposes to use during the examination, along with any proposed changes to be validated during the preparation week. During the review of each scenario that the facility selected for the examination, the chief examiner or designee will consider the quantitative and qualitative factors described in Appendix D, as summarized on Form ES 604-1.

   b. If the proposed scenarios require major changes to meet the guidance provided on Form ES-604-1, the chief examiner will inform the regional managers and determine the appropriate course of action. The NRC may revise the scenarios, as appropriate, or develop new scenarios to add to the facility's existing scenarios, if required. The NRC will communicate all scenario changes to the appropriate facility representative in sufficient time to allow for scenario validation prior to the preparation week. During the preparation week, the examiners may make minor changes to ensure that the scenario objectives are properly accomplished. The NRC staff will review the final scenarios with the facility's examination team representatives before the examination is administered. The NRC has the final authority to determine the content of the scenarios and decide whether a task is critical for evaluating the competency of the crew.

   c. A key element of the examination team's resolution of concerns regarding the scope, depth, and complexity of simulator scenarios involves a senior utility manager observing the proposed examination scenarios (subject to signing an appropriate examination security agreement) during examination preparation. If necessary, this executive would raise specific concerns to appropriate NRC regional management for resolution before the examination is administered.

D. Examination Administration

1. Administrative Requirements

   a. A facility manager or representative with responsibilities for conducting plant operations (as a minimum, a manager at the first level above shift supervisor) should be present while the simulator examinations are administered. The NRC's chief examiner or a designee will also be present during the administration of each dynamic simulator examination. The chief examiner is the principal point of contact between the facility manager and the NRC.

   b. The examination team briefs the operating crews before the start of the simulator scenarios, using the information in Parts A, C, and E of Appendix E.

c. Crews should be given adequate time to respond to all planned and unplanned events. A scenario's contact time should be approximately 45 to 60 minutes. Contact time means the actual time the operators spend in the scenario; it does not include time spent on briefings, simulator setup, or investigating simulator performance problems.

d. Under no circumstance will any member of the examination team modify the sequence of events and transients during the scenario. If the scenario is not properly administered as a result of a simulator operator error or an unexpected simulator response, the examination team will confer immediately after the scenario set to determine whether the crew has performed a sufficient number of transients and events to justify an evaluation of the required competencies. If necessary, the examination team can run an additional scenario to ensure that the required competencies are addressed.

e. Crew rotation practices shall be discussed and agreed to during the preparation week, and any problems shall be resolved before the administration of the operating test.

f. The members of the operating crew should maintain the same operating positions as during facility requalification evaluations. The crew members should rotate between positions in the manner identical to the facility's rotation practices for evaluations specified in the facility's requalification program.

g. Senior reactor operators (SROs) must be evaluated in at least one scenario in an SRO-licensed crew position. More than two simulator scenarios may be required to examine crews that consist of more than four SROs.

2. **Post-Scenario Activities**

a. If the NRC examiners and facility evaluators observe actions that are unclear during the simulator scenario, they should question the crew members as necessary to develop complete documentation of the crew's performance during the scenario. Questions should be factual and should clarify performance related to observations.

b. If an examiner observes an individual who demonstrates significant deficiencies in performing a CT, the NRC examiner and the facility evaluator will discuss those deficiencies at the completion of the scenario. If they determine that the operator's performance deficiencies cannot be assessed because of a lack of information, the examination team has the option to conduct an additional scenario or a job performance measure (JPM) to obtain the necessary information.

During the post-scenario discussion, the facility evaluator is expected to describe the operator's deficiencies to the NRC examiner and suggest a series of followup questions designed to identify the cause of the deficiency. The NRC examiner will assess the facility evaluator's ability to diagnose the operator's deficiency and document it in the examination report, if applicable. The NRC examiner has the option to augment the followup questions proposed by the facility evaluator, if necessary.

The examination team should minimize the time needed to conduct this review of crew and individual performance to minimize the impact on the operators. However, it is the examination team's responsibility to ensure that the review is thorough and complete.

The facility evaluator will conduct an individual evaluation of the operator in accordance with Section E.2 of this examination standard. The NRC examiner has the option to ask additional followup questions.

c. Upon completing any followup questioning, the NRC examiners and facility evaluators will dismiss the crew to await the next scenario and inform the crew that they may discuss the completed scenario among themselves.

d. The NRC examiners and facility evaluators will meet separately to compare observations and determine whether the crew omitted or incorrectly performed any CTs.

e. The facility evaluators will discuss the crew's performance with the NRC examiners after each scenario to clarify any performance deficiencies that have been noted. The examination team will determine whether the as-run scenario has invalidated any predesignated CTs or whether any new CTs should be designated to evaluate unpredicted events or actions taken by the crew during the scenario. The examination team will then revalidate the CTs in each scenario, using the methodology presented in Appendix D.

f. After the crew completes the last scenario, the NRC examiners and facility evaluators will independently complete Form ES-604-2, "Simulator Crew Evaluation Form," as discussed in Section E. The facility evaluators will also evaluate individual operator performance in accordance with their requalification program requirements and Section E.2. In addition, the NRC examiners will review the facility's evaluations of individual operator's performance after completing each crew evaluation.

E. Performance Evaluations

Two separate evaluations will be conducted based on the information obtained during the dynamic simulator examination. The first is a crew simulator evaluation. For the second, the examination team uses individual simulator performance to determine whether followup questioning of the operator is necessary. The examination team may conclude that, after observing the operator's performance in the dynamic simulator and evaluating the responses to followup questions, additional performance information about the operator must be obtained to make an individual evaluation. In this case, an additional scenario or JPM would be conducted. The individual followup would then be documented along with the individual's crew evaluation on Form ES-601-5, "Individual Requalification Examination Report."

Each operator will be subject to failure based on a competency evaluation of his or her performance on the dynamic simulator and the required followup evaluation, if he or she exhibited deficient performance in executing a crew CT.

1.   **Crew Simulator Evaluations**

After administering the dynamic simulator scenario set as discussed in Section D, the NRC examiners and facility evaluators will independently evaluate the crew's performance by completing Form ES-604-2. The facility is expected to provide its final crew evaluations to the NRC examiners before the crew members return to licensed duties or the end of the examination week, whichever is sooner. Specific guidance for completing Form ES-604-2 appears on the first page of the form.

The results of the crew evaluations will be factored into each individual's examination results and the facility requalification program evaluation. Members of a crew that receive an unsatisfactory crew evaluation are expected to receive remedial training from the facility licensee and to be reevaluated in accordance with the facility licensee's NRC-approved requalification program before returning to licensed duties. Although operators are not required to take an NRC-conducted requalification examination for purposes of license renewal, those who fail to pass (individually or as a member of a crew) an examination conducted by the NRC must be reevaluated by the NRC before their license will be renewed. The level of NRC involvement during the reevaluation will be determined on a case-by-case basis. (Refer to Section F.1 of ES-601, "Conducting NRC Requalification Examinations.")

NRC examiners will document the results of each operator's crew performance in the "Simulator Examination Results" section of Form ES-601-5.

2.   **Individual Operating Evaluations**

Individual operating evaluations on the dynamic simulator examination and the resulting remedial training are primarily the responsibility of the facility licensee. Unsatisfactory operator performance of a crew CT will be followed up after the simulator scenario and documented on Form ES-601-5.

Facility evaluators are expected to document and grade individual operator performance during the dynamic simulator examination in accordance with the requirements of the facility licensee's requalification program. The NRC expects the facility's grading methodology to identify operator deficiencies. The NRC also expects the facility evaluators to discuss those deficiencies with the NRC examiners during the meetings following the scenarios as described in Section D, and to document the deficiencies and remediate and retest the operators for the identified deficiencies in accordance with the facility licensee's requalification training program. At a minimum, the NRC expects the facility evaluators to identify any operator on the crew who was directly responsible for the omission or incorrect performance of validated CTs.

Individual followup is conducted if an operator has significant performance deficiencies linked to a CT. As described in Section D.2.b of this examination standard, the NRC examiner will assist in developing and administering followup questions specific to the deficiencies that the operator displayed in performing the CT. The examination team will determine the number and scope of the followup questions that will be asked based on a review of the operator's deficiencies at the completion of the scenario. The examination team has the option to gather additional information about an operator who displays performance deficiencies while attempting CTs, by either running an additional scenario or using JPMs, if the dynamic simulator examination and followup questioning are inconclusive.

Upon completion of the individual followup questions, the NRC examiner will complete an individual competency evaluation using the appropriate sections of Form ES-604-2 or the facility's equivalent form. Only those competencies that deal with the operator's individual performance deficiencies should be filled out. If the NRC examiner gives the operator a rating factor score of "1" in either of the following cases, the individual fails this portion of the examination:

- any two rating factors in any one competency

- any one rating factor in any one competency if, in the judgement of the examination team, the operator's performance deficiency jeopardizes the safety of the plant or has significant safety impact on the public. (NRC management will make the final decision concerning all operator failures resulting from a single rating factor evaluation of "1.")

When conducting the evaluation described herein, NRC examiners will not assign rating factor scores of "1" based solely on performance in the dynamic simulator. Followup questions will be asked and the operator's responses will be recorded to evaluate and document the knowledge or ability deficiency linked to the performance of a CT.

The NRC examiner will then apply the individual's responses to the questions asked to evaluate and justify individual performance deficiencies that warrant a rating factor score of "1." The examiner will document and include the followup questions asked and the responses given by the operator. Written comments describing the operator's performance and the as-run simulator scenario set will be included with the results of the operator's simulator examination.

The NRC examiner will document the pass or fail determination for each operator's individual followup under "Individual Followup" in the "Simulator Examination Results" section of Form ES-601-5, "Individual Requalification Examination Report."

If an operator demonstrates no performance deficiencies and, therefore, does not require any additional followup questioning, regardless of whether the crew passes or fails the dynamic simulator examination, the NRC examiner will record an "N/A" for "Individual Followup" in the "Simulator Examination Results" section of Form ES-601-5.

F.   Attachments/Forms

Form ES-604-1,   "Simulator Scenario Review Checklist"
Form ES-604-2,   "Simulator Crew Evaluation"

Note: *Attach a separate copy of this form to each scenario reviewed. The examination team uses this form as guidance as they conduct their review of the proposed scenarios.*

SCENARIO IDENTIFIER: _____   REVIEWER: _____

Qualitative Attributes

__ 1.   The scenario summary clearly states the objectives of the scenario.

__ 2.   The initial conditions are realistic, in that some equipment and/or instrumentation may be out of service, but it does not cue the crew to expected events.

__ 3.   The scenario consists mostly of related events.

__ 4.   Each event description consists of—
  •   the point in the scenario when it is to be initiated
  •   the malfunction(s) that are entered to initiate the event
  •   the symptoms/cues that will be visible to the crew
  •   the expected operator actions (by shift position)
  •   the event termination point

__ 5.   No more than one non-mechanistic failure (e.g., pipe break) is incorporated into the scenario without a credible preceding incident such as a seismic event.

__ 6.   The events are valid with regard to physics and thermodynamics.

__ 7.   Sequencing/timing of events is reasonable, and allows for the examination team to obtain complete evaluation results commensurate with the scenario objectives.

__ 8.   If time compression techniques are used, the scenario summary clearly so indicates. Operators have sufficient time to carry out expected activities without undue time constraints. Cues are given.

__ 9.   The simulator modeling is not altered.

__ 10. All crew competencies can be evaluated.

__ 11. The scenario has been validated.

__ 12. If the sampling plan indicates that the scenario was used for training during the requalification cycle, the need to modify or replace the scenario has been evaluated.

Simulator Scenario Review Checklist (Continued)

Note: The following criteria address scenario traits that are numerical in nature. A second set of numbers indicates a range to be met for a set of two scenarios. Therefore, to complete this part of the review, the set of scenarios must be available. This page should be completed once per scenario set.

Scenario Set Consists of Scenario _____ and Scenario _____

Quantitative Attributes

___ 13. total malfunctions inserted: 4 to 8/10 to 14

___ 14. malfunctions that occur after EOP entry: 1 to 4/3 to 6

___ 15. abnormal events: 1 to 2/2 to 3

___ 16. major transients: 1 to 2/2 to 3

___ 17. EOPs used beyond primary scram response EOP: 1 to 3/3 to 5

___ 18. EOP contingency procedures used: 0 to 3/1 to 3

___ 19. approximate scenario run time: 45 to 60 minutes
          (one scenario may approach 90 minutes)

___ 20. EOP run time: 40 to 70 percent of scenario run time

___ 21. crew critical tasks: 2 to 5/5 to 8

___ 22. technical specifications are exercised during the test

COMMENTS: _____

_____

_____

_____

The examination team should use this evaluation form during the dynamic simulator component of the requalification examination. The rating scales on this form are for evaluating the crew as a whole, rather than the individual operators. Use the following instructions when rating team performance on the simulator examination:

1. Review the rating scales before the simulator examination so that you are familiar with each competency to be evaluated.

2. Use Form ES-D-2, "Required Operator Actions," or an equivalent facility form to make notes during the examination, as described in Appendix D and ES-302, "Administering Operating Tests to Initial License Applicants."

3. Complete this form immediately after the simulator examination. Evaluate the crew's performance on each applicable rating factor by comparing the actions of the crew against the associated behavioral anchors and selecting the appropriate grade. The tasks planned and performed during the crew's scenario set may not permit you to evaluate every rating factor for every crew. Annotate those rating factors that are not used in the evaluation.

    The examination team should pay particular attention to the completion of tasks that they identified as critical to plant safety. The crew may compensate for actions that individual operators performed incorrectly, as long as the critical task was completed satisfactorily. Other less-significant deficiencies should also be accounted for in the rating factor evaluations to provide a source of information for crew remedial training during subsequent requalification training.

4. Justify all rating factor grades of "1," and document each justification in the space for "Comments" on the form. Rating factor grades of "1" must be linked to the performance of at least one critical task.

5. Complete the examination summary sheet, recording for each scenario, the scenario name (or identifier), and the critical tasks performed by the crew. Annotate whether the critical task was performed satisfactorily or unsatisfactorily. Complete the crew's overall evaluation using the criteria listed in the next paragraph. Space is provided for additional comments about the crew's performance.

6. The threshold for failing the simulator portion of the examination is to receive a (behavioral anchor) score of "1" in either of the following:

    a. any two rating factors in any one competency

    b. any one rating factor in any one competency if, in the judgement of the examination team, the crew's performance deficiency jeopardizes the safety of the plant or has significant safety impact on the public. (NRC management will make the final decision concerning all crew failures resulting from a single rating factor evaluation of "1.")

Simulator Examination Summary Sheet

Facility: _____ Examination Date:_____

Overall Dynamic Simulator Crew Evaluation: **SAT or UNSAT**

Crew Members	Docket No.	Scenario #1 Position	Scenario #2 Position
1. _____	55-_____	_____	_____
2. _____	55-_____	_____	_____
3. _____	55-_____	_____	_____
4. _____	55-_____	_____	_____
5. _____	55-_____	_____	_____
6. _____	55-_____	_____	_____

Scenario #1: [Enter scenario descriptor]		
Crew Critical Tasks	**SAT**	**UNSAT**
1. [Enter critical task descriptor]		
2.		
3.		
4.		
5.		

Scenario #2:		
Crew Critical Tasks	**SAT**	**UNSAT**
1.		
2.		
3.		
4.		
5.		

Comments:

_____

_____

Diagnosis of Events and Conditions Based on Signals or Readings

Did the crew—

(a) recognize off-normal trends and status?

3	2	1
Recognized status and trends quickly and accurately.	Recognized status and trends at the time of, but not before, exceeding established limits.	Did not recognize adverse status and trends, even after alarms and annunciators sounded.

(b) use information and reference material (prints, books, charts, emergency plan implementation procedures) to aid in diagnosing and classifying events and conditions?

3	2	1
Made accurate diagnosis by using information and reference material correctly and in a timely manner.	Committed minor errors in using or interpreting information and reference material.	Failed to use, or misused, or misinterpreted information or reference material that resulted in improper diagnosis.

(c) correctly diagnose plant conditions based on control room indications?

3	2	1
Performed timely and accurate diagnosis.	Committed minor errors or had minor difficulties in making diagnosis.	Made incorrect diagnosis, which resulted in incorrect manipulation of any safety control.

Grade for diagnosis of events and conditions based on signals and readings: SAT or UNSAT

Comments:_____

_____

_____

_____

Understanding of Plant and System Responses

Did the crew—

(a)   locate and interpret control room indicators correctly and efficiently to ascertain and verify the status/operation of plant systems?

3	2	1
Each crew member located and interpreted instruments accurately and efficiently.	Some crew members committed minor errors in locating or interpreting instruments or displays. Some crew members required assistance.	The crew members made serious omissions, delays, or errors in interpreting safety-related parameters.

(b)   demonstrate an understanding of the manner in which the plant, systems, and components operate, including setpoints, interlocks, and automatic actions?

3	2	1
Crew members demonstrated thorough understanding of how systems and components operate.	The crew committed minor errors because of incomplete knowledge of the operation of the system or component. Some crew members required assistance.	Inadequate knowledge of safety system or component operation resulted in serious mistakes or plant degradation.

(c)   demonstrate an understanding of how their actions (or inaction) affected systems and plant conditions?

3	2	1
All members understood the effect that actions or directives had on the plant and systems.	Actions or directives indicated minor inaccuracies in individuals' understanding, but the crew corrected the actions.	The crew appeared to act without knowledge of or with disregard for the effects on plant safety.

Grade on understanding of the response of plant and systems: SAT or UNSAT

Comments:_____

_____

_____

Adherence to and Use of Procedures

Did the crew—

(a)   refer to the appropriate procedures in a timely manner?

3	2	1
The crew used procedures as required and knew what conditions were covered by procedures and where to find them.	The crew committed minor failures to refer to procedures without prompting, which affected the plant's status.	The crew failed to correctly refer to procedure(s) when required, resulting in faulty safety system operation.

(b)   correctly implement procedures, including following procedural steps in correct sequence, abiding by cautions and limitations, selecting correct paths on decision blocks, and transitioning between procedures when required?

3	2	1
The crew followed the procedural steps accurately and in a timely manner, demonstrating a thorough understanding of the procedural purposes and bases.	The crew misapplied procedures in minor instances, but made corrections in sufficient time to avoid adverse effects.	The crew failed to follow procedures correctly, which impeded recovery from events or caused unnecessary degradation in the safety of the plant.

(c)   recognize EOP entry conditions and perform appropriate actions without the aid of references or other forms of assistance?

3	2	1
The crew recognized plant conditions and implemented EOPs consistently, accurately, and in a timely manner.	The crew had minor lapses or errors. Individual crew members needed assistance from others to implement procedures.	The crew failed to accurately recognize degraded plant condition(s) or execute efficient mitigating action(s), even with the use of aids.

Grade on adherence to and use of procedures: SAT or UNSAT

Comments:_____
_____
_____
_____

Control Board Operations

Did the crew—

(a)   locate controls efficiently and accurately?

3	2	1
Individual operators located controls and indicators without hesitation.	One or more operators hesitated or had difficulty in locating controls.	The crew failed to locate control(s), which jeopardized system(s) important to safety.

(b)   manipulate controls in an accurate and timely manner?

3	2	1
The crew manipulated plant controls smoothly and maintained parameters within specified bounds.	The crew demonstrated minor shortcomings in manipulating controls, but recovered from errors without causing problems.	The crew made mistakes manipulating control(s) that caused safety system transients and related problems.

(c)   take manual control of automatic functions, when appropriate?

3	2	1
All operators took control and smoothly operated automatic systems manually, without assistance, thereby averting adverse events.	Some operators delayed or required prompting before overriding or operating automatic functions, but avoided plant transients where possible.	The crew failed to manually control automatic systems important to safety, even when ample time and indications existed.

Grade on control board operations: SAT or UNSAT

Comments:_____

_____

_____

_____

Crew Operations

Did the crew members—

(a)   maintain a command role?

3	2	1
The crew took early remedial action when necessary.	In minor instances, the crew failed to take action within a reasonable period of time.	The crew failed to take timely action, which resulted in the deterioration of plant conditions.

(b)   provide timely, well-planned directions to each other that facilitated their performance and demonstrated appropriate concern for the safety of the plant, staff, and public?

3	2	1
Supervisor's directives allowed for safe and integrated performance by all crew members.	In minor instances, the supervisors gave orders that were incorrect, trivial, or difficult to implement.	The supervisor's directive(s) inhibited safe crew performance. Crew members had to explain why order(s) could not or should not be followed.

(c)   maintain control during the scenario with an appropriate amount of direction and guidance from the crew's supervisors?

3	2	1
Crew members stayed involved without creating a distraction, the crew members anticipated each other's needs, and the supervisors provided guidance when necessary.	Crew members had to solicit assistance from supervisors or each other, interfering with their ability to carry out critical action(s).	Crew members had to repeatedly request guidance. The crew failed to verify successful accomplishment of orders.

Crew Operations Continued on Next Page

Crew Operations (Continued)

Did the crew members—

(d)   use a team approach to problem solving and decision making by soliciting and incorporating relevant information from all crew members?

3	2	1
Crew members were involved in the problem solving and decision making processes for effective team decision making.	At times, crew members failed to get involved in the decision making process when they should have, detracting from the team-oriented approach.	The crew was not involved in making decision(s). The crew was divided over the scenario's progress, and this behavior was counter-productive.

Grade on crew operations: SAT or UNSAT

Comments:_____

_____

_____

_____

_____

_____

_____

_____

_____

_____

_____

_____

Communications

Did the crew—

(a)   exchange complete and relevant information in a clear, accurate, and attentive manner?

3	2	1
Crew members provided relevant and accurate information to each other.	Crew communications were generally complete and accurate, but sometimes needed prompting, or the crew failed to acknowledge the completion of evolutions, or to respond to information from others.	Crew members did not inform each other of abnormal indication(s) or action(s). Crew members were inattentive when important information was requested.

(b)   keep key personnel outside the control room informed of plant status?

3	2	1
Crew members provided key personnel outside the control room with accurate, relevant information throughout the scenarios.	In minor instances, the crew needed to be prompted for information and/or provided some incomplete/inaccurate information.	The crew failed to provide needed information.

(c)   ensure receipt of clear, easily understood communications from the crew and others?

3	2	1
The crew requested information/clarification when necessary and understood communications from others.	In minor instances, the crew failed to request or acknowledge information from others.	The crew failed to request needed information, or was inattentive when information was provided; serious misunderstandings occurred among crew members.

Grade on communications: SAT or UNSAT

Comments:_____

_____

_____

_____

ES-605
LICENSE MAINTENANCE, LICENSE RENEWAL APPLICATIONS, AND REQUESTS FOR ADMINISTRATIVE REVIEWS AND HEARINGS

A.   Purpose

This standard describes the requirements for maintaining an NRC operator's license and the procedures for processing license renewal applications, licensed operators' requests for administrative reviews and hearings in connection with failures of NRC-conducted requalification examinations, and denials of applications for license renewal.

B.   Background

The renewal license application differs in some respects from the initial license application. The staff developed this standard to establish the procedures for processing operators' renewal applications and requests for administrative reviews and hearings regarding the denial of renewal applications resulting from failures of NRC-conducted requalification examinations.

C.   License Maintenance

1.   Requalification Training and Testing

a.   Title 10, Section 55.53(h), of the *Code of Federal Regulations* [10 CFR 55.53(h)] imposes a condition that requires licensed operators to complete a requalification program as described by 10 CFR 55.59; the requirement applies to all operators, even if they do not maintain watch-standing proficiency pursuant to 10 CFR 55.53(e). 10 CFR 55.59(a)(1) requires licensed operators to successfully complete a requalification program not to exceed 24 months in duration, and 10 CFR 55.59(c)(1) requires the requalification program to be conducted for a continuous period not to exceed two years. To keep from exceeding the 24 month / 2 year duration requirement, a requalification program must be completed within the anniversary month of the second year. For example, if a licensed operator requalification program was started on June 1, 2004, the facility licensee would have until June 30, 2006, to complete their program to ensure compliance with 10 CFR 55.59(a)(1) and (c)(1).

Under 10 CFR 55.59(a)(2), licensed operators must pass a comprehensive requalification written examination as part of a 24-month requalification program, therefore, the exam must occur *during* the requalification program, rather than after its completion. Although the comprehensive written examinations are generally conducted on the same 24-month frequency, their timing can be adjusted somewhat near the end of the 24-month program to account for outages and other events, thereby resulting in some longer testing intervals if an examination is advanced during one 24-month program cycle and returned to its normal timing during the following cycle. Thus, the interval between the administration of successive comprehensive written requalification examinations may exceed 24 months for individual licensed operators. As long as a licensed operator successfully completes the facility licensee's Commission-approved requalification program, including its required comprehensive written examination, within 24 months, as required per 10 CFR 55.59 (a)(1), the operator's comprehensive written requalification examination can be administered more than 24 calendar months from the administration of his or her last comprehensive written examination without requesting an exemption in accordance with 10 CFR 55.11.

For example, consider a licensed operator who took a comprehensive requalification written examination on August 18, 2002, for a facility licensee requalification training program that ran for 24 months according to 10 CFR 55.59(a)(1) and 10 CFR 55.59(c) from October 1, 2000, through September 30, 2002. If that operator's next comprehensive requalification written examination is scheduled for September 27, 2004, the operator will exceed 24 calendar months between successive comprehensive requalification written examinations, but the facility licensee's requalification program and the licensed operator are still in compliance with 10 CFR 55.59(a)(1) and (2) and 55.59(c). The licensed operator will have successfully completed two consecutive requalification training programs, including comprehensive written examinations, within the 24-month requalification program time limit according to 10 CFR 55.59(a)(1) and 55.59(c).

b. Newly licensed operators are expected to enter the requalification training and examination program promptly upon receiving their licenses. Because they just passed the initial licensing examination, operators may be excused from taking any annual operating test or comprehensive written examination that is scheduled to be administered during the first requalification training cycle (nominally, lasting about 6 weeks) in which the operator participates. However, operators who complete one or more training cycles before the scheduled annual test or comprehensive examination should take the test and/or examination to ensure that they do not exceed the allowed testing intervals.

c. Under extenuating circumstances, a facility licensee may invoke the provisions of 10 CFR 55.59(b), "Additional Training," and request in writing that an operator temporarily suspend participation in the facility licensee's requalification training program. The NRC's regional office may authorize the operator to temporarily suspend participation in the requalification training program under the following circumstances:

   •   The operator will be reassigned to full-time, career-enhancing duties at another location, making it impractical to participate in the training program (e.g., assignment to the Institute of Nuclear Power Operations or a foreign interchange program, or college attendance).

   •   The duration of the assignment will not exceed 24 months; if the assignment extends beyond the date of license expiration, the operator may apply for timely license renewal in accordance with 10 CFR 55.55(b) and 55.57(a).

- The facility licensee's plan for ensuring the operator's qualifications and status is acceptable (i.e., the operator must be retrained, tested, reactivated, and medically fit for duty).

If the regional office approves the temporary suspension, it will amend the operator's license to prohibit the performance of licensed duties during the reassignment. The regional office will also confirm its expectations regarding the operator's return to licensed duties and the need for the facility licensee to certify when the actions have been completed. These expectations will be documented in a letter to the facility licensee with a copy to the operator.

The regional office shall refer situations outside the specified parameters to the NRR operator licensing program office for evaluation.

2. **Proficiency Watches**

a. In accordance with 10 CFR 55.53(e), licensed operators are required to maintain their proficiency by "actively performing the functions of an operator or senior operator" on at least seven 8-hour or five 12-hour shifts per calendar quarter. This requirement may be completed with a combination of complete 8- and 12-hour shifts (in a position appropriately credited for watch-standing proficiency as discussed below) at sites having a mixed shift schedule, and watches shall not be truncated when the operator satisfies the minimum quarterly requirement (56 hours). Overtime may be credited if the overtime work is in a position appropriately credited for watch-standing proficiency. Overtime as an extra "helper" after the official watch has been turned over to another watchstander does not count toward proficiency time.

b. In accordance with 10 CFR 55.4, "actively performing the functions of an operator [RO] or senior operator [SRO]" means that an individual has a position on a shift crew that requires an individual to be licensed as defined in the facility's technical specifications. Watch-standing proficiency credit may also be appropriate for certain licensed RO or SRO shift crew positions that are in excess of those required by a facility's technical specifications. However, in order to credit watch-standing proficiency for such excess positions, the facility licensee shall have in place the following procedural administrative controls:

(1) A list of all the licensed shift crew positions, including title, description of duties, and indication of which positions are required by technical specifications.

(2) For shift crew positions in excess of those required by technical specifications, a description of how the position is *meaningfully and fully* engaged in the functions and duties of the analogous minimum licensed position(s) required by technical specifications. For example, a dual unit facility with a common control room where technical specifications require two SROs per shift, could credit watch-standing proficiency for three SROs per shift, with one SRO responsible for overall plant operation, and the other two SROs each responsible for the command and control of a

single unit. In this case, the third SRO would be entitled to watch-standing proficiency credit, because he or she is performing duties analogous to the second SRO (who is required by technical specifications). Similarly, a dual unit facility with a common control room could credit watch-standing proficiency for four ROs (two per unit) per shift, at a facility where technical specifications require only three ROs, if the fourth RO is performing duties analogous to the third RO (who is required by technical specifications).

If a facility cannot justify, as explained above, crediting watch-standing proficiency for shift crew positions in excess of technical specifications, or does not implement administrative controls as described above, then an individual who stands watch in an excess position shall not receive proficiency credit. In order to maintain an active license under such circumstances, each licensed individual would have to rotate into a licensed shift crew position required by technical specifications for the minimum of seven 8-hour or five 12-hour shifts per calendar quarter, with sufficient administrative controls to document those activities.

Facility licensees that are uncertain if shift crew positions in excess of those required by technical specifications qualify for watch-standing proficiency credit should contact their NRC regional office.

c.   It is permissible for an individual with an SRO license to maintain only the RO portion of his or her license in an active state by performing the functions of an RO for a minimum of seven 8-hour or five 12-hour shifts per calendar quarter pursuant to 10 CFR 55.53(e). Moreover, an inactive SRO may reactivate only the RO portion of his or her license, pursuant to 10 CFR 55.53(f)(2), by completing a minimum of 40 hours of shift functions, including a plant tour, under the direction of an operator and in the position to which the individual will be assigned. However, the fact that an SRO license holder is routinely standing watches only as an RO does *not* maintain his or her proficiency as an SRO. Therefore, before such an SRO can resume duties that require an SRO license, he or she must reactivate that portion of the license, pursuant to 10 CFR 55.53(f)(2), by completing a minimum of 40 hours of shift functions, including a plant tour, under the direction of a senior operator and in the SRO position to which the individual will be assigned.

d.   To maintain the supervisory portion of a SRO license active, a SRO must stand at least **one** complete watch (8- or 12-hour shift) per calendar quarter in a shift crew position credited for SRO-only supervisory licensed duties. The remainder of complete watches (to meet the required minimum of seven 8-hour or five 12-hour shifts per calendar quarter) may be performed in either a credited SRO or RO position. A SRO may stand all of his or her required watches in credited SRO-only supervisory positions, and the RO portion of the license will still be considered active. Similarly, for a SRO to reactivate the supervisory portion of his or her license SRO, pursuant to 10 CFR 55.53(f)(2), a SRO must complete a minimum of 40 hours of shift functions, including a plant tour, under the direction of a SRO in a credited SRO-only supervisory position. A SRO who reactivates his or her license in this manner automatically reactivates the RO portion of the license; an additional 40 hours of under-direction watches in a credited RO position is not required.

e.   Individuals who are licensed on two (or more) similar units at a facility are not required to establish proficiency on each of the similar units unless they hold a separate license on each unit. Performing the required seven 8-hour or five 12-hour shifts of watch-standing per calendar quarter on a single unit maintains the license active for all similar units identified on the license. Similarly, individuals who are licensed on two (or more) similar units at a facility are not required to reactivate their license on each of the similar units identified on the license. Performing the required 40 hours of under-direction watches on a single unit reactivates the license for all similar units at a facility.

f.   In addition to the under-direction watch requirements discussed above, the following also apply to license reactivation pursuant to 10 CFR 55.53(f):

   •   The 40 hours of under-direction watches required by 10 CFR 55.53(f)(2) shall only be credited for standing watches in a RO or SRO position appropriately credited for maintaining license proficiency. It is not appropriate to credit reactivation watch hours while under the direction of an active license holder who is standing watch in an "extra" or non-credited position.

   •   When performing under-direction watches, only **one** under-direction watchstander shall be assigned to an active license holder. Given that the inactive operator is required to complete (not just observe) 40 hours of shift functions, it would not be appropriate to divide under-direction watch functions among multiple individuals.

   •   The 40 hours of under-direction watches for license reactivation do not need to occur in complete shifts or be completed on consecutive days. All 40 hours should occur within a reasonable time frame (e.g., 30 days), and at least one complete on-coming shift turnover and one complete off-going shift turnover must be performed while under the direction of the active license holder. Once all the requirements for license reactivation have been completed, the license is considered active for the remainder of the current calendar quarter, with proficiency watches (i.e., seven 8-hour or five 12-hour shifts) required to maintain the license active during subsequent calendar quarters.

   •   The 40 hours of under-direction watches do not need to occur in the control room; they may performed wherever the duties of the credited licensed position are performed.

   •   The 40 hours of under-direction watches must include at least one complete plant tour. Since it is a part of the 40 hours of under-direction watches, the plant tour must be performed under the direction of an active license holder. Although the regulations do not define the scope of a complete plant tour, the NRC expects that this tour will include all readily accessible major areas of the plant that are routinely toured by in-plant operators that contain safety related equipment. If a facility has developed a checklist of areas to tour, it is generally inappropriate to skip plant areas and mark the items as "non-applicable," unless there is sufficient justification (e.g., personnel or radiation hazard).

g.   The regulations do not include provisions for SROs who are limited to fuel handling (i.e., LSROs) to maintain proficiency between refueling outages. Consequently, unless such LSROs are licensed on multiple units that have refueling outages during successive calendar quarters, they would generally have to reestablish proficiency by standing an "under-direction" watch pursuant to 10 CFR 55.53(f)(2). Ideally, such a watch should be performed primarily in the fuel handling area during refueling operations [i.e., at a time when the presence of a senior operator is required pursuant to 10 CFR 50.54(m)(2)(iv)]. This would clearly meet the requirements of 10 CFR 55.53(f)(2), which mandates that the licensee must complete one shift of shift functions under the direction of a senior operator in the position to which the licensee will be assigned, as well as the definition of *actively perform the functions of a senior operator* (in 10 CFR 55.4), which requires that the licensee must fill a position on the shift crew that requires the individual to be licensed and to carry out and be responsible for the duties covered by that position. This also ensures that the trainee's activities are adequately supervised.

However, given the infrequency and short duration of shift functions that require the presence of an LSRO on the refueling floor, it may not always be practical for a facility licensee to delay its LSRO reactivations until those shift functions are actually underway. In such instances, the facility licensee can satisfy the intent of the regulation by implementing a reactivation program that specifies, in detail, the refueling tasks, activities, and procedures that an LSRO must satisfactorily complete or simulate in order to demonstrate watch-standing proficiency. Moreover, such a program shall exercise positive control to ensure that the LSRO completes the required tasks, activities, and procedures within a reasonable period of time (ideally, no more than 1 week) before he or she is assigned to supervise refueling shift functions.

To properly reactivate an LSRO license in accordance with 10 CFR 55.53(f), the individual should stand a watch under the direction and *in the presence of* an active SRO or LSRO, who will directly oversee the trainee's activities, provide feedback as appropriate, and enable an authorized representative of the facility licensee to certify that the operator's qualifications are current and valid, as required by 10 CFR 55.53(f)(1). Permitting trainees to perform self-directed activities on the refueling floor eliminates the opportunity for meaningful feedback, thereby casting doubt on the validity of the resulting certification. The NRC's requirements regarding the conduct of under-instruction or training watches are reflected in 10 CFR 55.13, which allows trainees to manipulate the controls of a facility "under the direction and in the presence of a licensed operator or senior operator..." This position is also evident in the responses to Questions 252 and 276 in NUREG-1262, "Answers to Questions at Public Meetings Regarding Implementation of Title 10, *Code of Federal Regulations*, Part 55 on Operators' Licenses," which indicate that a trainee's activities are to be closely monitored by the responsible person.

If a facility licensee needs to reactivate a regular SRO license for the purpose of supervising refueling activities, the operator must complete one shift under direction on the refueling floor, as discussed above, and the facility licensee must ensure that the operator is administratively restricted from performing full SRO duties.

If a facility licensee is unable to comply with the LSRO license reactivation requirements in 10 CFR 55.53(f)(2) despite the clarifications discussed above, the licensee may, pursuant to 10 CFR 55.11, "Specific Exemptions," request an exemption from the requirements in 10 CFR 55.53(e) and propose alternative criteria for maintaining active LSRO licenses. The Commission may grant such exemptions from the regulatory requirements as it determines are authorized by law and will not endanger life or property and are otherwise in the public interest. Such requests should provide the following information:

- the reason why the facility licensee is unable to comply with the requirements of 10 CFR 55.53(f)(2), as clarified above, for reactivating its (L)SRO licenses to supervise fuel handling

- the nature of the fuel handling activities that a licensee will have to complete in order to remain "active" and an explanation how those activities would maintain proficiency to supervise actual core alterations (identify those activities that must be performed and those that may be simulated and how the simulation will be accomplished)

- the minimum duration and frequency of the fuel handling activities required to remain "active"

- the nature, duration, and frequency of the training related to fuel handling that is given to its licensed fuel handlers

3.   Medical Standards

In accordance with Subpart C, "Medical Requirements," and Section 55.33(a)(1) of 10 CFR Part 55, the medical condition and general health of licensed operators must be such that it will not adversely affect the performance of assigned operator duties or cause operational errors that might endanger public health and safety. Therefore, licensed operators must be examined by a physician and determined to be fit every 2 years (measured from the date of the last physical examination, rather than the date of licensing), and, pursuant to 10 CFR 55.57(a)(6), their fitness must be certified on NRC Form 396 every time the license is renewed. As noted on NRC Form 396, the physician and facility licensee may use either the 1983 or the 1996 version of ANSI/ANS-3.4, "Medical Certification and Monitoring of Personnel Requiring Operator Licenses for Nuclear Power Plants," when making their fitness determinations. Both versions of the standard include provisions for medical waivers in those cases when the operator can demonstrate complete capacity to perform licensed duties, and conditional licenses in those cases when compensatory measures may be required to ensure public health and safety (refer to Section C.3.c below). However, in both cases, the examining physician and facility licensee must submit a recommendation and supporting evidence, on or with NRC Form 396, to enable the NRC to make a licensing decision.

a. If, during the term of the license, an operator is *temporarily* unable to meet medical standards but is expected to meet those standards again in the future, the facility licensee may administratively classify that operator's license as "inactive" or require compensatory measures, such as taking any medications as prescribed during the temporary period to maintain medical qualifications, or impose other operating restrictions to accommodate the operator's medical condition until the operator is once again certified to meet all medical standards by the facility licensee. Similarly, if the operator's medical condition precludes the operator from completing the requalification training program pursuant to 10 CFR 55.59(a), the facility licensee shall administratively control the operator's activities until he or she completes the requirements of 10 CFR 55.59(b), "Additional Training," including notification of the NRC.

    The facility licensee need not notify the NRC nor request a conditional license concerning an operator's temporary disability, including the temporary use of prescribed medications, provided that the facility licensee administratively prevents the operator from performing licensed duties or otherwise compensates or restricts the operator, as appropriate, throughout the period of his or her temporary disability. If the disability extends beyond the date of license expiration, the operator may apply for timely license renewal in accordance with 10 CFR 55.55(b) and 10 CFR 55.57(a). In that event, the facility licensee should document the nature of the operator's temporary disability on the medical certificate and submit a revised certificate to the NRC after the physician determines that the operator meets the requirements of 10 CFR 55.33(a)(1). The NRC will not renew the operator's license until the staff finds that all of the conditions specified in 10 CFR 55.57(b) are satisfied.

b. If the facility licensee determines that an operator's medical condition is *permanently* disqualifying in accordance with ANSI/ANS-3.4, the facility licensee shall notify the NRC within 30 days of learning of the diagnosis (see 10 CFR 50.74 and 55.25). If an operator develops a permanent medical condition that is not identified in ANSI/ANS-3.4, but the examining physician believes that it could affect the operator's performance or cause operator errors, then it would be prudent to report it to the NRC or at least contact the NRC to inquire whether it should be reported.

    While most of the medical conditions/disabilities identified in ANSI/ANS-3.4, including those that result in failure to meet the minimum requirements for medical qualification, are likely to be permanent, the examining physician is responsible for evaluating each operator's medical condition on a case-by-case basis and assessing whether the operator will be capable of meeting medical standards in the foreseeable future. For example, the facility licensee should report to the NRC a condition for an operator who takes medication to meet the minimum standard for blood pressure (i.e., less than or equal to 160/100 mmHg), unless the physician has reasonably determined that the condition will be controllable without medication in the foreseeable future. In addition, many physicians prescribe blood pressure medication prior to an individual reaching the 160/100 mmHg limit, and facility licensees should consider reporting this to the NRC as well.

When reporting a permanent disqualifying medical condition, if a conditional license is requested, the facility licensee shall provide medical certification and evidence on NRC Form 396 and recommend the exact wording of any license restriction that might be necessary. A permanent disqualifying condition is always reportable, even if it is being controlled and regardless whether the compensatory measures are recognized in the applicable version of ANSI/ANS-3.4.

c.   In accordance with 10 CFR 55.33(b), if an operator's medical condition does not meet the minimum standards under 10 CFR 55.33(a)(1), the NRC may condition the license to accommodate the medical defect. The NRC will consider the recommendations and supporting evidence provided on or with NRC Form 396 in determining the appropriate license condition. The following medical restrictions/conditions are illustrative but not all-inclusive:

- An operator may be required to wear **corrective lenses** while performing licensed duties if his or her vision does not meet medical standards.

- An operator may be required to wear a **hearing aid** while performing licensed duties if his or her hearing does not meet medical standards.

- An RO who is at risk of sudden incapacitation may have a ***no-solo*** restriction that requires another licensed operator to be in view when the restricted operator is performing control manipulations, and someone capable of summoning assistance to be present at all other times while the restricted operator is performing licensed duties. The analogous SRO restriction would require another licensed operator to be in view when the restricted operator is performing control manipulations, and another senior operator to be present on site at all other times while the restricted operator is performing SRO licensed duties or someone capable of summoning assistance to be present at all other times while the restricted operator is performing RO licensed duties. For LSROs, the no-solo restriction would require someone capable of summoning assistance to be in view when the restricted LSRO is performing licensed LSRO duties.

- An operator may be required to take ***medication as prescribed***, if an operator's medical qualification is contingent on taking a prescription medication.

- An operator whose medical condition is acceptable but unstable may be required to submit followup ***medical status reports*** (i.e., prognosis, treatment, and ability to perform licensed duties) at 3-, 6-, or 12-month intervals.

- An operator with ***respiratory problems*** may be restricted from performing licensed activities that require the use of a respirator.

d.   With regard to prescription medications, it is important that the examining physician understand what medical conditions are contained in the applicable version of ANSI/ANS-3.4. For example, the fact that a licensed operator is diagnosed with gastroesophageal (acid) reflux disease and placed on the appropriate prescription medication would, in all likelihood, not be reportable to the NRC, since this condition is not addressed in ANSI/ANS-3.4. However, when assessing *any* prescription medication, the examining physician needs to consider: (1) the possible side effects of the medication, to ensure that they will not cause operational errors or affect the operator's capacity to safely perform licensed duties; and (2) any delay in taking a medication that might be expected to result in the incapacity of the operator.

In addition, the actual wording of the license condition regarding medication will **not** specify a particular medical condition or medication, but it will simply state that the operator must "take medication as prescribed." Therefore, physician-prescribed changes in medication or dosing for an existing medical condition are not required to be reported to the NRC, unless the examining physician believes the operator's medical condition has become unstable (therefore requiring followup medical status reports to the NRC) or that operator requires a no-solo license restriction. However, any new permanently disqualifying medical condition(s), requiring new medication(s), must be reported to the NRC.

Facility licensees do not need to submit a revised NRC Form 396 for operators with existing medical conditions, simply because NRC Form 396 has recently been changed to include medications. If an operator is currently prescribed medication for an existing medical condition, the revised NRC Form 396 should be submitted the next time that operator's license is due for renewal, and marked to indicate that the license will be conditioned to require taking medication as prescribed. A new NRC Form 396 would also be required if the operator develops a new permanent physical or mental condition reportable under 10 CFR 55.25.

4.   Downgrading an SRO License

If a facility licensee desires to permanently downgrade the license of a senior operator at the facility, it may do so by submitting a written request to the NRC regional office. In such instances, the NRC regional office will (1) amend the license to restrict the operator's activities to those authorized for a reactor operator under 10 CFR Part 55; (2) condition the license to prohibit the operator from directing the licensed activities of licensed operators; and (3) inform the operator and facility in writing that the license will not be subject to renewal under 10 CFR 55.57 and that a new application (NRC Form 398) will be required pursuant to 10 CFR 55.31 if the operator desires to maintain an RO license upon expiration of the amended SRO license. The expiration date of the original license will not change, and the operator may transition to the RO requalification program upon receipt of the amended license.

D.   License Renewal

1.   An operator who wishes to renew a license must comply with the requirements of 10 CFR 55.57(a), as follows:

a.	The operator will complete NRC Form 398, including the operator's experience under the current license, the approximate number of hours that the operator spent on operating shifts, and the date and results of the applicant's most recent requalification written examination and annual operating test. The senior management representative on site shall provide evidence that the operator has safely and competently discharged his or her license responsibilities and satisfactorily completed the facility's approved requalification program by checking Item 19.c and signing in the designated space on Form 398.

b.	The facility licensee must certify on NRC Form 396 that a physician has performed a medical examination within the previous 2 years, as required by 10 CFR 55.21 and submit that form along with NRC Form 398.

c.	The operator must submit NRC Forms 396 and 398 not less than 30 days before the expiration date of the license. In accordance with 10 CFR 55.55(b), if the operator files a proper application for renewal at least 30 days before the date of expiration, the license shall not expire until the NRC has denied the application for renewal or issued a new license. If the application is received more than 60 days in advance, the regional office should contact the facility licensee to determine whether it would prefer to have the license renewed immediately with a new effective date (the license will not be predated, nor will it exceed a 6-year license term) or to resubmit the application within the 60- to 30-day window preceding the expiration date.

If an operator is waiting to be given a reexamination after failing a requalification examination, the operator *should still make timely application* for license renewal under the provisions of 10 CFR 55.55(b).

The NRC's regional office may allow for transit time and accept a license renewal application that is received 25 days before the license expiration date, provided that all signatures on NRC Forms 398 and 396 are dated before the 30-day timely renewal cutoff date. The submittal will not be considered timely if it is received less than 25 days before the date of license expiration unless positive evidence of receipt (e.g., postmark or docketing stamp) from the U.S. Postal Service or the NRC is available. If the application is received less than 25 days before the date of license expiration and too late for processing in the regional office, the license shall expire on the expiration date. The regional office may then issue a new license when it has finished processing the application.

d.	If the license for a RO expires while he or she is participating in the facility licensee's SRO upgrade training program, NRC Forms 396 and 398 should still be submitted for timely renewal of the RO license. However, if the RO is not current in the facility's requalification training and testing program, because he or she is attending SRO upgrade training, NRC Form 398 must note the exception in block 17, "Comments," and the operator must be administratively restricted from performing licensed duties until the individual is up-to-date in the requalification program.

e. Pursuant to 10 CFR 55.5, "Communications," facility licensees may submit these forms to the NRC by mail, in person, or, where practicable, via electronic information exchange (EIE) or on CD-ROM. Electronic submissions must be made in a manner that enables the NRC to receive, read, authenticate, distribute, and archive the submission, and process and retrieve it one page at a time. Detailed guidance on making electronic submissions can be obtained by visiting the NRC's public Web site at http://www.nrc.gov/site-help/eie.html, calling (301) 415-6030, sending an email message to EIE@nrc.gov, or writing to the Office of the Chief Information Officer, U.S. Nuclear Regulatory Commission, Washington, DC 20555-0001. Forms that have only a single signature, such as NRC Form 396, may be submitted electronically using an electronic digital signature. However, forms with multiple signatures, such as NRC Form 398, must rely on handwritten optically scanned signatures, because of the limited digital signature capability of the EIE system. For any textual documents submitted in an optically scanned format, please note that Searchable Image (Exact) PDF is required, to preclude optical character recognition errors. When sending forms via EIE, facility licensees are encouraged to follow up with a phone call or e-mail message to the operator licensing assistant in the regional office to ensure the forms are received.

f. After reviewing the renewal application, the NRC's regional office may ask the operator or facility to provide supplemental information. The operator or facility must forward the requested supplemental information to the regional office within 20 days.

If an applicant for renewal declines to provide the supplemental information requested by the NRC's regional office, or if the regional office concludes, after reviewing any additional information supplied by the operator, that the application is still inadequate for license renewal, the regional office will notify the operator in writing that the renewal application is denied. The operator may then exercise one of the following options within 20 days after the date of the proposed denial letter from the regional office:

(1) Do nothing. The denial will become final 20 days after the date of issuance, and the regional office will inform the facility licensee and the operator in writing that the license has been terminated.

(2) Request reconsideration of the application denial. Applicants must submit such requests in writing to the Director, Division of Inspection and Regional Support, U.S. Nuclear Regulatory Commission, Washington, DC 20555-0001. If submitting via private courier (e.g., FedEx, UPS), send your request to 11555 Rockville Pike, Rockville, Maryland 20852, instead of using the Washington, DC, address. Requests for informal reviews by the NRC shall list the items for which the applicant is requesting additional review and include documentation supporting the contentions made by the operator. The package containing the review request and supporting documentation must be mailed or delivered within 20 days of the date of denial.

(3) Request a hearing pursuant to 10 CFR 2.103(b)(2). Applicants must submit such requests to the Office of the Secretary, U.S. Nuclear

Regulatory Commission, Washington, DC 20555-0001, Attention: Rulemakings and Adjudications Staff, with a copy to the Associate General Counsel for Hearings, Enforcement, and Administration, Office of the General Counsel, at the same address. (Refer to 10 CFR 2.302 for additional filing options and instructions.) If submitting via private courier (e.g., FedEx, UPS), send your request to 11555 Rockville Pike, Rockville, Maryland 20852, instead of using the Washington, DC, address.

2.   Upon receipt of a renewal application, the NRC regional office may take the following actions, as appropriate:

   a.   Review the application and issue the license renewal if the staff finds that the applicant satisfies the conditions in 10 CFR 55.57(b). There is no minimum number of hours that the operator has to operate the facility in order to qualify for license renewal (i.e., inactive licenses are also renewable). However, the regional office should take the applicant's operating history into consideration as an additional piece of information if any of the requirements of 10 CFR 55.57(b) are not met.

   A RO license renewal application received from a RO who has not completed the facility licensee's requalification program, as required by 10 CFR 55.57(b)(2)(ii) and (iii), because he or she is participating in the facility's SRO upgrade training program, will be renewed if all other requirements are met. No waiver is required. Although attending SRO upgrade training will largely fulfill the requalification training requirements, the facility licensee would, nevertheless, be expected to administratively restrict the operator's duties until his or her qualifications and status are certified to be current. The NRC regional office will follow up on the facility licensee's administrative controls during periodic requalification program inspections.

   b.   If the renewal applicant does not meet the requirements of 10 CFR 55.57, the regional office shall inform the facility licensee of the deficiencies and request any supplemental information that the staff might require to make a relicensing decision. If, after evaluating the supplemental information, the regional office still concludes that the applicant does not meet the requirements for license renewal, the staff will issue a proposed denial letter to the operator (with a copy to the facility licensee).

   c.   If the operator requests informal reconsideration of the application denial or a hearing, the regional office will review the operator's request as directed by the NRR operator licensing program office. The NRR operator licensing program office will inform the operator, in writing, of the outcome of the review.

E.   NRC-Conducted Requalification Examination Results

1.   Passing an NRC-Conducted Requalification Examination

An operator who passes all portions of the requalification examination, including being a member of a crew that passes the dynamic simulator examination, will receive written notification from the NRC's regional office.

2. **Failing an NRC-Conducted Requalification Examination**

a. The NRC's regional office will notify the operator in writing of a failure of the requalification examination. On receiving the failure notification, the operator can request an informal review of the failed portion of the examination. The request must be made as described (for reconsideration of application denials) in Section D.1.d(2), above.

b. If an operator fails any part of an NRC-conducted requalification examination, the facility licensee is expected to remove the operator from licensed duty and take corrective action consistent with the provisions of its requalification program before returning the operator to licensed duty. If the facility licensee's requalification program is unsatisfactory, refer to Section F.2 of ES-601 for a list of other recommended actions to be taken, including those actions the facility licensee is expected to complete before attaining a "provisionally satisfactory" requalification program status.

c. Although the regulation [10 CFR 55.57(b)(2)(iv)] that required operators to pass an NRC-administered requalification examination as a prerequisite for license renewal was deleted effective March 11, 1994, the license of any operator who fails to pass any NRC-conducted requalification examination will not be renewed without some level of NRC involvement in the retesting process. The amount of NRC involvement may include conducting the retest in accordance with the applicable examination standard(s); inspecting the facility licensee in accordance with Inspection Procedure (IP) 71111.11, "Licensed Operator Requalification Program," as it retests the operator; or reviewing an examination prepared by the facility licensee. The NRC's regional office, in consultation with the NRR operator licensing program office, will determine the appropriate level of involvement on a case-by-case basis depending on the quality of the facility licensee's program. As long as the operator submits a timely renewal application, the term of the license will continue until the renewal requirements are satisfied or the operator fails three NRC-conducted examinations as discussed in Section E.2.e.

d. The NRC will normally administer a second (first retake) examination approximately 6 months after issuing the first failure notification in accordance with Section E.2.a of this standard. That examination will concentrate on the areas in which the operator exhibited deficiencies.

e. The NRC will normally administer a third (second retake) examination approximately 6 months after issuing the second failure notification in accordance with Section E.2.a of this standard. The third examination will be a *comprehensive* requalification examination.

Regardless of the status of the facility licensee's requalification program, if an operator fails a third requalification (second retake) examination, the NRC will thoroughly review the operator's examination performance and may conduct a complete review of the facility licensee's training program. The third failure may be grounds for suspending or revoking the operator's license. If an operator has an application pending for license renewal with the NRC at the time of a third requalification failure, that failure will provide the basis for denying the application. Notification of the operator will be handled on a case-by-case basis and coordinated through the NRR operator licensing program office.

ES-701
ADMINISTRATION OF INITIAL EXAMINATIONS
FOR SENIOR OPERATORS LIMITED TO FUEL HANDLING

A.   Purpose

This standard provides specific instructions for use in preparing, administering, grading, and documenting initial examinations for senior operators who are limited to fuel handling (LSROs).

B.   Background

Pursuant to Title 10, Sections 55.41 and 55.43, of the *Code of Federal Regulations* (10 CFR 55.41 and 55.43), the NRC's written LSRO examinations must contain a representative selection of questions concerning the specific knowledge, skills, and abilities needed to perform licensed fuel handling duties. Similarly, to the extent applicable, the operating tests must require the applicant to demonstrate an understanding of and the ability to perform the actions necessary to accomplish a representative sample of the items in 10 CFR 55.45. The regulations also stipulate that the content of the examinations and tests will be identified, in part, from learning objectives derived from a systematic analysis of the operators' duties performed by the facility licensee. Therefore, the facility licensee's job task analysis (JTA) for fuel handlers would provide an excellent source of information for developing the written examination and operating test.

Except as noted herein, the guidance in Examination Standards (ESs) 201, 202, 204, 301, 302, 303, 401, 402, 403, 501, and 502 for administering unrestricted initial licensing examinations at power reactors also applies to the LSRO examination. However, the "Procedure for Administering the Generic Fundamentals Examination [GFE] Program" (described in ES-205) does not apply to LSRO applicants.

C.   Responsibilities

1.   Facility Licensee

The facility licensee is responsible for the same activities specified in the unrestricted ESs, with the following exceptions and modifications:

a.   As an exception to ES-202, "Preparing and Reviewing Operator License Applications," the facility licensee may request LSRO licenses that are valid for more than one site. To do so, the facility licensee shall provide documentation that describes the differences in the design, procedures, technical data, and administrative controls of the separate facilities for which the license is being sought.

b.   The scope, content, administration, and grading of the written examination and operating test shall be as described in Sections D and E, below.

c.   In accordance with 10 CFR 55.46(b), the facility licensee shall request
       the Commission's approval to use the plant or a simulation facility, other than
       a plant-referenced simulator, in administering the operating test under
       10 CFR 55.45(b)(1) or (3).

2.   NRC Regional Office

The NRC's regional office is responsible for the same activities specified
in the unrestricted ESs, with the following exceptions and modifications:

a.   The regional office should generally conduct the LSRO examinations during
       a time when the fuel handling equipment will be available for the operating tests.

b.   With the concurrence of the NRR operator licensing program office,
       the regional office may issue LSRO licenses that are valid for units
       at more than one site, provided that the units are manufactured by the same vendor
       and are of similar design. The applicant must pass an examination that addresses
       the differences in the design, procedures, technical data, and administrative
       controls of the separate facilities for which the license is being sought.

c.   The scope, content, administration, and grading of the written examination
       and operating test shall be as described in Sections D and E, below.

d.   The regional office shall coordinate with the NRR operator licensing program office
       regarding approval to use the plant or a simulation facility, other than a plant-
       referenced simulator, in administering the operating test under 10 CFR 55.45(b)(1) or (3).

D.   Written Examination Instructions

1.   Preparation

The NRC's written LSRO examination should meet all of the guidelines and requirements
for question construction, quality, and facility reviews specified in ES-401, "Preparing
Initial Site-Specific Written Examinations," and Appendix B, "Written Examination
Guidelines," except as noted below:

a.   Develop the examination outline as described in Section D.1 of ES-401,
       with the following exceptions and clarifications:

       •   Instead of using the RO and SRO models in ES-401, use Form ES-701-1
               or Form ES-701-2, as applicable to the facility, and Form ES-701-3
               to develop the examination outline. As with the unrestricted examinations,
               topics that are not applicable to LSROs at the subject facility should be
               eliminated in accordance with Section D.1 of ES-401. Given the large
               number of knowledge and ability (K/A) statements that will not apply
               to LSROs, it may be advantageous to pre-screen the K/As as discussed
               in Item 4 of that Attachment. When reviewing K/As for elimination,
               do not focus only on the fuel handling equipment; rather, focus more broadly

on the knowledge and abilities that an LSRO would need to support safe operation during fuel handling. If the facility licensee's JTA identified other LSRO-relevant components, systems, and evolutions that are not included on Form ES-701-1 or ES-701-2, those items must be added to the appropriate tier of the outline before beginning the random selection process. Additional instructions are noted on the forms.

- Section D.1.c of ES-401 is not applicable to the LSRO examination.

- Use Form ES-701-5, "LSRO Examination Outline Quality Checklist," instead of Form ES-201-2 when reviewing the examination outline.

b. Select and develop questions as described in Section D.2 of ES-401, with the following exceptions:

- Construct the LSRO written examination so that a competent applicant can complete the examination in 2.5 hours. (The applicants will be allowed 4 hours to complete and review the examination.)

- Between 50 and 60 percent (20 to 24) of the LSRO examination questions shall be written at the comprehension/analysis level.

- Reactor theory, component, and thermodynamic questions that directly relate to the LSRO JTA may be selected from prior GFE examinations.

- Section D.2.d of ES-401 is not applicable to the LSRO examination.

- Limit the use of bank questions to no more than 30 and include at least 4 new questions on every examination; the remaining 6 examination questions may be new or significantly modified from the facility licensee's or *any* other bank. All questions developed must be relevant to the LSRO function. To be considered a significantly modified question, at least one pertinent condition in the stem and at least one distractor must be changed from the original bank question. Changing the conditions in the stem such that one of the three distractors in the original question becomes the correct answer would also be considered a significant modification.

- If the examination will be used to license the applicants at more than one facility, ensure that it adequately covers all of the applicable units. An examination developed for the purpose of cross-qualifying a licensed LSRO at another similar facility may focus exclusively on the differences between the facilities.

c. Review and assemble the examination as described in Sections D.3, D.4, and E of ES-401, using Forms ES-701-6 and ES-701-8 instead of the equivalent forms in ES-401.

2.   Administration and Grading

The NRC's written LSRO examination shall be administered and graded in accordance with ES-402, "Administering Initial Written Examinations," and ES-403, "Grading Initial Site-Specific Written Examinations." The examination may be administered concurrently and in the same room with full-scope, initial license examinations. However, in such instances, the proctor should minimize any disturbance to those applicants taking the longer examination.

E.   Operating Test Instructions

The LSRO operating test shall generally be prepared, administered, and documented in accordance with ES-301, "Preparing Initial Operating Tests"; ES-302, "Administering Operating Tests to Initial License Applicants"; and ES-303, "Documenting and Grading Initial Operating Tests," except as noted below and in the specific criteria at the bottom of Form ES-701-4, "LSRO Operating Test Outline."

The operating test shall be performance-based to the maximum extent possible; however, given the nature of the LSROs' duties, it is neither practical nor appropriate to administer the test on the plant-referenced simulator. Therefore, pursuant to 10 CFR 55.45(b), the test shall be administered in a plant walk-through and in either the plant or a simulation facility, as approved by the Commission under 10 CFR 55.46(b). The facility licensee is encouraged to permit the actual use of equipment to handle dummy fuel elements, assemblies, or modules during the operating test whenever feasible. This may require careful coordination with the facility licensee to establish a schedule and to make sure that a licensed SRO is available, if needed. When actual equipment is not available or accessible (e.g., because of high radiation), administer the test using walk-through methods near the actual equipment or by using mockup equipment. If the facility licensee has a refueling machine simulator, use it to the extent possible during the administration of the operating test.

The operating test shall assess the applicant's ability to execute normal, abnormal, and emergency procedures associated with fuel handling. Each applicant will be required to simulate or perform tasks related to fuel handling and, if necessary based on their performance, to answer questions associated with the refueling equipment and associated systems. The applicant shall not be held accountable for duties that are performed exclusively by the control room staff or shift supervisor.

1.   Preparation

The operating test shall consist entirely of job performance measures (JPMs) covering those administrative topics, systems, and emergency/abnormal plant evolutions (E/APEs) related to refueling. No distinction between control room and facility systems/evolutions is required, because most (if not all) of the test will be conducted outside the control room. The dynamic simulator operating test requirements and guidelines in Section D.5 of ES-301 do not apply to the LSRO license examination.

Part of the operating test may be conducted in the control room so that those controls, instruments, and other materials or equipment related to fuel handling (e.g., procedures and diagrams) are available for reference. Although LSROs will not operate any systems from the control room, they must be aware of the effects (e.g., alarms) that fuel handling operations will have in the control room. They must also be familiar with the methods and requirements for communicating with the control room staff and shift supervisor. At least two of the JPMs must require the applicant to use the facility's technical specifications.

The following additional guidelines clarify the expectations for each part of the LSRO operating test.

a.   Develop the Administrative portion of the operating test in accordance with Section D.3 of ES-301; however, given the reduced scope of the LSROs' responsibilities, the required number of tasks is reduced from five to three, distributed among the four administrative topics. Note that some "Conduct of Operations" subjects (e.g., reactor plant startup requirements) may not apply; however, most can be adapted for use during the LSRO operating test. The "Equipment Control" subjects all lend themselves to evaluating the required refueling maintenance and surveillance actions that the LSRO should be able to supervise or perform. All of the "Radiation Control" subjects apply to refueling operations and should be evaluated on a sampling basis. The "Emergency Plan" topic shall be evaluated to the extent that the applicant is required to respond to a declared event and the knowledge required of a radiation worker.

b.   Develop the Systems portion of the operating test as follows:

•   Develop two JPMs that require the applicant to manipulate the facility's fuel handling equipment.

•   Develop two JPMs related to systems other than fuel handling equipment (i.e., 234000 or 034) listed in Tier 2 of the appropriate written examination outline (i.e., Form ES-701-1 or ES-701-2, as modified in Section D.1.a, above).

•   The specific criteria in Sections D.4.a and b of ES-301 do not apply. Two of the tasks shall require the applicant to execute alternative paths within the facility's operating procedures.

c.   Develop the E/APE portion of the operating test as follows:

•   Develop three JPMs based on the evolutions listed in Tier 1 of the appropriate written examination outline (i.e., Form ES-701-1 or ES-701-2, as modified in Section D.1.a, above); one of the JPMs must involve a refueling accident.

•   One of the tasks shall require the applicant to execute alternative paths within the facility's operating procedures.

d.   The operating test should normally take between 4 and 6 hours, depending on whether the LSRO actually operates refueling equipment.

e. Use Form ES-701-4, "LSRO Operating Test Outline," to document the selection of Administrative, System, and E/APE JPMs to be performed (instead of using Forms ES-301-1 and ES-301-2); insert the applicable type codes and adhere to the specific criteria noted at the bottom of the form. Review the outline using Form ES-701-5, "LSRO Examination Outline Quality Checklist" (instead of Form ES-201-2).

f. Review the final operating test in accordance with Section E of ES-301, as applicable, using Form ES-701-7, "LSRO Operating Test Quality Checklist" (instead of Form ES-301-3).

2. **Administration**

Administer the operating test in accordance with Sections D.1 and D.2 of ES-302, as applicable; Section D.3 (in its entirety) does not apply to the LSRO operating test.

3. **Grading**

Grade and document the applicant's performance on the operating test in accordance with Sections D.1, D.2.a, D.3, and D.4 of ES-303, as applicable, with the following specific exceptions and clarifications:

a. Substitute Form ES-701-4 for Pages 2 and 3.b of Form ES-303-1 and determine a grade for each Administrative, System, and E/APE JPM as described in Section D.2.a of ES-303. "N/A" the "Simulator Operating Test" in the Summary section on page 1 of Form ES-303-1.

b. The applicant must achieve a satisfactory grade on at least 80 percent of the JPMs (8/10) overall and at least 60 percent (2/3) of the administrative JPMs (i.e., the same criteria as in ES-303).

F. **Attachments/Forms**

Form ES-701-1, "LSRO BWR Written Examination Outline"
Form ES-701-2, "LSRO PWR Written Examination Outline"
Form ES-701-3, "LSRO Generic Knowledge and Abilities Outline (Tier 3)"
Form ES-701-4, "LSRO Operating Test Outline"
Form ES-701-5, "LSRO Examination Outline Quality Checklist"
Form ES-701-6, "LSRO Written Examination Quality Checklist"
Form ES-701-7, "LSRO Operating Test Quality Checklist"
Form ES-701-8, "LSRO Written Examination Cover Sheet"

Facility:												Date of Exam:	

Tier	K/A Category Points												
	K1	K2	K3	K4	K5	K6	A1	A2	A3	A4	G*	Total	
1. Emergency & Abnormal Plant Evolutions												10	
2. Plant Systems												20	

3. Generic Knowledge and Abilities Categories	1	2	3	4	GFE	
						10

Note: 1.   Ensure that at least one topic from every K/A category is sampled within each tier.

2.   The point total for each tier in the proposed outline must match that specified in the table. The final point total for each tier may deviate by ±1 from that specified in the table based on NRC revisions. The final exam must total 40 points.

3.   Select topics from many systems and evolutions; avoid selecting more than two K/A topics from a given system (except fuel handling equipment) or evolution (except refueling accident).

4.   The shaded areas are not applicable to the category/tier.

5.*   The generic (G) K/As in Tiers 1 and 2 shall be selected from Section 2 of the K/A Catalog, but the topics must be relevant to the applicable evolution or system.

6.   If the applicants have not previously taken the GFE, Tier 3 shall include basic reactor theory, component, and thermodynamic topics that apply to fuel handling operations.

7.   Systems/evolutions within each tier are identified on the associated outline. Enter the K/A numbers, a brief description of each topic, the topics' importance ratings (IR) for the SRO license level, and the point totals (#) for each system and category. Enter the tier totals for each category in the table above.

8.   For Tier 3, select topics from Section 2 of the K/A catalog, and enter the K/A numbers, descriptions, importance ratings, and point totals (#) on Form ES-701-3.

9.   Refer to ES-401, Section D.1, for guidance regarding the elimination of inappropriate K/A statements. The facility licensee's JTA for fuel handlers should be used as the basis for eliminating or adding testable topics.

	K 1	K 2	K 3	A 1	A 2	G	K/A Topic(s)	IR	#
295003 Partial or Complete Loss of AC									
295004 Partial or Total Loss of DC									
295014 Inadvertent Reactivity Addition									
295018 Partial or Total Loss of CCW									
295021 Loss of Shutdown Cooling									
295023 Refueling Accidents									
295033 High Secondary Containment Area Radiation Levels									
295034 Secondary Containment Ventilation High Radiation									
295006 SCRAM									
295008 High Reactor Water Level									
295009 / 295031 Reactor Low Water Level									
295017 / 295038 High Offsite Release Rate									
295019 Partial or Total Loss of Inst. Air									
295020 Inadvertent Cont. Isolation									
295030 Low Suppression Pool Wtr Lvl									
295035 Secondary Containment High Differential Pressure									
600000 Plant Fire On Site									
K/A Category Totals:							Tier Point Total:		10

	K 1	K 2	K 3	K 4	K 5	K 6	A 1	A 2	A 3	A 4	G	K/A Topic(s)	IR	#
205000 Shutdown Cooling														
215004 Source Range Monitor														
233000 Fuel Pool Cooling/Cleanup														
234000 Fuel Handling Equipment														
262001 AC Electrical Dist.														
263000 DC Electrical Dist.														
290002 Reactor Vessel Internals														
201002 RMCS														
201003 Control Rod and Drive Mechanism														
203000 RHR/LPCI: Injection Mode														
204000 RWCU														
211000 SLC														
212000 RPS														
214000 RPIS														
215001 Traversing In-Core Probe														
215003 IRM														
215005 APRM / LPRM														
223001 Primary CTMT and Aux.														
223002 PCIS/Nuclear Steam Supply Shutoff														
261000 SGTS														
264000 EDGs														
272000 Radiation Monitoring														
286000 Fire Protection														
288000 Plant Ventilation														
290001 Secondary CTMT														
300000 Instrument Air														
400000 Component Cooling Water														
K/A Category Totals:												Tier Point Total:		20

Facility:									Date of Exam:				
Tier	K/A Category Points												Total
	K1	K2	K3	K4	K5	K6	A1	A2	A3	A4	G*		
1. Emergency & Abnormal Plant Evolutions													10
2. Plant Systems													20
3. Generic Knowledge and Abilities Categories	1		2		3		4		GFE				10

Note:
1. Ensure that at least one topic from every K/A category is sampled within each tier .
2. The point total for each tier in the proposed outline must match that specified in the table. The final point total for each tier may deviate by ±1 from that specified in the table based on NRC revisions. The final exam must total 40 points.
3. Select topics from many systems and evolutions; avoid selecting more than two K/A topics from a given system (except fuel handling equipment) or evolution (except refueling accident).
4. The shaded areas are not applicable to the category/tier.
5.* The generic (G) K/As in Tiers 1 and 2 shall be selected from Section 2 of the K/A Catalog, but the topics must be relevant to the applicable evolution or system.
6. If the applicants have not previously taken the GFE, Tier 3 shall include basic reactor theory, component, and thermodynamic topics that apply to fuel handling operations.
7. Systems/evolutions within each tier are identified on the associated outline. Enter the K/A numbers, a brief description of each topic, the topics' importance ratings (IR) for the SRO license level, and the point totals (#) for each system and category. Enter the tier totals for each category in the table above.
8. For Tier 3, select topics from Section 2 of the K/A catalog, and enter the K/A numbers, descriptions, importance ratings, and point totals (#) on Form ES-701-3.
9. Refer to ES-401, Section D.1, for guidance regarding the elimination of inappropriate K/A statements. The facility licensee's JTA for fuel handlers should be used as the basis for eliminating or adding testable topics.

	K 1	K 2	K 3	A 1	A 2	G	K/A Topic(s)	IR	#
000025 Loss of RHR System									
000026 Loss of Component Cooling Water									
000032 Loss of Source Range NI									
000036 (BW/A08) Fuel Handling Accident									
000061 ARM System Alarms									
000033 Loss of Intermediate Range NI									
000055 Station Blackout									
000056 Loss of Offsite Power									
000057 Loss of Vital AC Inst. Bus									
000058 Loss of DC Power									
000062 Loss of Nuclear Svc Water									
000065 Loss of Instrument Air									
000067 Plant Fire On Site									
000069 (W/E14) Loss of CTMT Integrity									
W/E16 High Containment Radiation									
K/A Category Totals:							Tier Point Total:		10

	K 1	K 2	K 3	K 4	K 5	K 6	A 1	A 2	A 3	A 4	G	K/A Topic(s)	IR	#
005 Residual Heat Removal														
015 Nuclear Instrumentation														
033 Spent Fuel Pool Cooling														
034 Fuel Handling Equipment														
103 Containment														
062 AC Electrical Distribution														
063 DC Electrical Distribution														
002 Reactor Coolant														
004 Chemical and Volume Control														
008 Component Cooling Water														
013 Engineered Safety Features Actuation														
064 Emergency Diesel Generator														
072 Area Radiation Monitoring														
076 Service Water														
078 Instrument Air														
079 Station Air														
086 Fire Protection														
K/A Category Totals:												Tier Point Total:		20

ES-701		LSRO Generic Knowledge and Abilities Outline (Tier 3)		Form ES-701-3	
Facility:			Date of Exam:		
Category	K/A #	Topic		IR	#
1. Conduct of Operations	2.1.				
	2.1.				
	2.1.				
	2.1.				
	Subtotal				
2. Equipment Control	2.2.				
	2.2.				
	2.2.				
	2.2.				
	Subtotal				
3. Radiation Control	2.3.				
	2.3.				
	2.3.				
	2.3.				
	Subtotal				
4. Emergency Procedures / Plan	2.4.				
	2.4.				
	2.4.				
	2.4.				
	Subtotal				
5. Generic Fundamentals					
	Subtotal				
Tier 3 Point Total					10

Applicant Docket Number: 55- Facility:		Date of Examination:	Page 2 of
Title / Description of Tasks (JPMs)	**Type Codes***	**Evaluation (S or U)**	**Comment Page Number**
Administrative			
1.			
2.			
3.			
Systems			
1.			
2.			
3.			
4.			
Emergency/Abnormal Plant Evolutions			
1.			
2.			
3.			

Type Codes & Criteria: (A)lternative path (2 systems; 1 E/APE))
                                  (C)ontrol room
                                  (D)irect from bank (≤ 7)
                                  (I)n-plant
                                  (N)ew or (M)odified from bank including 1(A) (≥ 1 / section)
                                  (L)ast NRC exam (≤ 1 / section)
                                  (R)efueling accident (1)
                                  (T)echnical specification (≥ 2)

	Facility:	Date of Examination:			
				Initials	
Item		Task Description	a	b*	c#
1. W R I T T E N	a.	Verify that the outline fits the model in accordance with ES-701.			
	b.	Assess whether the outline was systematically and randomly prepared in accordance with Section D.1 of ES-401 and whether all K/A categories are sampled at least once.			
	c.	Assess whether the outline over-emphasizes any systems, evolutions, or generic topics.			
	d.	Assess whether the justifications for deselected or rejected K/A statements are appropriate.			
2. O P E R A T I N G	a.	Verify that the overall operating test: (1) includes at least two tasks that require the use of technical specifications (2) does not duplicate any tasks from the applicants' audit test(s)			
	b.	Verify that the administrative tasks: (1) are distributed among the four administrative topics described in ES-301 (2) include no more than one repeat from the last NRC licensing examination (3) include at least one task that is new or significantly modified			
	c.	Verify that the systems walk-through includes: (1) two tasks requiring the manipulation of fuel handling equipment (2) two additional tasks related to Tier 2 systems other than fuel handling equipment (3) two tasks requiring implementation of alternative path procedures (4) no more than one repeat from the last NRC licensing examination (5) at least one task that is new or significantly modified			
	d.	Verify that the E/APE walk-through includes: (1) three JPMs based on the Tier 1evolutions, including a refueling accident (2) one task requiring implementation of an alternative path procedure (3) no more than one repeat from the last NRC licensing examination (4) at least one task that is new or significantly modified			
	e.	Determine whether there are enough different outlines to test the projected number of applicants and ensure that no items are duplicated on subsequent days.			
3. G E N E R A L	a.	Assess whether plant-specific priorities (including PRA and IPE insights) are covered inthe appropriate exam section.			
	b.	Assess whether the 10 CFR 55.41/43 and 55.45 sampling is appropriate.			
	c.	Assess whether the sampling process adequately considered plant-specific refueling components, systems, and procedures that are not included in the generic models.			
	d.	Ensure that K/A importance ratings (except for plant-specific priorities) are at least 2.5.			
	e.	Check for duplication and overlap among exam sections.			
	f.	Check the entire exam for balance of coverage.			
	g.	Assess whether the proposed sample is consistent with the LSRO's job responsibilities.			

		Printed Name / Signature	Date
a.	Author	_____	_____
b.	Facility Reviewer (*)	_____	_____
c.	NRC Chief Examiner (#)	_____	_____
d.	NRC Supervisor	_____	_____

Note: * The facility reviewer's initials/signature are not applicable for NRC-developed examinations.
Independent NRC reviewer initial items in Column "c"; chief examiner concurrence required.

		Initial		
	Item Description	a	b*	c#
1.	Questions and answers are technically accurate and applicable to the facility.			
2.	a. NRC K/As are referenced for all questions (as applicable). b. Facility learning objectives are referenced as available.			
3.	Questions are appropriate for LSRO applicants.			
4.	The sampling process was random and systematic (If more than 3 questions were repeated from the last 2 NRC licensing exams, consult the NRR OL program office).			
5.	Question duplication from the license screening/audit exam was controlled as indicated below (check the item that applies) and appears appropriate: __ the audit exam was systematically and randomly developed, or __ the audit exam was completed before the license exam was started, or __ the examinations were developed independently, or __ the licensee certifies that there is no duplication, or __ other (explain)			
6.	Bank use meets limits (no more than 30 questions from the bank, at least 4 new, and the rest modified); enter the actual question distribution at right. Bank | Modified | New			
7.	Between 50 and 60 percent (20 and 24) of the questions on the exam are written at the comprehension/analysis level; enter the actual question distribution at right. Memory | C/A			
8.	References/handouts provided do not give away answers or aid in eliminating distractors.			
9.	Question content conforms with specific K/A statements in the previously approved examination outline and is appropriate for the Tier to which they are assigned; deviations are justified.			
10.	Question psychometric quality and format meet guidelines in ES Appendix B.			
11.	The exam contains 40 one-point, multiple choice items; the total is correct and agrees with value on cover sheet.			

	Printed Name / Signature	Date
a. Author	_____	_____
b. Facility Reviewer (*)	_____	_____
c. NRC Chief Examiner (#)	_____	_____
d. NRC Regional Supervisor	_____	_____

Facility: Date of Exam:

Note: * The facility reviewer's initials/signature are not applicable for NRC-developed examinations.
        # Independent NRC reviewer initial items in Column "c"; chief examiner concurrence required.

Facility:	Date of Examination:	Operating Test Number:		

Item Description	Initials		
	a	b*	c#
1. The operating test conforms with the LSRO's job responsibilities and the previously approved outline (Form ES-701-4).			
2. Any changes from the previously approved outline have not caused the test to deviate from any of the acceptance criteria (e.g., item distribution, bank use, repetition from the last two NRC examinations) specified on the outline.			
3. There is no day-to-day repetition between this and other operating tests to be administered during this examination.			
4. The operating test does not duplicate items from the applicants' audit test(s). (See Section D.1.a of ES-301).			
5. Overlap between the written examination and the operating test is within acceptable limits.			
6. It appears that the operating test will differentiate between competent and less-than-competent applicants.			

7.   Each JPM includes the following, as applicable:
• initial conditions
• initiating cues
• references and tools, including associated procedures
• reasonable and validated time limits (average time allowed for completion) and specific designation if deemed to be time-critical by the facility licensee
• specific performance criteria that include:
   – detailed expected actions with exact criteria and nomenclature
   – system response and other examiner cues
   – statements describing important observations to be made by the applicant
   – criteria for successful completion of the task
   – identification of critical steps and their associated performance standards
   – restrictions on the sequence of steps, if applicable			

Printed Name / Signature	Date
a. Author	
b. Facility Reviewer(*)	
c. NRC Chief Examiner (#)	
d. NRC Supervisor	

NOTE: *   The facility signature is not applicable for NRC-developed tests.
           #   Independent NRC reviewer initial items in Column "c"; chief examiner concurrence required.

U.S. Nuclear Regulatory Commission

LSRO Written Examination

Applicant Information

Name:

Date: | Region: I ☐   II ☐   III ☐   IV ☐

Facility/Unit: | Reactor Type: W ☐   CE ☐   BW ☐   GE ☐

Start Time: | Stop Time:

Instructions

Use the answer sheets provided to document your answers. Staple this cover sheet on top of the answer sheets. The passing grade requires a final grade of at least 80.00 percent. Examination papers will be picked up 4 hours after the examination begins.

Applicant Certification

All work done on this examination is my own. I have neither given nor received aid.

_____
Operator's Signature

Results

Test Value	_____ Points
Applicant's Score	_____ Points
Applicant's Grade	_____ Percent

ES-702
ADMINISTRATION OF REQUALIFICATION EXAMINATIONS
FOR SENIOR REACTOR OPERATORS LIMITED TO FUEL HANDLING

A.   Purpose

The NRC's requalification examinations for senior reactor operators limited to fuel handling (LSROs) are administered under this standard in accordance with the provisions of Title 10, Section 55.59(a)(2)(iii), of the *Code of Federal Regulations* (10 CFR 55.59(a)(2)(iii)).

B.   Background

In conjunction with ES-601 through ES-603, this examination standard provides general guidance for facility licensees and requirements for NRC examiners to use in preparing, administering, grading, and documenting NRC requalification examinations for LSROs. Except as noted herein, the methodology and guidance presented in ES-601 through ES-603 also applies to LSRO requalification examinations, as they relate to administering full-scope requalification examinations at power reactors.

C.   General Differences

The LSRO examinations will be conducted in accordance with the methodology outlined in ES-601, "Conducting NRC Requalification Examinations," with the following exceptions:

1.   The NRC will coordinate with the facility licensee to schedule the NRC's LSRO examinations concurrent with the facility licensee's LSRO requalification examination schedule. If practical, the examination team will conduct the LSRO examination shortly before or after an outage to facilitate access to refueling equipment because some of the equipment is not accessible during plant operations.

      The NRC may administer LSRO requalification examinations concurrent with full-scope initial license or operator requalification examinations.

2.   The facility licensee's LSRO requalification program, LSRO job task analysis, and associated learning objectives will provide the basis for the examination if they are of sufficient scope and depth. The items in 10 CFR 55.43 and 55.45 will be sampled as appropriate to the LSRO's limited responsibilities.

3.   The LSRO requalification examination will consist of a written examination and a walk-through operating test, which is administered and evaluated individually. References to the crew-based dynamic simulator test and the associated crew evaluation criteria and forms do not apply to LSROs.

4.   Whenever possible, the facility licensee should include an LSRO on the examination team.

5.   The requirement to examine at least 12 operators to arrive at a program evaluation is not applicable to LSRO examinations. The region and the NRR operator licensing program office will determine the appropriate sample size based on the number of LSROs licensed at the facility.

6.   The "Corporate Notification Letter" (Attachment 2 to ES-601) shall be revised as necessary to reflect the examination arrangements and to specify a modified list of reference material requirements associated with LSRO fuel handling activities. The NRC's regional office will review the reference material using the applicable portions of Form ES-601-2.

7.   The NRC staff expects the facility licensee to maintain JPM and written examination banks for use in evaluating LSROs. Facility licensees should periodically update these examination banks to reflect areas of emphasis in training and to ensure that they represent all applicable knowledge and skills. There is no minimum threshold or ceiling for these banks.

8.   The NRC's regional office will document the agency's LSRO requalification examination results using Forms ES-702-1, "LSRO Requalification Examination Report," and ES-702-2, "LSRO Requalification Results Summary," instead of Forms ES-601-3, 4, and 5.

9.   This standard does *not* provide for a formal LSRO requalification program evaluation; however if more than one-third of the examined LSROs at a facility fail, the NRC may need to inspect the LSRO requalification program. The regional staff is responsible for determining whether such an inspection should be conducted. If an inspection is performed, the staff should assess at least the following considerations:

      a.   the content of the training program, the development of examination materials, and quality controls

      b.   the administrative controls for maintaining training material current with procedural revisions and design changes

      c.   the training and evaluation techniques of the facility licensee's evaluators

      d.   the evaluation techniques that the facility licensee uses to determine if it has effectively implemented and assessed its training

      e.   the frequency, scope, and depth of the training provided to the operators

Section D discusses specific exceptions related to each category of the examination. Any questions regarding the program office's expectations regarding the conduct of LSRO requalification examinations shall be referred to the NRR operator licensing program office for resolution.

D. Examination Differences

1. Written Examination

The written examination will be developed, administered, and evaluated as described in ES-602, "Requalification Written Examinations," with the following exceptions:

a. The written examination will be "open reference" and will contain a minimum of 25 points in a single section; static simulator scenarios do not apply to the LSRO examination. The time limit for completing the examination shall be 2 hours, but the examination should be constructed so that a competent LSRO can complete it in 1.5 hours. The examination should emphasize refueling procedures, administrative controls, and abnormal and emergency procedures. The examination should include questions associated with industry and licensee event reports (LERs) and recent plant modifications that affected refueling operations and systems that apply to the facility.

b. Form ES-702-3 will be used as a cover sheet rather than Form ES-602-1.

2. Walk-Through Operating Test

The walk-through operating test will be developed, administered, and evaluated as described in ES-603, "Requalification Walk-Through Examinations," with the following exceptions:

a. Each LSRO will be administered an operating test consisting of five tasks/JPMs. Whenever possible, these tasks/JPMs should include the use of refueling equipment to manipulate *dummy fuel only,* or the use of a refueling machine simulator if one is available at the facility. If dummy fuel manipulation or the use of a simulator is not possible, the refueling tasks should be simulated. The requirement to conduct a minimum number of JPMs in the control room/simulator is not applicable to LSRO examinations.

b. Each JPM will consist of a task that is normally performed by fuel handling personnel and will include tasks performed both before and after refueling and for maintenance, surveillance, or testing of systems or equipment. The examination team may evaluate the LSRO's ability to perform normal fuel handling administrative tasks including documenting clearances, maintenance activities, and surveillances. The operating test should also evaluate the LSRO's response to abnormal or emergency events associated with fuel handling.

c. If sufficient facility-developed JPMs are not available, the NRC can conduct a walk-through examination of the type administered to an initial LSRO applicant, as discussed in ES-701, "Administration of Initial Examinations for Senior Operators Limited to Fuel Handling."

3.   **Dynamic Simulator Operating Test**

The dynamic simulator operating test described in ES-604, "Dynamic Simulator Requalification Examinations," is not applicable to LSRO requalification examinations.

E.   Attachments/Forms

Form ES-702-1   "Individual LSRO Requalification Examination Report"
Form ES-702-2,   "Power Plant LSRO Requalification Results Summary"
Form ES-702-3,   "LSRO Written Requalification Examination Cover Sheet"

PRIVACY ACT INFORMATION — FOR OFFICIAL USE ONLY

<table>
<tr><td colspan="4" align="center">**U.S. Nuclear Regulatory Commission**
Individual LSRO Requalification Examination Report</td></tr>
<tr><td colspan="2">Operator's Name:</td><td colspan="2">Facility:</td></tr>
<tr><td colspan="2">Docket No.: 55-</td><td>Retake Exam: 1st/2nd/#</td><td>Date of Last Exam:</td></tr>
<tr><td colspan="2">License No.: SOP -</td><td colspan="2">Expiration Date:</td></tr>
<tr><td colspan="4" align="center">**Written Examination Results**</td></tr>
<tr><td colspan="2">Date of Exam:</td><td>NRC Examiner:</td><td>Facility Evaluator:</td></tr>
<tr><td rowspan="2" colspan="2" align="center">Overall
Grade (%) >>>></td><td align="center">NRC</td><td align="center">Facility</td></tr>
<tr><td align="center">%</td><td align="center">%</td></tr>
<tr><td colspan="4" align="center">**Operating Test Results**</td></tr>
<tr><td colspan="2">Date of Test:</td><td>NRC Examiner:</td><td>Facility Evaluator:</td></tr>
<tr><td colspan="2" align="center">No. of JPMs Correct</td><td align="center">of</td><td align="center">of</td></tr>
<tr><td colspan="2" align="center">Final Grade (%) >>>></td><td align="center">%</td><td align="center">%</td></tr>
<tr><td colspan="4" align="center">**NRC Examiner Recommendations**</td></tr>
<tr><td colspan="2">Category</td><td align="center">Results</td><td align="center">Signature</td></tr>
<tr><td colspan="2" align="center">Written</td><td align="center">Pass / Fail</td><td></td></tr>
<tr><td colspan="2" align="center">Operating</td><td align="center">Pass / Fail</td><td></td></tr>
<tr><td colspan="4" align="center">**NRC Supervisor Review**</td></tr>
<tr><td colspan="2">Date:</td><td align="center">Pass / Fail</td><td></td></tr>
</table>

PRIVACY ACT INFORMATION — FOR OFFICIAL USE ONLY

PRIVACY ACT INFORMATION — FOR OFFICIAL USE ONLY

Power Plant LSRO Requalification Results Summary						
Facility:				Exam Date:		
Examiners:						

Overall Results --->	Total # of Operators		Passed (# / %)		Failed (# / %)	

Individual Results

Operator's Name	Docket 55-	Grader	JPM % Overall	Written (%)	Results (P/F)	
					Written	Operating
		NRC				
		Fac				
		NRC				
		Fac				
		NRC				
		Fac				
		NRC				
		Fac				
		NRC				
		Fac				
		NRC				
		Fac				
		NRC				
		Fac				

PRIVACY ACT INFORMATION — FOR OFFICIAL USE ONLY

U.S. Nuclear Regulatory Commission

LSRO Written Requalification Examination

Operator Information

Name:	
Date:	Region: I / II / III / IV
Facility/Unit:	Reactor Type: W / CE / BW / GE
Start Time:	Stop Time:

Instructions

Use the answer sheets provided to document your answers. Staple this cover sheet on top of the answer sheets. Points for each question are indicated in parentheses after each question. The passing grade requires a final grade of at least 80.00 percent. Examination papers will be picked up 2 hours after the examination begins.

Operator Certification

All work done on this examination is my own. I have neither given nor received aid.

_____
Operator's Signature

Results

Test Value	_____ Points
Operator's Score	_____ Points
Operator's Grade	_____ Percent

APPENDIX A
OVERVIEW OF GENERIC EXAMINATION CONCEPTS

A.   Purpose

This appendix provides an overview of two fundamental examination concepts — validity and reliability — as they apply to the development of NRC operator licensing and requalification examinations. Specifically, this appendix discusses the following topics:

*   the rationale for providing guidance for the construction, review, and approval of NRC examinations (Section B)

*   the various aspects of validity and how the NRC establishes the validity of its examinations (Section C)

*   the concept of reliability and how it is maintained on NRC examinations (Section D)

B.   Background

The fact that the NRC's operator licensing examinations are prepared and administered by many different individuals working in various locations makes it imperative to establish and follow a defined set of administrative structures and protocols to ensure that the examinations are administered successfully and consistently. External attributes, such as the number and types of items, the length of the examination, security procedures, proctoring instructions, and other administrative details are essential to the orderly conduct of an examination. These factors have a significant effect on the reliability and validity of an examination — the cornerstones that allow the NRC to make confident licensing decisions.

The internal attributes of the examination, such as its level of knowledge, level of difficulty, and use of item banks, also impact the operational and discriminatory validity of the examination, which, in turn, can affect its consistency and reliability. If the internal and external attributes of examinations are allowed to vary significantly, the *uniform conditions* that are required by Section 107 of the *Atomic Energy Act of 1954*, as amended, and the basis upon which the NRC's licensing decisions rest are challenged. The NRC must reasonably control and structure the examination processes to ensure the integrity of the licenses it issues.

Acceptable levels of examination consistency, uniformity, and fairness would be impossible to achieve without quantitative and qualitative acceptance criteria. The examination standards identify many of the quantitative criteria necessary for a well-balanced and consistent examination. Although the NRC's Knowledge and Abilities Catalogs for pressurized- and boiling-water reactors (NUREG-1122 and 1123, respectively) have brought a degree of consistency to the qualitative issue of safety-significance, there is no comparable mechanism to aid in determining an examination's level of knowledge or difficulty before it is administered. In the end, the validity and consistency of the NRC's examinations depend largely on the individual and collective judgments of the people who write and review the examinations. The discussions herein clarify the intent of the NRC's examination criteria, thereby decreasing the likelihood that inconsistencies among examinations, particularly with regard to the level of knowledge and difficulty, will jeopardize the validity of the NRC's licensing decisions.

C.   Validity

For a test to be considered valid, it must be shown to measure that which it is intended to measure. In the case of the NRC examinations, the intent is to measure the examinee's knowledge and ability, such that those who pass will be able to perform the duties of a reactor operator (RO) or senior reactor operator (SRO) to ensure the safe operation of the plant. The following subsections outline the three principal facets of test validity and the techniques that are used to establish the validity of NRC examinations.

1.   Content Validity

a.   *Establish a Link to Job Duties*

In order to develop valid examinations, the knowledge and abilities (K/As) selected for testing must be linked to and based upon a description of the most important job duties. This is accomplished by conducting a job task analysis (JTA), focusing on the delineation of essential K/As.

The testing industry endorsed this approach to the development of content valid licensing examinations in the 1985 revision of the "Standards for Educational and Psychological Testing" published by the American Educational Research Association, the American Psychological Association, and the National Council on Measurement in Education. Those standards treat licensing examinations in a separate section in recognition of their importance and uniqueness. Accordingly, those seeking additional technical guidance are encouraged to consult Chapter 11 of the Standards for further clarification.

To ensure content validity in the NRC's examinations, the JTA performed on the licensed operator and senior operator positions by the Institute of Nuclear Power Operations (INPO) served as the initial source of information. The INPO JTA identified more than 28,000 K/As and nearly 800 tasks. The extensive number of tasks and K/A statements is attributable, in part, to the specific purpose of the analysis, which was to provide an information base to be used in developing training programs that would be applicable to all pressurized- and boiling-water facilities. Accordingly, many of the individual statements were too specific and/or too elementary for use as the basis for development of the NRC's examinations. The job content of special interest to the NRC is that subset of K/As that are required for the safe operation of the nuclear plant. Although *safe* performance and *efficient* performance may have considerable overlap, any K/A that contributes to efficiency but not safety is an inappropriate focus for the NRC's examinations.

NUREG-1122, "Knowledge and Abilities Catalog for Nuclear Power Plant Operators: Pressurized-Water Reactors," and NUREG-1123, "Knowledge and

Abilities Catalog for Nuclear Power Plant Operators: Boiling-Water Reactors," provide the basis for the development of content valid examinations for ROs and SROs, consistent with the testing industry standards described above. The fact that the K/As from which test items are developed are drawn or sampled from the same universe regardless of who develops the examination ensures that the examinations are consistently content valid. Furthermore, developing the examinations using the appropriate K/A catalog in conjunction with the applicable examination standards and related appendices will ensure that the examinations cover a representative sample of the topics listed under Title 10, Part 55, of the *Code of Federal Regulations* (10 CFR Part 55).

The NRC's K/A catalogs were developed on the basis of the INPO JTA and were reviewed by licensed ROs and SROs, as well as the NRC's own license examiners. These experts reviewed the K/A statements for accuracy and completeness, and then rated each statement with respect to its importance to safe plant operation. Further explanation of the content of the K/A catalogs is provided in Section 1 of each catalog.

In addition to the NRC's K/A catalogs, learning objectives from the facility licensee's training program often provide a supportive reference for test items to be included in the NRC's examination. Since facility learning objectives are specific to the job requirements at a given site, they should provide an excellent basis for test item development. However, because they are not always stated at the comprehension or analysis levels of knowledge (the preferred focus for NRC examinations) they should be referenced only to the extent that they support a test item that is being developed.

b.   *Use a Sample Plan*

Once the essential K/As have been identified through the conduct of the JTA, test specifications must be developed. The test specifications consist of a content outline or sample plan indicating what proportion of items or questions shall deal with each K/A. Because a single test cannot measure every knowledge or ability required to be a licensed operator, it must sample the required knowledge or performance in a manner that allows inferences to be made regarding the examinees' performance on the broader population of knowledge, even though it was not tested. The sample must be evenly distributed and soundly based so that the NRC can confidently assume that the untested knowledge is proportionately known or not known in relation to the score on the sample. In other words, by testing performance on the sample, it is possible to make inferences concerning the broader area of knowledge not tested. This is referred to as a "validity inference."

The sample plan is at the heart of making a validity inference. Research indicates that when samples are not chosen systematically and according to the sample plan, the sample is biased and, therefore, its validity is reduced. When the sample is biased or skewed in a particular direction, it introduces some degree of sampling error, which makes it impractical to infer or generalize that the examinees have mastered the larger population of untested knowledge from which the sample was drawn.

Test items selected for inclusion in an NRC examination should be based on K/As contained in the appropriate K/A catalog. Testing outside the documented K/As can jeopardize the content validity of the examination. Content validity can also be reduced if important K/As are omitted from the examination. Therefore, the sample of K/As that are tested should cover all of the K/A categories in the catalog in a fashion that is consistent with their contribution to the public protection function of the examination. Not all categories are equal in this regard. This conclusion is based on the analysis of ratings on importance and testing emphasis collected from licensed SROs and NRC license examiners. The specific examination standards provide additional guidance on how to develop test outlines that will ensure adequate content coverage.

It is important to note that there is a difference in the testing demands for an initial examination versus a requalification examination. The requalification examination is based upon the plant's systems approach to training during the requalification cycle and will more closely parallel the training received in the requalification program. Consequently, the instructional and testing processes are more closely linked. The initial examination, on the other hand, covers all instruction related to safety-significant K/As that either were or should have been taught during the training program. The examination standards ensure that the K/As are sampled in a relatively uniform process that would likely include content and instruction that occurred from the beginning to the end of the program and *not* be focused upon any particular segment of instruction.

2.   Operational Validity

The second facet of validity is operational validity. To the extent possible, test items should address an actual or conceivable mental or psychomotor activity performed on the job. In this regard, the more operationally oriented a test item is, the more valid the test item. Since operationally valid items involve skills central to job performance (i.e., analysis, prediction of events or system responses, or problem-solving), the items should be written at the comprehension or analysis level, rather than the level of simple fundamental knowledge. The theoretical knowledge classification system upon which the NRC bases its operational validity estimates is Bloom's Taxonomy.

Bloom's taxonomy suggests that testing knowledge at higher cognitive levels (i.e., comprehension and analysis) is more efficient and operationally valid because those higher levels include the fundamental knowledge required, in part, to answer the higher-level question. Furthermore, the higher the level tested in the test item, generally the more operationally valid that test item will be, since it is at the higher levels that questions invoke problem-solving, diagnosis, prediction, and analysis of conditions, events, and responses.

Designing test items that test the *application* of knowledge in different content situations (i.e., process testing) is at the heart of designing good, discriminatory test items. Just as a mathematics teacher would not design a test to ask multiplication questions that were identical to practice questions, so too should the examination author minimize asking questions that are identical to those previously rehearsed or tested. Test items should attempt to assess similar knowledge applications in different contexts, thereby assessing students' problem solving skills in new and different applications. These applications should be item substitutions of comparable difficulty, neither harder nor easier than those practiced. This practice provides assurance that the examination is valid and discriminatory, since the process (rather than the specific content) is primarily measured.

The NRC cannot make confident and consistent validity inferences (i.e., licensing decisions) if one examination assesses knowledge at lower cognitive levels, and another assesses knowledge at higher levels (greater depth). While each examination may meet sample plan coverage guidelines, they test different levels of knowledge and, consequently, they are different and inconsistent measuring instruments. Therefore, they yield different validity inferences regarding minimally safe operator performance. Refer to Section D for a more detailed discussion of consistency and reliability and to Appendix B for a more detailed discussion of the various levels of knowledge as they relate to the development of written test questions.

3. **Discrimination Validity**

The third facet of validity concerns the examination's ability to discriminate, or make some distinction along a continuum of examinee performance. In that regard, the primary objective of the NRC's examinations is to determine whether or not the examinees have sufficiently "mastered" the knowledge, skills, abilities, and other attributes to perform the job of an RO or SRO at a specific plant. The NRC's examinations are not intended to distinguish among levels of competency or to identify the most qualified individuals, but to make reliable and valid distinctions at the minimum level of competency that the agency has selected in the interests of public protection.

a. *Criterion-Referenced Testing*

The NRC's initial and requalification examinations, like most licensing examinations, are criterion- rather than norm-referenced tests. This means that there is a pass/fail or minimal cut score or grade that the examinee must achieve to demonstrate sufficient knowledge and ability to safely operate the power plant. If the examination does not intend to discriminate at an agreed-upon minimal measure of knowledge or performance, there is little reason to administer the examination. For a criterion-referenced test to be effective, both the individual test items and the overall examination must discriminate between applicants who have and have not mastered the required knowledge, skills, and abilities.

b. *Cut Scores*

For NRC examinations, the overall cut scores (on the written examination and walk-through) are fixed at 80 percent (although lower cut scores apply to subparts of the examination); it is the content of the examination that varies from occasion to occasion because of the plant-specific character of the test material. As discussed below, there are several reasons why the cut score must be fixed, including the uniqueness of each examination, consistency, and public confidence.

In the writing, reviewing, setting of scoring standards, and grading of any particular NRC examination, both the examination author and the reviewer are well aware of the NRC-established passing score of 80 percent. They may also have knowledge of how prior examinees have performed on questions similar to those being used on the examination under construction and expectations as to how a qualified or unqualified applicant should perform on the examination. They must use this knowledge to control the nature and difficulty of the examination, such that an examinee who is deemed to be qualified scores above the passing grade, while an examinee who is deemed to be unqualified scores below that grade.

The traditional cut score on the examination should not be viewed as arbitrary. Rather, it reflects a point on the test at which author and reviewer judgment separates the qualified from the unqualified. Nonetheless, the judgment is probably similar to other methodologies for determining passing test scores. For example, rather than *explicitly* judging the probability that a minimally qualified applicant will pass an item, the author is *implicitly* being asked to write an examination on which, *in the author's judgment*, the minimally qualified applicant will obtain a score of at least 80 percent. Achieving this objective requires the author and reviewer to integrate their content and process skills.

c.   *Level of Knowledge Versus Level of Difficulty*

As further discussed in Appendix B, the NRC uses Bloom's Taxonomy as the basis for classifying the level of knowledge of its test items [i.e., written examination questions, job performance measures (JPMs), and simulator events]. Simply stated, level of knowledge represents the range of mental demands required to answer a question or perform a task; in other words, level of knowledge is a continuum of mental rigor that ranges from retrieving fundamental knowledge (low level) to retrieving that knowledge and also understanding, analyzing, and synthesizing that knowledge with other knowledge (high level).

The accurate classification of knowledge as low- or high-level requires the application of objective criteria. While different reviewers can arrive at different conclusions regarding the knowledge level of individual test items, a common set of criteria can make the classification an informed process, thereby minimizing the differences among reviewers. Consistency among reviewers is important because this NUREG establishes specific criteria relative to the number of higher cognitive level (HCL) test items on the site-specific written licensing examination. Keep in mind that classifying a test item's level of knowledge is not equivalent to determining its level of difficulty, which is discussed as a separate issue below.

When evaluating level of knowledge, two key elements must be considered. Specifically, those elements are (1) the number and type of mental steps necessary to process the given data and arrive at the correct answer, and (2) the training and experience level of the target test group.

Generally, an HCL test item will require at least two mental steps, one of which requires the recall of acquired knowledge, and the other requires associating two or more pieces of data. The number and types of mental steps that must be considered are those necessary to *rule out* plausible incorrect distractors as well as the steps needed to *identify* the correct answer. Distractors can contain knowledge that the applicant might need to manipulate with other information contained in the question in order to answer the question and this, in turn, may raise the level of knowledge of the question. However, it is largely the *stem* of the question that drives the mental thought required to answer the question.

An HCL test item will have at least *two data points* that must be associated. These data points may be provided in the test item, or they may have to be recalled from memory by the examinee. For example, the examinee may be given one plant operating parameter in the stem of a question and have to recall a setpoint to evaluate whether a particular action should have occurred. This is considered HCL because it requires the examinee to (1) recall a setpoint beyond the information given in the stem and (2) compare the setpoint to a given data point. Since more than one mental step was necessary to answer this question, and two data points had to be associated or compared, it should be classified as HCL.

Similarly, if a test item elicits a mental demand that requires a "why" or "how" response such that the examinee must derive the correct explanation, prediction, or action, the item is testing at the comprehension or application level. Comprehension/application level test items require the examinee to recall stored knowledge and understand the relationship between *two or more* pieces of data (such as events or conditions) given in the stem of the test item. In sum, HCL test items require multiple mental processing steps, which are usually the recall and integration of two or more pieces of data. Good HCL test items are operational in nature and require demonstration of understanding and problem solving.

Test items that simply ask examinees to provide a single answer that requires a "who, what, when, or where" response are typically fundamental knowledge (low-level) questions because they involve recalling or recognizing a single answer or chunk of information. The examinee is not required to understand cause-effect relationships or system responses. Therefore, if a test item simply asks for a reactor trip setpoint, and does not require a comparison with an operating parameter value, it would be considered a lower cognitive level question because only one mental step, with no data association, is necessary to arrive at the answer.

With regard to the operating test items (i.e., JPMs and simulator events), the regulations [10 CFR 55.45(a) and 55.59(a)(2)(ii)] specifically require an assessment of the examinees' understanding of and ability to perform the actions specified in the regulation. Alternate path JPMs are used to assess such understanding during the walk-through because they require examinees to evaluate unplanned conditions or events while executing procedures and to implement acceptable, alternative methods of accomplishing the assigned tasks.

As previously noted, the training and experience of the target test group must also be considered when evaluating level of knowledge. A reviewer can approach the classification from the perspective of an "expert," with a predetermined belief about the mental processes required to answer a given question, and incorrectly assume that the novice applicant will use the same processes. This is a form of perceptual bias that can affect level of knowledge, as well as level of difficulty, classifications. When examining new license applicants, it is expected that the typical applicant will need to mentally analyze, or figure out, the answers to HCL questions. Whereas the expert is able to answer a test item quickly and easily, the novice may have to eliminate plausible distractors to arrive at the correct answer, an indication of an HCL question. Therefore, when making the level of knowledge determination, examination writers and reviewers should place themselves in the context of the "novice applicant" and assess the components of the test item that the novice must manipulate to answer the test item.

Keep in mind that many test takers may easily arrive at the answer; however, ease of answering a question is a relative concept and should be clearly separated from the mental processes, or level of knowledge, required to answer the test item.

In summary:

- Level of knowledge is a taxonomy to determine the mental processes used to answer a question. Those processes are classified as either lower or higher cognitive level and should not be confused with level of difficulty.

- An HCL test item requires at least two mental steps, one involving the recall of acquired knowledge, and the other requiring the association of two or more pieces of data. The number and type of mental steps that must be considered include those necessary to rule out plausible incorrect distractors as well as the steps needed to identify the correct answer. If there is doubt concerning the number of associations, err on the side of classifying the item as HCL. As a tip, attempt to answer the question in an unaided recall manner (i.e., if the question were in the completion or short answer format, cover the distractors and attempt to complete the answer). Then, analyze the mental process needed to answer the question using the "who, what, when, or where" (fundamental) or the "how or why" (comprehension/analysis) criteria discussed above.

- When assessing level of knowledge, the examination writers/reviewers must use the perspective of the test taker in the target group (i.e., novice versus expert) to avoid perceptual bias. The reviewer has seen the item, knows the answer, and may not appreciate the mental processes that an examinee may use to answer the question.

Level of difficulty is a separate concept, but is often influenced by the test item's level of knowledge. Although HCL test items are generally more difficult, this may not always be true. A fundamental knowledge question may be easy (e.g., How many inches are in a foot?) or difficult (e.g., In what year was the printing press invented?).

The NRC evaluates a test item's level of difficulty to ensure that the item can help discriminate between safe and unsafe operators. The examination's overall level of difficulty, as well as that of its individual test items, should center around the 80 percent cut score. (See additional guidance below.)

Assigning a level of difficulty rating to an individual test item is a somewhat subjective process. As when assessing the level of knowledge, examination authors and reviewers must "detach themselves" as subject matter experts (SMEs), place themselves in the position of the novice applicant, and apply what they know about previous applicants' performance on similar test items. For example, if a particular item was "missed" by 10 to 20 percent of past license applicants, the item would be considered moderately discriminating, with a difficulty rating of 3 on a 5-point scale. It would be reasonable to expect that a similar item will perform similarly with a comparable test group. Conversely, if 95 percent or more of license applicants typically answer a particular test item correctly, it is likely that future use of a comparable item will yield a similar result, so a difficulty rating of 1 would be justified.

d.   *Cut Scores and the Level of Difficulty*

For the cut score of 80 percent to be meaningful requires that individual test items must be written "near" that level. A target level of difficulty range of 70 to 90 percent is recommended for individual test items. Test items that are so difficult that few (if any) of the examinees are expected to answer correctly do not discriminate and should not be used on an NRC examination. Similarly, test items that are so easy or fundamental that even those examinees who are known to have performance problems will be able to answer correctly should be used with discretion. It is expected that every examination will contain some test items that all or most of the examinees will answer correctly or incorrectly. This does not necessarily mean that the test items or the examination are invalid.

It should be stressed that the intent is not for everyone to get a score of 80 percent. In fact, historically about 90 percent of examinees score 80 percent or above on the NRC examinations. A score of 80 percent is the minimal pass score that the author and reviewer must keep in mind as a functional level of discrimination for setting item difficulty. In order to achieve this, the test author must keep in mind and integrate the following concepts:

- the level of knowledge required of examinees taking the examination

- the operational validity of the questions (i.e., are they expressed as a conceivable job behavior?)

- the ability of the distractors to distract the examinees

- the examinees' past performance on items of similar difficulty

e. *Use of Item Banks*

Test item banks are a valuable resource for learning and represent one fundamental basis for training and testing. However, it would be inappropriate to copy all or a significant portion of the items for an examination directly from the bank if the same items were previously used for testing or training. Test item banks must be used properly to maintain the validity, reliability, and consistency of the examinations. Previously administered test items reduce examination integrity because examination discrimination is reduced.

Discrimination is reduced because the cognitive level at which the examinees are tested could decrease to the simple recognition level if the item bank is small and available for the examinees to study. The comprehension and analysis levels of knowledge may not be assessable because mental thought has been reduced to a recognition level, and decision-making is absent because test items, JPMs, or scenario events have been rehearsed and are anticipated. In short, challenge and mental analysis are lost and the examinees are tested at a rote-rehearsal level. An examination cannot assess higher cognitive and analytical abilities if a significant portion of the items within the examination have already been seen.

Furthermore, when the bank of items from which the examination is drawn is known to the examinees prior to the examination, the examination is said to be highly predictable. Predictable examinations tend not to discriminate because what is being tested is simple recognition of the answer. Although studying past examinations can have a positive learning value, total predictability of examination coverage through over-reliance upon examination banks reduces examination integrity. When the examinees know the precise and limited pool from which test items will be drawn, they will tend only to study from that pool (i.e., studying to the test) and may likely exclude from study the larger domain of job knowledge. When this occurs, it decreases the confidence in the validity inferences that are made from performance on the test to that of the larger realm of knowledge or skill to be mastered.

Therefore, the NRC has placed limits on the use of facility item banks or other such available banks or resources that have been published, reviewed, or used as the basis for training; the specific limits are discussed in the examination standards. The NRC appreciates the amount of resources required to develop new test items that are appropriate for use on an NRC examination, and realizes that existing test items are a valuable resource that should not be wasted.

Therefore, the NRC has elected to strike a balance in setting limits on the mix of previously used bank items, modified bank items, and newly developed (i.e., not previously seen) items. Additional limits have been placed on the repetition of test items from prior quizzes and examinations given at the facility.

D. Reliability

Reliability is the second fundamental testing concept that has played a decisive role in the development of the NRC's initial and requalification examination programs. Whereas the notion of validity emphasizes the appropriateness of the content of the NRC examinations, reliability stresses consistency, repeatability, and the degree of confidence that the examination process will result in valid pass/fail decisions. The reliability of an examination is as important as its validity; if an examination is not reliable, it cannot be valid.

The importance of examination consistency (reliability) cannot be overstated. In fact, test reliability represents the consistency among examinations which, in turn, gives the NRC the confidence that all examinations are valid measures from which to make confident and valid licensing decisions. The combined effects of item bank use, the level of knowledge tested in the individual test items, and the expected discriminatory (difficulty) level of the items play an important role in determining the reliability of the examination.

The higher the reliability of a test, the fewer errors will be made in determining whether the examinees have mastered the job requirements. Examinations should differ only in the specific content covered, not in their developmental processes, manner of sampling, item construction criteria, level of item bank use, or their levels of knowledge and difficulty. The standardization of the process creates consistency of measurement. Ideally, any two examinations that are written in accordance with these procedures and guidelines and administered to the same group of examinees should produce comparable results; likewise, the results of any examination administered to different but similarly trained and qualified examinees should also yield comparable results.

The standardized examination development, administration, and grading procedures described in this NUREG have evolved over a period of years in an effort to enhance the reliability and, hence, the validity of the NRC's licensing decisions. The importance of having these procedures and complying with their intent has grown in proportion with the number of individuals and organizations that have become involved in the examination process.

Section 107 of the *Atomic Energy Act of 1954*, as amended, requires the Commission to prescribe uniform licensing conditions for operators. Therefore, facility licensees are expected to develop and submit their proposed examinations based on the guidelines and instructions contained herein. The NRC discourages facility licensees from using testing methodologies that do not conform to the policies, procedures, and practices defined in this NUREG. Nonetheless, facility licensees may propose alternatives to specific guidance in NUREG-1021, and the NRC will review and rule on the acceptability of the alternatives.

APPENDIX B
WRITTEN EXAMINATION GUIDELINES

A.   Purpose

This appendix provides background information concerning the principles and practices for developing multiple-choice written test questions for NRC initial and requalification examinations. Examination authors and reviewers should use the guidance herein when selecting, constructing, and reviewing questions for use in NRC written examinations. Specifically, this appendix addresses the following topics:

- written examination background (Section B)

- the basic psychometric principles (i.e., low level of knowledge, low operational validity, low discriminatory validity, implausible distractors, confusing language or ambiguous questions, confusing or inappropriate negatives, collection of true/false statements, backward logic) and other guidelines applicable to the question development process (Section C)

- a checklist for reviewing multiple-choice questions (Attachment 1)

- examples of questions that illustrate the psychometric principles (Attachment 2)

- a list of references that provide additional information concerning the development of written examinations (Attachment 3)

For a discussion of the specific written examination criteria that apply to NRC initial and requalification examinations, refer to ES-401, "Preparing Initial Site-Specific Written Examinations," and ES-602, "Requalification Written Examinations," respectively.

B.   Background

1.   The Importance of the Written Examination

Our society has institutionalized written examinations as an accepted and important facet of performance testing, and they are routinely used as an integral factor in measuring human performance in nearly every field of study. Educational institutions from elementary through graduate schools use written examinations, in part or in whole, to measure intended competencies. Moreover, many fields of business, including the legal, medical, educational, and accounting professions, use written examinations for licensing and credentialing activities.

The importance of knowledge testing should not be underestimated, because knowledge is the underpinning of professional performance. The objectives of knowledge testing are varied; they may include assessing fundamental understanding, as well as testing more advanced levels of expertise. The most effective tests of knowledge include questions and test items that measure the application of knowledge that directly relates to an individual's job. In the case of operator licensing, the NRC's written examination yields a key measure that allows the agency to make a confident decision regarding the safety-significant performance of the individual seeking a license.

De-emphasizing or sidestepping knowledge testing through careless or simplistic testing processes, or treating it secondarily to other portions of the examination that are more operationally oriented could affect subsequent job performance. Failing to focus on testing the individual operator's cognitive abilities (i.e., comprehension, problem-solving, and decision-making), or paying insufficient attention to the operator's fundamental understanding of job content (e.g., systems, components, and procedures), may ultimately place job performance at risk of gradual degradation. When the demand for disciplined learning and study declines or the level of knowledge (depth of application) required for the job is reduced, it could lead to less time spent in training preparation, less mental review and practice, more forgetting of factual details, less reinforcement and application of job concepts, and a gradual decline in performance.

Moreover, without a solid fundamental knowledge base, operators may not perform acceptably in situations that are not specifically addressed in procedures. Since every performance has an underlying knowledge component, that knowledge and its depth need development and assessment to ensure the operators' competence on the job. Recent studies assessing mental performance in cognitively demanding emergencies point out that higher-level cognitive thought (such as event diagnosis and response planning) are important in responding to safety-related events.

2.   **Objective Versus Subjective Test Items**

Traditionally, questions that require the examinee to supply an answer (e.g., short answer and essay) have been considered "subjective," while questions requiring the examinee to select an answer (e.g., multiple-choice) have been considered "objective." These terms arose from the scoring of the items. If graders require subject matter expertise to interpret the answers, the question is considered subjective. By contrast, if graders can score the examination by verifying a single letter or number, the question is considered objective.

Multiple-choice items are the most common and most popular of the select-type items. For reasons of consistency and reliability, they are currently the only type of items acceptable for use on NRC initial licensing examinations. Although multiple-choice items are not as easy to construct as other forms, they are very versatile, can be used to test for all levels and types of knowledge, and minimize the likelihood that the examinee will obtain the correct answer by guessing. Scoring multiple-choice examinations is also considerably more reliable and less time-consuming than scoring open-ended response items. Furthermore, since each item requires less time to answer, more items can be used to test a larger sample of K/As. This provides better content coverage, which also increases test reliability.

For purposes of NRC requalification examinations and initial operating tests, the definition of "objective" differs from the traditional definition discussed above. In this case, an objective test item is one for which (1) there is only one correct answer, and (2) all qualified graders would agree on the amount of credit allowed for any answer.

Therefore, all questions on NRC examinations shall be objectively gradable, regardless of the item format. Questions with no single correct answer or for which the credit given can vary, depending on who graded it or when it was graded, have *no* place on an NRC examination.

C.   Question Development

Examination authors and reviewers should observe the following generic principles and question construction guidelines when preparing NRC written examinations. The guidance is based upon psychometrics — the process of applying sound qualitative processes to mental measurements. The generic principles apply to all question formats, including multiple-choice, while the guidelines in Section C.2 apply strictly to the multiple-choice format. It is important to minimize the number of psychometric errors in NRC examinations because test items that are free of psychometric errors yield greater measurement validity.

The following principles and guidelines are summarized on Attachment 1, which can be used as a desk-reference during the process of developing and reviewing questions. The list appears to be long, but the concepts become internalized with practice, and the process becomes less difficult. Many of the principles are accompanied by examples that illustrate the psychometric errors that should be avoided. Attachment 2 provides additional examples.

1.   Generic Principles

    a.   Ensure that the concept being measured has a direct, important relationship to the ability to perform the job. Although Appendix A stresses the importance of relevant knowledge and abilities (K/As) and testing objectives, it is equally important that the construction of the question itself clearly reflects the importance of the topic. Phrase the question so that is has "face validity," as well as underlying content validity. That is, make sure that the question would be considered reasonable to other subject matter experts utilizing the same reference materials.

            It is not always necessary to establish a direct, word-for-word match between a question and a facility learning objective. A broadly stated learning objective may support any number of related questions.

            Similarly, the absence of a facility learning objective does not preclude the development of a valid, K/A-based question. This is consistent with the concept of the NRC examination as providing checks and balances on the facility licensee's training program, thereby alerting the licensee that it may need to develop such a learning objective.

            Although it is appropriate to develop questions regarding knowledge that is embedded in or covered by procedures, the knowledge tested should not be trivial in nature.

b.   Make sure that the question matches the intent of the K/A.

It is very easy to wind up with a question that tests a relatively trivial aspect of an important K/A topic. When reviewing your draft question, ask yourself whether it is likely that someone could answer the question correctly and still not meet the objective or intent of the K/A or perform the responsibilities or tasks for which the K/A is needed.

If you are having difficulty translating a K/A into a test question, ask yourself the following questions to help you generate ideas for potential test questions:

(1)   What are the common misconceptions about this topic?
(2)   Why is this topic important to satisfactory job performance?
(3)   In what sort of circumstances might it be important to understand this topic?
(4)   What might an individual do if he or she does not understand this topic?
(5)   What might be the consequences of a lack of knowledge about this topic?
(6)   How can the individual demonstrate his or her knowledge of this topic?

c.   State the question unambiguously, precisely, and as concisely as possible, but provide all necessary information.

Often the individuals who develop a question assume that certain stipulations or conditions are inherent in the question when, in fact, they are not. It is very difficult for the person who wrote a question to review it impartially or through the eyes of a new reader. Therefore, it is very important to have others review your questions to ensure that it includes all necessary information, and excludes all extraneous or superfluous information. For example, it is not necessary to provide a status for each annunciated parameter that is in its normal (non-alarming) state. Refer to Section C.3 for additional guidance regarding examination reviews and to Part B of Appendix E for the instructions provided to applicants regarding question clarity and assumptions.

However, as discussed in Appendix A, keep in mind that the key purpose of any test item is to assess important knowledge and abilities at a level that distinguishes between safe and unsafe applicants. A test item's ability to make that distinction is referred to as its discrimination validity. For a question to discriminate at the appropriate level, the test author must exercise judgement in establishing the initial conditions posed in the stem of the question. Providing too much information may "lead the applicant to the answer" and decrease the discrimination validity of the question because the answer is obvious to all applicants.

For closed-reference questions related to a specific plant procedure, it is generally desirable for the question to cite the number and title of the subject procedure, thereby limiting the possibility of an alternative correct answer if another procedure happens to relate to the same activity. For open-reference questions, use caution to ensure that the question does not become a direct look-up, with a pointer to help the applicant find the answer.

d.   Write the question at the highest level of knowledge reflected in the testing objective.

One of the most challenging aspects of question development is attaining the appropriate level of knowledge. The reference benchmark that the NRC uses to classify the levels of knowledge of test items is Bloom's Taxonomy, a classification scheme that permits the grouping of items by the level (depth) of mental thought and performance required to answer the items. (Refer to Attachment 3 for references related to Bloom's Taxonomy). Although Bloom's Taxonomy is most pertinent to written examination questions, it can also be applied to simulator scenarios and JPM items. In ascending order, the three levels (depths) of mental thought and performance are as follows (refer to Section A of Attachment 2 for examples of each level):

- Level 1 (i.e., fundamental knowledge or simple memory) tests the recall or recognition of discrete bits of information. Examples include knowledge of terminology, definitions, set points, patterns, structures, procedural steps and cautions, or other specific facts.

- Level 2 (i.e., comprehension) involves the mental process of understanding the material by relating it to its own parts or to some other material. Examples include rephrasing information in different words, describing or recognizing relationships, showing similarities and differences among parts or wholes, and recognizing how systems interact, including consequences or implications.

- Level 3 (i.e., analysis, synthesis, or application) testing is a more active and product-oriented testing approach, which involves the multi-faceted mental process of assembling, sorting, or integrating the parts (information bits and their relationships) to predict an event or outcome, solve a problem, or create something new. This level requires mentally using the knowledge and its meaning to solve problems.

Although test questions should be written to reflect the level of knowledge that is most appropriate for a specific K/A, it is best to avoid high percentages of fundamental knowledge-level questions on the examination. (Refer to ES-401 for specific limits.) When there is a choice between two levels of knowledge, try to write the question to reflect the higher level. In general, test items at the comprehension and analysis levels are the most operationally oriented and, therefore, tend to be the most valid and discriminatory measure of operator knowledge and safe performance. Questions that require only memorization or recall are not acceptable for use on open-reference examinations.

e.   Avoid questions that are unnecessarily difficult or irrelevant.

As discussed conceptually in Appendix A, both the level of knowledge and the difficulty of an item are at the heart of examination discrimination. Authors should develop examinations that are estimated to center around the 80-percent cut score level, with individual item difficulty estimated to fall in the 70- to 90-percent difficulty range. (These parameters should not be viewed as precise benchmarks, but rather as approximate end points.) Examination authors should consider the results of past examinations when preparing a new one. Past performance on individual test questions may provide a basis for generating new questions and for estimating the level of difficulty of the examination. For example, questions that everyone answered incorrectly may indicate that the topic did not receive sufficient emphasis in training or that the item was poorly worded. Conversely, questions that everyone answered correctly may indicate that the item was written at too low a level or the distractors were not very plausible.

Since item difficulty can usually be decreased or increased through revision, the examination author need not be overly preoccupied with difficulty when writing the items. Instead, the author should focus on achieving a valid measure of the concept he or she is attempting to evaluate.

When attempting to determine the appropriate level of difficulty, it may be helpful to think of two groups of individuals, one composed of experienced operators and the other of typical applicants, and evaluate the likelihood that each group of individuals will be able to answer the question. If at least 80 percent of the job incumbents or license applicants should be able to answer the question as written based on the expected knowledge levels for the position (operator or senior operator), the item is likely written at an appropriate discriminatory level. Examination authors and reviewers may also consider the following factors in an effort to identify questions that are unnecessarily difficult or irrelevant:

•   Could someone do the job safely and effectively without being able to answer the question? If so, is it because (1) the content is inappropriate, (2) the wording is unclear, or (3) the level of understanding is too great?

•   What aspects of the item or option might cause the most difficulty? Has the item been made artificially difficult? Can a person understand the principle being tested and still miss the item?

Estimates of difficulty made by the examination author and reviewers may vary somewhat, but should not vary widely. Unless there is some reason to doubt the estimates of some reviewers, the average estimate may be taken as a basis for assessing the suitability of item difficulty for the examination. Items should be revised if estimates fall well below or above the 70 to 90 percent target range.

Research has shown that when authors write test items in their own area of specialization, they have a tendency to underestimate the difficulty of a concept or principle being tested. This tendency can manifest itself in two ways. Specifically, (1) the author will view items of average difficulty as being easy, or (2) in an effort to include plausible misleads among distractors in a multiple-choice test item, the author may make the item even more difficult. For this reason, an estimate of item difficulty made by the reviewers will probably be more accurate than one made by the author of the item.

Examination authors should take care not to develop an examination with wide swings of individual item difficulty. For example, writing half the items at a 60-percent difficulty level with the other half at a 100-percent difficulty level would yield an average of 80 percent; however, this approach has numerous flaws. The items at the 100-percent level, by design, would be meaningless, since they would fail to discriminate at any level because the expectation is that nearly everyone would answer the question correctly. On the other hand, those written at the 60-percent difficulty level, by design, would also not discriminate and would likewise be unfair because 40 percent of the examinees would not be expected to answerable those items correctly.

f. Limit the question to one concept or topic, unless a synthesis of concepts is being tested.

There is a common misconception that testing for multiple K/A topics in one question is a time-efficient way to examine. However, questions containing a variety of topics and issues only serve to confuse the examinee about the purpose of the question and what is expected in terms of a correct response. Each individual question should test one K/A topic, and that topic (as well as the intent of the question) should be clear to both the reviewer and the examinee.

g. Avoid copying text directly from training or other reference material.

Another common tendency among examination developers is to copy sentences directly from reference material and turn them into test questions. Unfortunately, questions written in this way generally encourage rote memorization. Further, copying from reference material can cause ambiguity or deficiency in questions because the replicated material often draws its meaning and importance from its surrounding context. Therefore, important assumptions or stipulations stated elsewhere in the material are often omitted from the test question. Finally, such questions can frequently be answered correctly by examinees who do not really understand the concept, but do remember the specific wording on a page of reference material. Conversely, examinees who understand the topic, but not in the exact way it was written in the material, may miss the question because of unstated assumptions or other missing information.

h.   Avoid "backward logic" questions that ask for what should be provided in the question, and provide what should be required in the examinee's response.

In addition to testing on valid topics, it is important to examine on those topics in a way that is consistent with how the K/A should be remembered and used. Do not test on the topic in a backward way. Section G of Attachment 1 provides examples of backward logic questions.

2.   **Other Question Construction Guidelines**

The following principles and guidelines apply specifically to multiple-choice questions:

a.   Use four answer options.

The four-distractor multiple-choice item with only one correct answer is the only style that is considered acceptable for NRC examinations. However, the use of test items with multiple correct answers from which examinees must select the "most correct" answer is not acceptable because such items significantly reduce the reliability of examination results by increasing the effect of examiner subjectivity in the examination development and grading processes.

The five-answer option contributes nothing to the question, and any format with fewer than four distractors makes guessing correctly more probable. The following four basic models are acceptable and may be used in combination with one another.

Model A: a.   correct answer
            b.   plausible incorrect answer
            c.   plausible incorrect answer
            d.   plausible incorrect answer

This model depicts the traditional multiple-choice design format with one correct single-word/phrase answer followed by three incorrect single-word/phrase options. Note that all options are of similar length.

Model B: a.   correct answer
            b.   plausible misconception
            c.   plausible incorrect answer
            d.   plausible incorrect answer

This variation of Model A uses a plausible misconception as one of the three incorrect answers. Again, note that all options are of similar length.

Model C: a.   correct answer with correct condition
                               (e.g., because, since, when, if, etc.)
                  b.   correct answer with plausible incorrect condition
                  c.   plausible incorrect answer with incorrect condition
                  d.   plausible incorrect answer with incorrect condition

Model C depicts an acceptable design that uses answers with conditions (i.e., a setting, event, cause/effect) that may make the answer correct or incorrect. Note that Model C shows only one correct answer with its correct condition, and all options are similar in length.

Model D: a.   correct answer
                  b.   plausible incorrect answer
                  c.   correct answer with plausible incorrect condition
                  d.   plausible incorrect answer with incorrect condition

Model D is useful when it is not possible to create four options of similar length. This model shows paired lengths (two long and two short options), which prevents any one option from standing apart (either too long or too short) from the remaining options.

When using Models C or D, it is particularly important to maximize the plausibility of any incorrect conditions that appear in multiple distractors, in order to minimize the chances that examinees will be able to eliminate those distractors by detecting one piece of implausible information.

b.   Do not use "all of the above" or "none of the above."

"All of the above" questions provide inadvertent clues to the examinee. When the "all of the above" option is the correct response, the examinee must simply recognize that two options are correct to answer the question correctly. Similarly, when "all of the above" is used as a distractor, the examinee only needs to be able to determine that one option is incorrect in order to eliminate this option. "None of the above" responses should not be used with "best answer" multiple-choice questions, since it may always be defensible as a response.

c.   Do not present a collection of true-false (T/F) statements as a multiple-choice item.

As previously discussed, each item should focus on one K/A topic that is determined by the stem of the question. A question containing answer options related to many separate issues does not increase the efficiency of the question. To the contrary, questions with multiple topics only confuse the examinee about the meaning and purpose of the question.

A way of determining if you have a test item that is a collection of T/F statements is to check whether the answer can be determined or the distractors can be rejected without the information contained in the stem. If so, you likely have a question that is a T/F collection. Refer to Section F of Attachment 2 for sample questions that illustrate this psychometric deficiency.

d.   Define the question, task, or problem in the stem of the question.

In designing multiple-choice questions that are operationally based and require an application/use scenario, one suggestion is to provide the conditions in the first part of the question, separated by a double space from the body of the question, and blocked to the left column with each condition bulleted, as in the following example:

"Given the following conditions:

- Both main feed pumps tripped
- Auxiliary feedwater (AFW) automatically started
- AFW valves reset to control steam generator water level
- AFW suction pressure decreased to 7 psig

Which ONE of the following describes the AFW pump response for the given conditions?

a.   Suction will automatically shift to the nuclear service water system.
b.   Suction will automatically shift to the upper surge tank.
c.   Trip when suction pressure decreases to 5 psig.
d.   Trip after a 6-second time delay."

Include as much necessary information as possible about the problem or situation in the stem, leaving only the solution, action, or effect for the answer options. Consider the following "poor" and "better" examples:

(Poor) "At 50% power—

a.   The equilibrium xenon reactivity worth is approximately equal to the equilibrium xenon worth at 100% power.
b.   The equilibrium xenon reactivity worth is approximately one-half the equilibrium xenon worth as 100% power.
c.   The equilibrium xenon reactivity worth is approximately two-thirds the equilibrium xenon worth at 100% power.
d.   The equilibrium xenon reactivity worth is approximately three-fourths the equilibrium xenon worth at 100% power."

(Better) "How does the equilibrium xenon reactivity worth compare to the equilibrium xenon reactivity worth at 100% power?

a.   equal to
b.   one-half
c.   two-thirds
d.   three-fourths"

e.   When possible, avoid using negatively stated stems. If a negative stem is necessary, highlight the negative word (e.g., *not, never, least*).

It is very tempting to write negatively stated questions, since they can be constructed by picking three true statements out of the reference material and changing a fourth statement to make it false. However, studies have shown that examinees do not do as well on negatively stated questions, because they overlook the negative word and/or because negatively stated questions require examinees to select an answer that is not true or characteristic, which can be somewhat confusing. In addition, these questions tend to emphasize negative learning. For example, consider the following stem of a multiple-choice question:

> "During 100% power operation, the feedwater heater 2A high level dump valve opens inadvertently. The condensate pumps will not do which of the following:"

This stem can be made to read positively:

> "During 100% power operation, the feedwater heater 2A high level dump valve opens inadvertently. The condensate pumps will:
>
> a.   increase flow to maintain feedwater flow rate
> b.   trip due to a runout condition
> c.   have no response
> d.   trip due to low suction pressure"

Although a negatively stated question is sometimes unavoidable, *never* use a negatively stated stem with a negatively stated answer option, as illustrated by example E.3 in Attachment 2.

f.   Provide sufficient counterbalance in questions with multi-part answers.

Multiple-choice questions can legitimately contain multi-part answer options. However, if the answers contain too many parts and/or too many options for each part, cues indicating the correct answer may be unavoidable. Consider the following example:

> "The reactor coolant system (RCS) is in hot standby with no reactor coolant pumps (RCPs) running. If the once-through steam generator (OTSG) pressure is decreased, according to the plant verification procedure, which of the following temperature responses indicates the presence of natural circulation?
>
> a.   T-H increases, T-C remains the same
> b.   T-H increases, T-C decreases
> c.   T-H decreases, T-C decreases
> d.   T-H remains the same, T-C decreases"

The examinee could choose the correct answer (c) without knowing about the T-C temperature response in this situation, since "T-H decreases" only occurs in option "c."

Notice that two-part answers, with each part containing a two-option response, provide complete counterbalance, since all contingencies can be covered in four responses, as in the following example:

"Which of the following is a definition of quadrant power tilt ratio (QPTR)?

a. minimum upper detector output divided by average upper detector output
b. maximum upper detector output divided by average upper detector output
c. minimum upper detector output divided by average lower detector output
d. maximum upper detector output divided by average lower detector output"

A highly recommended multi-part question format is one in which the two-part answer options consist of a two-level response (e.g., yes/no; off/on) and a reason, as in the following example:

"Which of the following describes the behavior of equilibrium xenon reactivity over core life?

a. It decreases because of the increased fuel burnup.
b. It decreases because of the decrease in plutonium-xenon yield.
c. It increases because of the increase in thermal flux.
d. It increases because of the decrease in boron concentration."

Sometimes, in an effort to improve their plausibility, distractors may include secondary pieces of information that have lower relative importance and discriminatory value than the key point of the distractor. However, those secondary pieces of information are not irrelevant; the value of the question should be considered as a whole and not discounted because the answer choices contain information of lower importance.

g. When possible, include common misconceptions as distractors. Since the purpose of the examination is to differentiate between competent and less-than-competent examinees, a good source of questions involves topics in which there are common misconceptions about important K/A topics. For example, the following question was based upon a common misconception about loss of subcooling margin:

"During a small-break loss-of-coolant accident (LOCA) with a resultant loss of subcooling margin, why are the RCPs secured?

a. to prevent pump damage resulting from operation under two-phase conditions
b. to prevent core damage resulting from rapid phase separation upon subsequent loss of RCS flow
c. to reduce RCS pressure by removing the pressure head developed by the RCPs
d. to remove the heat being added to the RCS by the operating RCPs"

h.   Make all answer options homogeneous and highly plausible.

Consider the following "poor" and "better" examples:

"On a loss of condenser circulating water intake canal, the upper surge tank, hotwell, and condensate storage tank will supply sufficient feedwater to allow decay heat removal for approximately:

	Poor	Better
a.	15 minutes	8 hours
b.	8 hours	24 hours
c.	48 hours	48 hours
d.	3 months	72 hours"

Notice how one method of changing the difficulty level of a question is to vary the similarity of the answer options. The distractors should be similar enough to be chosen by those who do not meet the testing objective, yet different enough so they do not test trivial issues or distinctions. Also note how the answer options in each case are listed in order of magnitude.

i.   If the answer options have a logical sequence, put them in order (as in "h," above).

j.   Avoid overlapping answer options, as in the following example:

"The self-powered neutron detector (SPND) uses rhodium, which decays with a half-life of 42 seconds. How long will it take for a detector to indicate approximately 95% of an instantaneous power level change?

	Poor	Better
a.	2 to 4 minutes	1 to 2 minutes
b.	4 to 6 minutes	3 to 4 minutes
c.	6 to 8 minutes	5 to 6 minutes
d.	8 to 10 minutes	7 to 8 minutes"

k.   Do not include trivial distractors with more important distractors.

In the search for distractors, it is very tempting to include relatively trivial facts along with options that focus on more important issues or concepts, as in the following example:

"Which of the following is true concerning the turbine?

a.   The turbine is rotated at low speed when shut down in order to prevent distortion of the turbine casing.
b.   Turbine eccentricity is the measure of turbine speed.
c.   The turbine blades are cooled by hydrogen gas.
d.   Technical specifications require that at least one turbine overspeed protection system must be operable in Mode 2."

Relative to the other options, option "c" could be considered a trivial distractor. Even if included as an incorrect answer, relatively unimportant information jeopardizes the content validity of the question. Also, note that this question consists of a collection of true/false statements as described in Section C.2.c.

l.   Vary the position of the correct answer; avoid a pattern.

Make sure the position of the correct answer is randomized throughout the examination. This means that options "a," "b," "c," and "d" should be correct about an equal number of times, but in no specific order.

m.   Avoid "specific determiners" that give clues as to the correct answer. Specific determiners include the following:

(1)   distractors that do not follow grammatically from the stem, as in the following example:

"During 100% normal power operation, a single steam flow element in the steam generator feedwater control system fails high. This will cause—

a.   the feedwater valves to increase steam generator level slightly before returning the level to normal
b.   before returning the level to slightly above normal, the feedwater valves to increase the steam generator level significantly
c.   the feedwater valves to increase the steam generator level to the level of a reactor trip
d.   the feedwater valves to increase the steam generator level slightly and maintain the increased level"

Note the improvement when distractor "b" is reworded as follows:

"b.   the feedwater valves to increase the steam generator level significantly before returning the level to slightly above normal"

(2)   options that can be judged correct or incorrect without reading the stem

(3)   equivalent and/or synonymous options, which rule out both options for an examinee who recognizes the equivalence

(4)   an option that includes another option [for example: (a) less than 5; (b) less than 3...]

(5)   implausible distractors

(6)   a correct answer that is longer than the distractors

(7)   qualifiers in the correct answer (e.g., probably and ordinarily) unless they are also used in the distractors

(8)   words such as "never" or "always," which suggest an incorrect option

(9)   a correct option that differs from the distractors in favorableness, style, or terminology, as in the following example:

> "Which action or occurrence is likely to cause water hammer?
>
>   a.   maintaining the discharge line from an auto starting pump filled with fluid
>   b.   water collecting in a steamline
>   c.   pre-warming of steam lines
>   d.   slowly closing the discharge valve of an operating pump"

In the above question, all options except for "b" (the correct answer) describe preventive actions, while option "b" describes a condition that occurs as a result of negligence or oversight. A test-wise examinee would only need to know that water hammer is not a desired occurrence to determine that "b" is the least favorable and, therefore, the correct answer.

n.   When appropriate, use distractors that are generically correct statements, but do not correctly answer the question, as illustrated in the following example:

> Preparations are being made for refueling, and the following plant conditions exist:
>
> –   The refueling cavity is filled with the transfer tube gate valve open.
> –   The SFP LO LEVEL and CTMT SUMP HI LEVEL annunciators are in alarm.
>
> Which ONE of the following is the required IMMEDIATE ACTION in response to these conditions?
>
>   a.   Verify alarms by checking the containment sump level recorder and spent fuel level indication.
>   b.   Sound the containment evacuation alarm.
>   c.   Initiate containment ventilation isolation.
>   d.   Initiate control room ventilation isolation.

Answer "a" is a generic good practice, but it is not responsive to the conditions specified in the stem of the question. It is not a required immediate action, nor is it an appropriate response in light of the mutually confirmatory annunciators that are in alarm.

3. **Reviewing Test Items**

Examination reviewers can assist examination authors by performing technical content, level of difficulty, psychometric, and editorial checks, and it is advantageous to consider each of these four areas separately and in this order. If there is a need to revise an item on the basis of one stage of the review, the changes should be made before going further because the changes at each stage could well affect the subsequent reviews. For example, a criticism that appears to affect only one distractor may ultimately lead to changes in other parts of the item, so time spent reviewing the item for grammar and punctuation may be wasted.

There are also some advantages associated with having the questions reviewed for clarity, grammar, expression, spelling, and punctuation by someone who is not familiar with the area being tested. Such a reviewer can determine whether an item can be correctly answered by a person without knowledge of the field.

The examination author and reviewers should ask themselves the following types of questions:

- Will the examinees clearly know what they are expected to do?
- Do they have all the information they need to work with?
- Does answering the question depend on certain assumptions that must be stated?

Attachment 1 presents a more thorough list of suggestions for examination authors and reviewers.

D. Attachments/Forms

Attachment 1,   "Question Development Checklist"
Attachment 2,   "Examples"
Attachment 3,   "References"

1. Does the concept being measured have a direct, important relationship to the ability to perform the job?

2. Does the question match the testing objective and intent of the K/A?

3. Is the question clear, concise, and easy to read? Could it be stated more simply and still provide the necessary information? Should it be reworded or split up into more than one question?

4. Is each question stated positively, unless the intent is to test knowledge of what not to do?

5. Does the question provide all necessary information, stipulations, and assumptions needed for a correct response? Does the stem include as much information as possible?

6. Is the question written at the highest appropriate level of knowledge or ability for the job position of the person being tested?

7. Is the question free of unnecessary difficulty, trickiness, or irrelevance?

8. Is the question limited to one concept or topic, making it something other than a collection of true-false items?

9. Does the question have face validity?

10. Are key points underlined or highlighted?

11. Is each question separate and independent of all other questions?

12. Are the answer options homogeneous and highly plausible? Are common misconceptions used as distractors? Is the question free of trivial distractors?

13. Are "none of the above" and "all of the above" avoided?

14. Are there four answer options for each question?

15. Are the answer options of the questions ordered sequentially?

16. Is the question free of "specific determiners" (e.g., logical or grammatical inconsistencies, incorrect answers that are consistently different, verbal associations between the stem and the answer options)?

A.   Levels of Knowledge

The first three examples illustrate how the level of knowledge tested can vary among
a series of questions that focus on the same pair of K/As. Even though the K/A statements
use verbs (identify, define) that elicit a fundamental or simple memory level of knowledge,
the item author can increase its operational validity by testing at a higher cognitive level.

- 191004K101 (PWR) or 291004K101 (BWR):   Identification, symptoms, and consequences of cavitation

- 193006K111 (PWR) or 293006K109 (BWR):   Define or explain cavitation

1.   Fundamental Knowledge/Simple Memory

"Which one of the following describes pump cavitation?

a.   Vapor bubbles form when the enthalpy difference between pump discharge
      and pump suction exceeds the latent heat of vaporization.
b.   Vapor bubbles form in the eye of the pump and collapse as they enter
      higher-pressure regions of the pump.
c.   Vapor bubbles are produced when the localized pressure exceeds
      the vapor pressure at the existing temperature.
d.   Vapor bubbles are discharged from the pump, where they impinge
      on downstream piping and cause a water hammer."

This question simply asks for a description of cavitation and, as such, is a "low cognitive
order" question that does not require any understanding, analysis, or problem-solving.
The examinee merely needs to recognize the correct description (b); the other options
appear plausible but are, nonetheless, incorrect.

2.   Comprehension

"Cavitation in an operating pump may be caused by—

a.   lowering the pump suction temperature
b.   throttling the pump suction valve
c.   increasing the pump back-pressure
d.   increasing the pump suction pressure"

This example requires the examinee to determine causation, which requires
an understanding of the correct answer and a recognition that the incorrect answers
are indeed, incorrect. As with any item, the quality of this item is determined by
the distractibility of the incorrect options.

3.   **Analysis**

"While on surveillance rounds, an operator notices that a centrifugal pump is making a great deal of noise (like marbles rattling inside the pump casing) and the discharge pressure is fluctuating. This set of conditions indicates pump—

a.   runout
<u>b</u>.   cavitation
c.   bearing deterioration
d.   packing deterioration"

This example requires the candidate to analyze multiple abnormal indications (multiple effects) for an operating centrifugal pump and determine the cause (complex cause-effect). All the distractors are initially plausible in that they have face validity (i.e., they have reasonable connections to centrifugal pump operation).

4.   **"Low Level of Knowledge" Questions**

The following four examples illustrate *"low level of knowledge"* questions, which should be used judiciously on NRC examinations.

"Which one of the following is powered from 4160 VAC bus 1A?

<u>a</u>.   Residual heat removal (RHR) pump A
b.   RHR pump B
c.   RHR pump C
d.   RHR pump D"

"Select the full core display indication of a drifting control rod.

<u>a</u>.   red light
b.   white light
c.   blue light
d.   amber light"

Although the above items have a high K/A value, they are written at a low level of knowledge and also have low operational validity and low discriminatory value. The following question tests at a low level of knowledge because it does not test the examinee's ability to recognize the class of fire or select the correct extinguisher. All the examinee has to know is that water is used for class A fires.

"Concerning use of water as a fire extinguishing agent, select the correct statement from the following:

a. It is the primary agent for extinguishing Class A fires and is also effective on Class B and C fires.
b. It is the primary agent for extinguishing Class B fires and is also effective on Class A and C fires.
c. It is the primary agent for extinguishing Class A and B fires but is not effective on Class C fires.
d. It is the primary agent for extinguishing Class B and C fires but is not effective on Class A fires."

The next question might be considered a fundamental knowledge level question that errs in the opposite direction. That is, it could be too difficult *unless* the operators are expected to memorize the correct time requirement in order to preclude damage to equipment. Moreover, this item may also have low discriminatory validity unless at least 80 percent of the examinees are expected to know the answer from memory.

"RCP 2A tripped after running for 50 minutes. The RCP was restarted, but tripped again within 15 seconds. Which ONE of the following is the minimum required interval before the next attempt to start RCP 2A?

a. 15 minutes
b. 30 minutes
c. 45 minutes
d. 60 minutes"

B.   Low Operational Validity

The next three questions illustrate another common psychometric deficiency, known as low operational validity, which should be avoided on NRC examinations.

1. "Under which one of the following conditions should the shift supervisor inform the shop steward?

   a. initiation of a directed overtime request
   b. disciplinary action against a supervisory employee
   c. medical injury of a contractor employee
   d. personnel error by a bargaining unit member"

While this question may be related to a shift supervisor's job, it has nothing to do with nuclear safety and should not be included on an NRC examination.

2.   "Which one of the following main steam line components is designed to limit the differential pressure across the steam dryer assembly?

       a.   main steam line flow elbows
       b.   main steam isolation valves
       c.   main steam shutoff valves
       <u>d</u>.   main steam line flow restrictors"

Knowing the purpose of a flow restrictor is not a good indicator of the operator's ability to operate the plant. Thus, knowing the answer to this question is not clearly job-related.

3.   "Given that all components controlled by the 'Locked Valve, Breaker, and Component Control' administrative procedure must be properly sealed and tagged, which one of the following is the correct location for the 'XXXX-XXXX' tag for an electrical breaker?

       a.   wired to the breaker handle
       <u>b</u>.   glued to the breaker cubicle
       c.   attached to the breaker cubicle with a magnetic clip
       d.   wired to the breaker cabinet door"

This question is likely unrelated to the reactor operator's job function and, therefore, would be unacceptable.

C.   Low Discriminatory Validity

The next three questions illustrate another common psychometric deficiency, known as low discriminatory validity, which should be avoided on NRC examinations.

1.   "Which one of the following reactor water levels will initiate the RHR pumps?

       <u>a</u>.   level 1 only
       b.   level 1 and 2 only
       c.   level 1 and 2 and 3 only
       d.   level 6 only"

The information in this question should be known by all operators at all times. Therefore, the question has low discriminatory value and also tests at a low level of knowledge.

2.   "The plant is recovering from a scram that resulted from a spurious Group I isolation. The cause of the isolation has been repaired and preparations are being made to reopen the main steam isolation valves (MSIVs). Reactor pressure is currently 825 psig and the main steam lines are being pressurized.

WHICH ONE (1) of the following is the LOWEST main steam line pressure that will allow the MSIVs to be opened in accordance with the procedure?

     a.   625 psig
     <u>b</u>.   675 psig
     c.   725 psig
     d.   775 psig"

This question does not discriminate and has low operational validity because, in real life, the applicant may not be expected to have memorized the procedure.

3.   "SG (corrected) = SG (uncorrected) + $\dfrac{(T - 77 \text{ degrees F})(.001)}{3}$ + (Level Mark)(.003)

Based on the above information, the specific gravity (SG) is ___?___, which ___?___ meet the Technical Specification (TS) Category A limit. Note: This question requires the use of TS 3.8.2.3.

     a.   1.198, does NOT
     b.   1.195, does NOT
     c.   1.207, does
     <u>d</u>.   1.201, does"

This question might appears to test the examinees' ability to understand and apply battery parameters to the determination of TS operability. However, the question really only tests their ability to substitute certain parameters into a given equation and perform an arithmetic calculation. Reference to the TS noted in the question is not required based on the three different values of SG (corrected) supplied as distractors. Therefore, the question has a low discriminatory value because any individual possessing adequate arithmetic knowledge will arrive at the correct answer.

D. Implausible Distractors

The next two questions illustrate the concept of implausible distractors, which is another common psychometric deficiency that should be avoided on NRC examinations.

1. "Which of the following will cause the RHR pumps to start during a design-basis LOCA?

    a. low drywell pressure
    b. high reactor water level
    c. high drywell pressure
    d. MSIVs in the NOT OPEN position"

    Distractors "a," "b," and "d" are implausible, considering minimal knowledge of the plant response to a LOCA.

2. "Which ONE of the following conditions will NOT result in a shutdown of the standby gas treatment system (SBGTS)?

    a. manual shutdown
    b. high-temperature (107 °C, 225 °F) charcoal bed
    c. high-temperature (82 °C, 180 °F) heater inlet
    d. overloads in the local control panel"

    Distractor "a" is very implausible, and distractor "d" is subjective. The question is also written from a negative perspective.

E. Confusing Language

The following questions illustrate how confusing language and inappropriate negatives in the stem of the question can mislead examinees. Such questions should be avoided on NRC examinations.

1. "Which one of the following parameters will start high-pressure coolant injection (HPCI), reactor core isolation cooling (RCIC), and the SBGTS?

    a. low reactor water level
    b. high primary containment pressure
    c. high reactor building exhaust radiation
    d. low reactor building differential pressure"

    This question could result in four correct answers, since the question could be interpreted individually or collectively.

2.   "Which ONE of the following most accurately describes the response to a static inverter failing.

    a.   The power supply will automatically transfer to the alternate 600 V Bus 2C / Vital AC Transformer 2A.

    b.   The 125 VDC battery will maintain power to the Vital AC Cabinet for up to 5 hours.

    c.   The power supply can be manually transferred to the alternate 600 V Bus 2C / Alternate Static Inverter by pressing a transfer pushbutton.

    <u>d.</u>   The power supply can be manually transferred to the alternate 600 V Bus 2C / Vital AC Transformer 2A by positioning the transfer switch to ALTERNATE."

This question implies an automatic response, but the listed correct answer and one distractor are operator actions, not responses to the loss of the static inverter.

3.   "Regarding temporary plant alterations (TPAs), technical reviews are NOT required for—

    a.   a TPA NOT installed using an approved procedure

    b.   TPAs installed on BOP systems BUT ARE required for safety related systems

    c.   a TPA that has NOT been directed by the shift supervisor to be an emergency TPA

    d.   all TPAs directed by the shift supervisor."

This question contains multiple problems. (1) While negative questions can be used, they should be used for good reason; there appears to be no good basis for asking this question negatively. (2) Two of the distractors ("a" and "c") also contain a negative, creating a double negative with readability confusion, a violation of good item writing practice. The question should more appropriately ask the conditions under which technical reviews are required, thereby eliminating the negative in the stem.

F.   Collections of True/False Statements

Collections of true/false statements typically only test simple rote memory; the examinee simply needs to recall a definition or condition. The questions elicit no comprehension or problem-solving; hence, they lack operational validity. This type of question allows an examinee to answer the question without referring to the stem of the question and should be avoided on NRC examinations.

1.   "Which ONE of the following is true?

    <u>a.</u>   High drywell pressure will auto-start the emergency diesel generators.

    b.   Low reactor water level will trip the main turbine.

    c.   High reactor pressure will initiate RCIC.

    d.   High reactor power with the mode switch in startup will NOT close the MSIVs."

2.   "Which one of the following describes pump cavitation?

       a.   Vapor bubbles form when the enthalpy difference between pump discharge and pump suction exceeds the latent heat of vaporization.
       <u>b</u>.   Vapor bubbles form in the eye of the pump and collapse as they enter higher-pressure regions of the pump.
       c.   Vapor bubbles are produced when the localized pressure exceeds the vapor pressure at the existing temperature.
       d.   Vapor bubbles are discharged from the pump, where they impinge on downstream piping and cause a water hammer."

G.   Backward Logic

Backward logic questions ask the examinee for information that is normally received, while providing the examinee with information that he or she normally has to supply. In an operational setting, operators are faced with conditions and required to know what procedure(s) to use. These questions ask them to do just the opposite and, therefore, should be avoided on NRC examinations.

1.   "Which of the following parameters will <u>simultaneously</u> start HPCI, RCIC, and SBGTS?

       a.   high RPV water level
       <u>b</u>.   high drywell pressure
       c.   low RPV water level
       d.   low drywell pressure"

     It would be better to select a parameter and then request the expected system response because that is more operationally relevant.

2.   "If it takes 0.354 cubic meters (12.5 cubic feet) of concrete to build a square loading pad that is 6 inches thick, what is the length of one side of the pad?"

     This question gives the examinees information they should be asked to calculate, while it requires them to provide information they would be given in an actual work situation.

1.   Bloom, Benjamin, *Taxonomy of Educational Objectives: The Classification of Educational Goals, Handbook I: Cognitive Domain*, David McKay Company, New York, 1956.

2.   Isaac, Stephen, *Handbook in Research and Evaluation, 2nd ed.*, Edits Publishers, San Diego, 1990.

3.   Novac, J., and D. Gowin, *Learning How to Learn*, Cambridge University Press, 1993.

APPENDIX C
JOB PERFORMANCE MEASURE GUIDELINES

A.   Purpose

This appendix provides a framework for preparing and evaluating job performance measures (JPMs) to ensure they are of appropriate substance and format for initial operator licensing and requalification examinations. The following elements are discussed in detail or attached for information:

- a basic procedure for developing new JPMs (Section B), including forms to document the JPMs and to assess the quality of the product (Form ES-C-1 and ES-C-2)

- guidelines for developing and using alternate path JPMs (Section C)

- a discussion of walk-through evaluation techniques (Section D)

Adhering to the concepts and guidelines discussed herein, in association with the specific operating test criteria cited in ES-301, "Preparing Initial Operating Tests," or ES-603, "Requalification Walk-Through Examinations," as applicable, will enhance the consistency and validity of the walk-through tests.

B.   Developing and Reviewing JPMs

This section addresses the major JPM components and instructions for their development. The instructions apply to both the initial and requalification examination programs, except as noted. Although they are written from the perspective of developing new JPMs, the instructions should also be referenced, as necessary, when modifying existing JPMs for reuse and when reviewing proposed JPMs for quality.

Select the systems and tasks to be evaluated during the walk-through portion of the operating test in accordance with the specific initial and requalification examination criteria in ES-301 and ES-603, respectively. If a JPM already exists for the selected task, it should be reviewed against the guidelines and criteria discussed herein to ensure that it is acceptable for use. If a new JPM is required to evaluate the selected system or task, prepare the JPM in accordance with the following basic steps and document the JPM using Form ES-C-1, "Job Performance Measure Worksheet," or equivalent. Form ES-C-2, "Job Performance Measure Quality Checklist," can be used to verify that the relevant criteria are satisfied.

1.   Specify Initial Conditions

Determine those system and plant conditions that would permit the task to be performed realistically. They should provide sufficient information regarding the status of the plant and system to facilitate task performance, without coaching the examinee. If the task is intended to be performed on the simulator, it is worthwhile to differentiate those specific initial conditions and system realignments that are necessary for the task to be performed as planned from those other general conditions that add realism and set the stage for performing the task but have no real bearing on its successful execution. Breaking down the initial conditions in such a manner will simplify the simultaneous administration of different tasks by two or more examinees.

All of the required operator actions preceding the starting point of the JPM should be completed unless a given action is purposely omitted as part of an alternate path JPM. If the JPM is intended to evaluate the examinee's ability to implement an alternate path (refer to Section C) within the facility licensee's procedural guidance, the initiating equipment or instrument failure should be reflected in the simulator initial condition specifications.

The JPM shall also include an *initiating cue* that provides the stimulus for the examinee to begin performing the task. When appropriate, the cue should clearly specify the desired endpoint for the task. For example, if it is desired for the examinee to start and load the emergency diesel generator, the cue should state the load at which the task will be considered complete. Alternate path tasks, as described in Section C, may have an actual endpoint different from that stated in the initiating cue.

The initial conditions and initiating cue may be duplicated on a separate sheet of paper so that they can be handed to the examinee. This is particularly helpful for tasks with detailed initial conditions or those that will be performed in high-noise areas. Take care to ensure that the initial conditions and initiating cue do not reveal the nature of any alternate path JPMs that are planned.

2. Identify References and Tools

The JPM shall identify those plant procedures that require task performance, as well as the procedures that provide guidance, directions, or standards for performing the task. When reviewing JPMs selected from the facility licensee's bank, it is important to ensure that the procedures identified in the JPM are still current.

The JPM shall also identify any special tools or equipment (e.g., a stop watch, wrench, fuse puller, or spool piece) that the examinee will need to perform the task. It is helpful to the examiner who will be administering the test if the JPM states the location(s) in which these items may be found. It is expected that any required tools will be readily available to the plant operators; they should not be staged specifically for the examination.

3. Develop Performance Criteria

The JPM should have meaningful performance requirements that will provide a legitimate basis for evaluating the examinee's ability to safely operate the system or the plant. Artificially subdividing existing tasks to generate new ones may dilute the value of the JPM to a point where it becomes meaningless.

In accordance with 10 CFR 55.45(a) and 55.59(a)(2)(ii), operating tests require operators and senior operators to demonstrate an understanding of and ability to perform necessary actions. Therefore, JPMs selected for the walk-through examination shall not test solely for simple recall or memorization. Although it was written in a style to address written examinations, refer to ES-602, Attachment 1, "Guidelines for Developing and Reviewing Open-Reference Examinations," when preparing JPMs as well. Although an operating test does not require every JPM to be alternate path or demonstrate detailed system understanding, simple one-step JPMs or JPMs that only require directly looking up the correct answer are not appropriate. JPMs that incorporate the testing of immediate actions steps from memory are acceptable. However, JPMs should not solely test immediate action steps, and should include testing additional steps or items that are not from memory.

The JPM shall identify specific *performance standards*, or check points, that will permit the examiner to evaluate successful progress toward completing the task in accordance with the procedural references. Detailed control and indication nomenclature and criteria (e.g., switch positions and meter readings) should be identified whenever possible, even if these criteria are not specified in the procedural step. The JPM should also note any *important observations* that the examinee should make while performing the task.

The JPM must clearly identify the *task standard* (i.e., the predetermined qualitative and/or quantitative outcome) against which task performance will be measured. Every procedural step that the examinee must perform correctly (i.e., accurately, in the proper sequence, and at the proper time) in order to accomplish the task standard shall be identified as a *critical step* and shall have an associated performance standard.

If there are any specific procedural restrictions on the sequence in which the steps are performed, they shall be clearly noted in the JPM.

4.   Develop Examiner Cues

The JPM shall identify appropriate *system response cues* so that the examiner can provide the examinee with specific feedback regarding the component and system reactions to the examinee's manipulations, especially those procedural steps that are identified as critical to task completion. The response cues are particularly important in the following situations:

- in-plant tasks that will be simulated because the examinee will not have available the normal indications (e.g., alarms, flow rates, temperatures, and pressures) that would be observed during actual task performance

- alternate path JPMs that require the examinee to perform auxiliary procedures when equipment or instrumentation fails during use

System response cues may not be necessary for those tasks that will be performed on the simulator.

To the extent that it is possible to anticipate incorrect actions that the examinees might take, it is beneficial to note the expected system response cues in the JPM as an aid to the examiner who will be administering and evaluating the task.

The JPM shall also identify any *additional cues or instructions* that the examiner might need to provide to the examinee in response to procedural steps for which the examinee will not be held accountable (i.e., those steps that have either already been performed or will be performed by other personnel in remote locations).

5. **Develop a Time Standard**

Every JPM shall identify an estimated average time for completing the task. The time should be measured from the moment that the examinee is read the initiating cue at the plant location in which an operator would normally be given the order to perform the specified task.

JPMs that are considered time-critical (i.e., those having a task standard that must be completed within a time period specified in a regulation or a facility commitment to the NRC) shall be uniquely identified and specifically validated. The facility licensee must agree that a failure to complete the task within the specified time will justify a failure of the given JPM.

C. Developing and Using Alternate Path JPMs

JPMs are intended to be tasks that an operator must be able to perform, which relate to the operator's particular job task analysis (JTA). Operators are frequently challenged to perform auxiliary procedures when equipment or instrumentation fails during use. Therefore, examinees are expected to be able to use alternative methods to perform tasks. Alternative paths are evaluated during an examination by incorporating malfunctions of instrumentation or components that require the examinee to perform actions other than those performed when a system responds normally.

JPMs in which malfunctions occur are used to provide a methodology to evaluate whether an examinee has the skills and knowledge at the level needed to safely operate the system. This type of JPM, called "alternate path," provides an excellent opportunity to observe how the examinees execute alternative paths within the wide spectrum of procedures under their cognizance that would not otherwise be examined. All alternate path JPMs should include the following five characteristics:

1. **Success Path**: Each JPM should have a valid, facility-endorsed success path. This path may require analyzing initial conditions to determine an alternative method for completing the task, mitigating a system-related problem that occurs during the task, or realigning the system.

2. **Procedurally Driven**: For each JPM, a procedure should address the actions that are required (i.e., if the JPM requires an alternative method to complete the task, the procedure would have an exit step that directs the use of that alternative method). The examinee may be required to use some common practices endorsed by the facility that are addressed through generic administrative procedures or policies (e.g., shifting controls to manual).

3. **Logical Sequence**: The sequence of procedurally driven actions should be logical. For example, an examinee performing a normal evaluation when a malfunction occurs should not be expected to enter emergency operating procedures (EOPs). More realistically, the examinee would attempt to correct the problem by referring to an annunciator response procedure (ARP) or abnormal operating procedure (AOP). However, an examinee performing a normal evolution may encounter a situation requiring a reactor trip. The JPM should not contain a cascading sequence of

malfunctions, for which several procedures must be used simultaneously, that occur while performing a task. This type of activity is better tested in the dynamic simulator portion of the examination.

4. **Independent of Crew Dynamics**: Each JPM should allow the examinee to complete the task or mitigate a problem that occurs during a task without having to rely on the actions of other control room operators. This provision does not prohibit simulator operators from acknowledging non-pertinent alarms or unexpected reactions of other systems that are not associated with the task. Also, the JPMs may still require the examinee to use the simulator operator to perform needed manipulations in the plant.

5. **Validated in Advance**: Each JPM should be validated before the examination begins and should not be changed thereafter. The JPM should not be a surprise to the examiners or simulator operators. Each JPM should be validated as early as possible before the examination is to be administered to allow time for changes to be made.

D. Walk-Through Evaluation Techniques

This guidance is intended to assist NRC examiners and facility evaluators in administering JPMs by illustrating good and bad examples of walk-through examination techniques.

1. **Providing Cues**

    Cuing refers to the information that an examiner provides to an examinee when conducting a JPM. When conducting JPMs on the simulator, the simulator provides most of the required cues. However, when conducting JPMs outside of the simulator, the examiner must provide realistic and timely information to the examinee.

    a. *Verbal Cues*

        Verbal cues are often required to provide relevant system information, such as valve position, meter deflection, or indicating light status. The examiner must be careful to provide the examinee with the indications that should be readily observed (e.g., "the red light just illuminated" or "the valve position indicator does not move"). An examiner can give too much information or inappropriate information (e.g., providing indications that are not visible or audible to the examinee) that could invalidate the JPM. The examiner must keep in mind what the examinee would see and hear while performing the JPM, and provide consistent cues.

    b. *Non-Verbal Cues*

        It is important to maintain a "poker face" when an examinee provides an incorrect response or performs the wrong procedural step. Voice inflections indicating something has been performed incorrectly, or changing the manner in which cues are given (e.g., talking more methodically, or rapidly) are examples of non-verbal communications that should be avoided.

Thorough preparation and familiarity with the JPM is vital to providing proper cuing. Knowledge of what indications will be available and how they will respond to the examinee's actions allow an examiner to give accurate and timely cues when an examinee is incorrectly performing the task.

2. Evaluation Skills

When evaluating an examinee, an examiner must have the ability to differentiate between what he or she knows or believes to be true about an examinee's ability and how the examinee actually performs on the JPM. As previously discussed, an examiner must be familiar with the JPM to be able to accurately evaluate performance. Errors made by the examinee performing the JPM may not be seen, or pertinent questions may not be asked, if the examiner has not prepared for the examination.

An examiner must remain attentive to the examinee's actions at all times. This will ensure that the examiner provides timely cues and detects errors in performance.

3. Exam Administration

While conducting the walk-through examination, the examiner must be aware of conduct that is appropriate for a trainer, but is inappropriate for an examiner. As a trainer, interacting with the examinee during the performance of the JPM to gain insight into what the examinee is thinking is a good practice. However as an examiner, this is distracting to the examinee and may inadvertently result in prompting or leading the examinee.

When conducting JPMs in the simulator, examiners should not manipulate any controls or silence/acknowledge any alarms. The examiner must take a "hands off" approach to maintain the proper testing environment.

The examiner must be careful to shield any notes or grading from the examinee to prevent giving an indication of performance, which may either provide a false sense of security or increase stress levels.

If an examinee's actions are not clear, the examiner must be prepared to ask appropriate followup or clarifying questions. Documenting these questions and the subsequent answers is important as they may have a bearing on an examinee's overall grade.

E. Attachments/Forms

Form ES-C-1,   "Job Performance Measure Worksheet"
Form ES-C-2,   "Job Performance Measure Quality Checklist"

Facility: _____   Task No: _____

Task Title: _____   Job Performance Measure No: _____

K/A Reference: _____

Examinee: _____   NRC Examiner: _____

Facility Evaluator: _____   Date: _____

Method of testing:

Simulated Performance _____   Actual Performance _____

Classroom _____   Simulator _____   Plant _____

Read to the examinee:

I will explain the initial conditions, which steps to simulate or discuss, and provide initiating cues. When you complete the task successfully, the objective for this job performance measure will be satisfied.

Initial Conditions:

Task Standard:

Required Materials:

General References:

Initiating Cue:

Time Critical Task: Yes/No

Validation Time:

Performance Information

Denote critical steps with a check mark

_____   Performance step:

Standard:

Comment:

_____   Performance step:

Standard:

Comment:

_____   Performance step:

Standard:

Comment:

Terminating cue:

Verification of Completion

Job Performance Measure No. _____

Examinee's Name:

Examiner's Name:

Date Performed:

Facility Evaluator:

Number of Attempts:

Time to Complete:

Question Documentation:

Question:_____
_____
_____

Response:_____
_____
_____
_____

Result: Satisfactory/Unsatisfactory

Examiner's signature and date: _____   _____

Every JPM should:

1. _____ be supported by the facility licensee's job task analysis.

2. _____ be operationally important (meet the NRC's K/A Catalog threshold criterion of 2.5 (3 for requalification exams) or as determined by the facility and agreed to by the NRC). JPMs shall not test only for simple recall or memorization (refer to ES-602 Attachment 1).

3. _____ be designed as either SRO only, RO/SRO or AO/RO/SRO.

4. include the following, as applicable:

    a. _____ initial conditions

    b. _____ initiating cues

    c. _____ references and tools, including associated procedures

    d. _____ validated time limits (average time allowed for completion) and specific designation of those JPMs that are deemed to be time-critical by the facility operations department

    e. _____ operationally important specific performance criteria that include:

        (1) _____ expected actions with exact control and indication nomenclature and criteria (switch position, meter reading), even if these criteria are not specified in the procedural step

        (2) _____ system response and other cues that are complete and correct so that the examiner can properly cue the examinee, if asked

        (3) _____ statements describing important observations that the examinee should make

        (4) _____ criteria for successful completion of the task

        (5) _____ identification of those steps that are considered critical

        (6) _____ restrictions on the sequence of steps

APPENDIX D
SIMULATOR TESTING GUIDELINES

A.   Purpose

This appendix provides a framework for preparing and evaluating simulator scenarios to ensure that they are of appropriate scope, depth, and complexity for the NRC's initial operator licensing and requalification examinations. Specifically, this appendix includes detailed discussions or attachments concerning the following elements:

- a basic procedure for developing new simulator scenarios (Section B), including a description of the associated qualitative and quantitative attributes (Section C) and the critical task (CT) methodology (Section D)

- the competencies in which reactor operators (ROs) and senior reactor operators (SROs) are expected to be proficient (Section E)

- the simulator security considerations that should be kept in mind during scenario validation and administration (Section F)

- selected examples of initial and requalification scenarios (Attachments 1 and 2)

Adhering to the concepts and guidelines discussed herein, in association with the specific criteria cited in ES-301, "Preparing Initial Operating Tests," or ES-604, "Dynamic Simulator Requalification Examinations," as applicable, will enhance the consistency and validity of the dynamic simulator operating tests.

B.   Integrated Scenario Development

This section summarizes the major activities that contribute to the development of dynamic simulator scenarios. The instructions apply to both initial and the requalification examination programs, except as noted. Although they are written from the perspective of new scenario development, the instructions should also be referenced, as necessary, when modifying existing scenarios for reuse and when assessing the quality of proposed scenarios.

1.   Identify Scenario Objectives

A scenario should have specific objectives. For a requalification examination, these should derive, in part, from the facility's requalification training program objectives. However, Title 10, Part 55, of the *Code of Federal Regulations* (10 CFR Part 55) requires that the initial licensing and annual requalification operating tests include a comprehensive sampling of items (2) through (13) in 10 CFR 55.45. Therefore, both tests should sample the various operating skills and abilities that the NRC requires for licensing an operator and the operating crew. Thus, it is not sufficient to limit a requalification examination to topics covered in the requalification cycle.

The basic objective of a scenario should be to evaluate the operators' ability to respond to events that are most appropriately tested in a dynamic simulator environment. Specifically, such events include those that require the operators to demonstrate their knowledge of integrated plant operations, as well as their ability to diagnose abnormal plant conditions and work together to mitigate plant transients that exercise their knowledge and use of abnormal and emergency operating procedures (AOPs and EOPs). Additionally, the scenario should require the operators (usually the SROs) to utilize technical specifications (TS) and, for requalification examinations, to implement the emergency plan. Section E of this appendix discusses the full range of competencies in which operators must demonstrate proficiency during the simulator test.

2.   Select Initial Conditions

The initial conditions established for a dynamic simulator operating test must allow each scenario to commence realistically. In other words, the initial conditions should be representative of a typical plant status, with various components, instruments, and annunciators out of service. It is also realistic to have maintenance or surveillance activities in progress. All, some, or even none of these initial conditions may have a bearing on subsequent scenario events. Initial conditions should also be frequently changed, to prevent future events from becoming predictable. In addition, initial conditions should be varied among the scenarios and should include startup, low-power, and full-power situations.

Briefly describe the initial conditions, including any items that should be addressed during the shift turnover, in the space provided at the top of Form ES-D-1, or equivalent.

3.   Select and Document Events

After establishing the initial conditions, select a sequence of events designed to achieve the stated objectives. Section C discusses a number of qualitative and quantitative criteria that should be considered when selecting events. The specific requirements for each quantitative criterion are enumerated in ES-301 and ES-604, as applicable.

Each event should have or contribute to an objective, whether it is to evaluate the operators' knowledge of a recent system modification, their ability to respond to a safety-significant event, or their use of the TS for a particular safety-related component. Uncomplicated events that require no operator action beyond the acknowledgment of alarms and verification of automatic actions provide little basis for evaluating the operators' competence and should not be included on the operating test unless they are necessary to set the stage for subsequent events.

The scenarios should be developed so that various systems are affected by each type of event (i.e., normal evolutions, instrument failures, component failures, and major plant transients). Having one equipment failure cause or exacerbate another can also be useful to evaluate the operators' understanding of system and component interactions. Balancing the severity of events and the demands they place on each operating position (e.g., RO and BOP) will allow each operator to demonstrate his or her competence across a range of conditions.

All events do not have to be linked; that is, one event need not occur for the next event to logically occur (although in many instances, such a relationship adds to the credibility of the scenario). However, the scenario should not consist of a series of totally unrelated events. A well-crafted scenario should flow from event to event, giving the operators sufficient time in each event to analyze what has happened, evaluate the consequences of their action (or inaction), assign a priority to the event given the existing plant conditions, and determine a course of action. Exercise care that one event does not fully mask the symptoms of another because the operators could overlook the malfunction and cause the event or competency coverage for the scenario set to be deficient.

Record each planned operation, malfunction, and transient on Form ES-D-1 and number them sequentially. Cross-reference each event to a simulator malfunction number, if applicable, or briefly describe the simulator instructions that must be entered.

For each event listed on Form ES-D-1, prepare a Form ES-D-2, "Required Operator Actions" (or equivalent), by entering the scenario, event, and page numbers and a brief description of the event at the top of the form. Each event description should include when it is to be initiated (e.g., by signal of the lead examiner/evaluator, time line, or plant parameter). The form shall also identify the symptoms or cues that the operators will be provided, the expected actions to be taken, communications to be made, the references to be used by each operating position (e.g., the SRO, RO, and BOP operators) on the crew, as well as the event terminus (i.e., the anticipated point at which the examiners or evaluators will have enough information on operator performance to move on to the next event).

Every required operator action should be included on Form ES-D-2; this is particularly important for the critical tasks (refer to Section D, "Critical Task Methodology") and other verifiable actions and behaviors that will provide a useful basis for evaluating the operators' competence. All CTs shall be flagged in a manner that makes them apparent to the individuals who will be administering the operating test (e.g., by using underlines, asterisks, or bold type), and the measurable performance indicators shall be identified. When possible, set points and other parameters should be included to provide an objective method for evaluating the operators' performance. Statements such as "Performs actions in accordance with Procedure XXXXX" generally do not provide sufficient guidance and are inadequate. However, the statement "Performs actions of steps XXX of Procedure XXX (attached)" is acceptable.

Although the expected actions should, to the extent possible, be listed in chronological order, certain actions may be required throughout the event (for instance, if a safety or relief valve fails open, the operators should continually monitor pressure and water level). Flag these actions to show that they are continuous.

The expected actions on Form ES-D-2 should be widely spaced to leave room for notes to document the operators' performance during the simulator test. The far-left column of the form should also be left blank so that it may be used to record the actual time at which key actions occurred while giving the test.

4. Determine the Scenario Endpoint

The last operator action sheet (Form ES-D-2) in the scenario should specify the endpoint of the scenario by identifying a particular plant condition, procedural step, or other point that is clearly recognizable. The scenario should not be terminated until the stated objectives have been achieved.

5. Validate the Scenario

Every scenario should be validated to ensure that it will run as intended. If a previously validated scenario is being modified slightly, real-time validation may not be necessary. However, if there are major changes or if someone questions the validity, revalidation in real-time is recommended.

C. Scenario Attributes

All valid scenarios contain common elements that make the scenarios useful as evaluation tools. A properly constructed scenario provides for an accurate test of each individual operator's skills and abilities as well as an opportunity to evaluate the crew members' team-dependent skills and abilities. Each scenario should be of sufficient scope and complexity to demonstrate the difference between competent operators and crews and those that are not performing at an acceptable level. It also should require the crew to demonstrate its ability as a team to adequately protect the public health and safety in emergency conditions, using the facility's EOPs.

Scenario attributes can be characterized as both qualitative and quantitative. No single qualitative or quantitative attribute or group of attributes can be used to determine the acceptability of a scenario. However, a trained examiner should be able to assess of the adequacy of a scenario or develop a new scenario, using both sets of attributes. This assessment, combined with validation of the scenario on a real-time basis, should be sufficient to determine whether a scenario is an acceptable tool for use in measuring the competency of a crew and/or its individual members.

1. Qualitative Attributes

a. *Realism/Credibility*

Introducing unrealistic or incredible events into a scenario can affect the validity of the scenario and provide negative training. Piping, component, and instrument failures often occur in such a way that deterioration can be tracked over a discrete time period (e.g., a small leak that propagates over time or a pump failure preceded by a high-vibration condition). Including such precursors in scenarios is important, where appropriate. A great deal of evaluative feedback can be obtained by observing how an operator or crew responds to a gradually worsening condition. A good technique inserts an event precursor (e.g., a small steam generator tube leak) and maintains the plant at a slightly degraded condition to observe how the crew incorporates that condition into its conduct of subsequent plant operations.

Although scenarios may include faults that occur with little or no warning (e.g., valve operators fail, fires occur in breakers or transformers, undetected pipe erosion results in piping failures), such faults often provide minimal evaluative benefits because they happen so suddenly that operators have little to do but watch the event unfold. These events are most useful when trying to establish a plant condition for subsequent evaluation goals or to assess the ability of an operator or crew to use procedures in a symptom-based (rather than an event-based) mode.

Mechanistic component failures are well-documented events that occur each year and often in multiple numbers. However, non-mechanistic failures (e.g., pipe breaks) generally occur singularly; therefore, unless there is a connective precursor, such as a seismic event, it would not be realistic or credible to have several piping systems fail during any one scenario.

Simulated events that appear to violate the laws of physics and thermodynamics contribute to negative training and are to be avoided. Time compression techniques, which are discussed later, may also contribute to negative training. However, if the intent of a scenario is to evaluate a crew's ability to execute procedural steps that may take a long time to reach during an event (e.g., hydrogen generation during a core uncovery event), such a technique may be useful. In such instances, the scenario must contain a cue that, when the crew detects the indications for such events, they are informed that the parameters are not responding as expected for the actual plant and that time is being compressed. This cue should be presented at the first opportunity that does not distract the crew from responding to available indications and before the crew challenges the validity of those indications. For example, in the first PWR scenario (Attachment 1), the cue should be given following the crew's determination that a reactor coolant system (RCS) feed and bleed may be necessary (in accordance with FR-H.1), but prior to steam generator levels requiring initiation.

b.   *Event Sequencing*

The sequence of events has a major effect in establishing the complexity of a simulator scenario. The pace at which malfunctions are entered can also adversely affect the way an operator or a crew responds.

Malfunctions may be entered simultaneously at separate control panel locations, provided that an individual applicant can handle each event without requiring extensive assistance.

Too short a time between malfunctions may mask the effects of a particular malfunction and divert the operators' attention. This cuts short the observers' ability to evaluate the operators' response to the earlier malfunction and may be prejudicial to a fair evaluation. Conversely, extending the time between malfunctions so that no operator activity is in progress may cause undue stress. During an examination, the operators expect something to occur; too much time between events should be avoided.

Therefore, the insertion of malfunctions in the scenario should be carefully timed. Rigorously following a planned time sequence of events is often less valid than initiating malfunctions on the basis of plant parameters or operator actions. The appropriate sequencing of events relates directly to the objectives of the scenario.

Event sequencing may involve time compression to speed up the response of key parameters so that the scenario can proceed to the next event within a reasonable time. Time compression may be accomplished by adjusting parameter indications or accelerating plant behavior characteristics so that plant indications trigger an event more quickly than would typically occur in reality (e.g., opening a drain path from a steam generator that is not noticeable to the operator so that the simulation reaches the entry conditions for a loss of heat sink.) This method is acceptable as long as the time compression allows the operators time to perform tasks that they would typically perform during the period in which time is compressed. To avoid wasting the operators' time determining the validity of their indications, the examiner should inform the crew before the scenario begins that time compression may be used during an event, and should debrief them after the scenario to minimize the potential for negative training.

Frequently, important evaluative benefit is gained in terms of safety significance by having key components or instruments fail after entering the EOPs. This process compels the operators to respond immediately to a safety-related situation by taking alternative actions to mitigate the event. This process also allows for a better evaluation of the operators' overall knowledge of plant procedures and systems because the event must be incorporated into the mitigation strategy for the remainder of the scenario. Conversely, instrument and component failures that are initiated after the major transient sometimes require little action and may provide little insight into the operators' competence.

c.   *Simulator Modeling*

The scenario should not exceed the limits of the facility licensee's configuration management system by altering a simulator model to obtain a desired effect. For example, it is not appropriate to increase the post-trip decay heat input in order to maximize internal core temperatures during a loss of cooling event; the simulator model should be allowed to perform as designed. The scenario may simulate events for which a simulator malfunction does not exist by using overrides or remote functions for local operator actions. An example would be failing indicators to simulate an inoperable component.

d.   *Evaluating Competencies*

Each scenario set shall ensure that all of the rating factors within each competency can be evaluated; moreover, the scenario must incorporate events that will allow an unsatisfactory evaluation of an operator or crew in a particular rating factor if the operator's or the crew's actions (or inaction) degrade the condition of the plant or threaten public health and safety. Scenarios that require little analysis or problem-solving and few operator actions may not provide an adequate basis to evaluate the required rating factors.

Section E describes the individual competencies that apply to the RO and SRO license levels during initial and requalification examinations. ES-303, "Documenting and Grading Initial Operating Tests," identifies the rating factors within each competency for the initial licensing examination (specifically, on Forms ES-303-3 and ES-303-4 for RO and SRO applicants, respectively), while ES-604 identifies the crew competencies that apply only to requalification examinations.

e.   *Level of Difficulty*

The dynamic simulator operating test must discriminate between those examinees who have and have not adequately mastered the knowledge, skills, and abilities required to be licensed operators. Simulator scenarios that are either too easy or too difficult are not effective discriminators.

In general, the level of difficulty of a scenario will increase with an increase in its quantitative attributes, such as the number of malfunctions or CTs (discussed below). However, the number of quantitative attributes in a scenario is not always indicative of the scenario's level of difficulty; that is, two scenarios having the same quantitative attributes can vary significantly in level of difficulty. There are no definitive minimum or maximum attribute values that can be used to identify inappropriate scenarios that will not discriminate because they are too easy or difficult.

The two most important determinants of the level of difficulty of a simulator scenario are the amount of analysis and problem-solving and the number of operator actions required to mitigate the events in the scenario. Malfunctions that require analysis or problem-solving increase the level of difficulty because they require the examinees to integrate a number of system conditions, evaluate their interrelationships, and take actions that demonstrate an understanding of the underlying concepts. Scenarios that consist of a number of unrelated malfunctions that require little or no operator analysis or response are generally less challenging.

2. Quantitative Attributes

Those traits discussed in the previous section provide for a qualitative assessment of the complexity of a simulator scenario. However, some characteristics of a scenario can be quantified and generally have a bearing on the complexity and level of difficulty of the scenario. The following discussion describes these characteristics, while ES-301 and ES-604 enumerate a target range for each trait that is applicable to the initial and requalification examination, respectively. The ranges are not absolute limitations; some scenarios may be an excellent evaluation tool but may not fit within the ranges. A scenario that does not fit into these ranges should be evaluated to ensure that it is appropriate.

a. *Normal Evolutions*

Normal evolutions include activities such as a feed pump startup, turbine loading, generator synchronization, and reactivity manipulations, which include evolutions such as a reactor startup or changing power with boron concentration, control rods, or core flow. Reactivity manipulations are considered significant if they produce a *clearly observable plant response*, such as bringing the reactor critical from a substantially subcritical state, raising power to the point at which reactivity feedback from nuclear heat addition is noticeable and a heatup rate is established, or changing reactor power manually with control rods or recirculation flow.

Normal evolutions can be used as a backdrop on which to stage the emergency or abnormal situations. For example, a main feedwater control valve may fail passively (i.e., as is) before the operators conduct a normal power change.

Time-consuming normal evolutions (such as a power escalation from low power) can provide an opportunity to evaluate the SRO's supervisory or resource management skills. Events such as component or instrument failures may be added to challenge the operators while continuing the power escalation.

Short surveillances (e.g., exercising safety rods or paralleling the emergency diesel generator with the grid) may be used to examine the operators' dexterity on the control panels or to involve operators who are not engaged in other activities.

b. *Total Malfunctions*

Total malfunctions are the number of instrument (e.g., nuclear, control, or process) and component failures (e.g., pump, motor, valve, or pipe) used to initiate the events that constitute a scenario, including those initiated after EOP entry (see Item C.2.c below). To count as a separate malfunction, they must involve a significant system response and require operator action to correct. For example, an anticipated transient without scram or trip (ATWS/ATWT) is a single malfunction, regardless of how many instructions a simulator operator must program to produce it.

Components that are placed out of service at the beginning of a scenario as part of the shift turnover conditions, and of which the crew is made aware, are not considered malfunctions. Component or instrument failures that require no operator actions or response do not count toward the recommended total number of malfunctions.

c.   *Malfunctions After EOP Entry*

Some malfunctions should result in vital instruments or components failing after the EOPs have been entered (these may have been inoperable at the beginning of the scenario or before EOP entry) and should influence the operators' choice of mitigation strategy. For example, failing a high head safety injection (SI) pump to start on a large-break loss-of-coolant accident (LOCA) does not affect the mitigation strategy; however, this would have an effect if it were the only available high head SI pump on a small-break LOCA.

d.   *Abnormal Events*

Each scenario should evaluate the operators' ability to implement AOPs. An abnormal event may or may not be a precursor to the major transient (see Item C.2.e below), although it can add to the credibility of a scenario, such as preceding a total loss of feed water with a single feed pump trip. However, certain events may cue the operators about subsequent events. Therefore, if a scenario is derived from the facility licensee's bank, it is wise to vary or modify the precursor events that lead to the major transient. It is also good to insert abnormal events that are not always predictive of the same major transient (e.g., a steam generator tube leak does not always lead to a subsequent tube rupture).

Some abnormal events for each scenario should require the operators to recognize and interpret technical specifications. This recognition and interpretation can also be incorporated into the scenario by designating TS-related equipment that is out of service at the start of the scenario.

Components or instrument failures that occur following EOP entry do not count toward the recommended total number of abnormal events.

e.   *Major Transients*

A major transient is one that has a significant effect on plant safety and leads to an automatic (or manual, if initiated by an operator) protective system actuation, such as a reactor trip or an engineered safety system actuation. A single major transient that actuates more than one automatic protective system actuation will be counted as a single major transient. Examples include loss of offsite power, LOCA, steam or feed line break, steam generator tube rupture, and loss of feedwater. A major transient should normally involve activation of the facility's emergency plan.

f.   *EOPs Used*

A scenario that requires the operators to refer to many different EOPs may not be as complex as a scenario for which only one EOP is used, but which requires use of alternative decision paths and prioritization of actions within the EOP to deal with the situation. Therefore, this attribute should reflect the EOPs that have measurable actions that the crew must take. Moreover, the primary scram response procedure that serves as the entry point for the EOPs is not counted.

For boiling-water reactors (BWRs), the number of "EOPs Used" should be counted consistent with the following four top-level Emergency Procedures Guidelines: (1) RPV control, (2) primary containment control, (3) secondary containment control, and (4) radioactivity release control. Use of multiple control sections of these guidelines do not count separately as "EOPs Used." For example, use of RPV level control and RPV pressure control should be counted as "one EOP Used — RPV control."

g.   *EOP Contingency Procedures Used*

Contingency procedures are used when there is a challenge to a critical safety function or if plant conditions have become severely degraded. Therefore, using them in a scenario provides an opportunity to observe the operators attempt to execute a mitigation strategy that clearly has safety significance to the plant and the public health and safety. Each scenario set should require the operators to enter and perform safety-related tasks within an EOP contingency procedure at least once.

The following list of contingency procedures is neither unique nor all-inclusive. Scenario developers and reviewers should consider it as a set of general guides that may not fully apply to all scenarios.

(1)   <u>Westinghouse</u>

Optimal Recovery Procedures designated as Emergency Contingency Action (ECA) procedures:

- Loss of All AC Power With or Without SI Required
- Loss of Emergency Coolant Recirculation
- LOCA Outside Containment
- Uncontrolled Depressurization of All Steam Generators
- Steam Generator Tube Rupture (SGTR) With Loss of Reactor Coolant Subcooled Recovery
- SGTR With Loss of Reactor Coolant-Saturated Recovery
- SGTR Without Pressurizer Pressure Control

Functional Recovery Procedures entered as a result of RED or ORANGE conditions on the Critical Safety Function Status Trees:

- Response to Nuclear Power Generation/ATWS
- Response to Inadequate Core Cooling
- Response to Degraded Core Cooling
- Response to Loss of Secondary Heat Sink
- Response to Imminent Pressurized Thermal Shock Conditions
- Response to High Containment Pressure
- Response to Containment Flooding

(2) Combustion Engineering

- Entry into Functional Recovery Procedures (FRPs)
- Transition among Functional Recovery Safety Function success paths
- Transition from one safety function to another within the FRPs

(3) Babcock and Wilcox (B&W)

The B&W EOP structure does not identify procedures that can be easily recognized as contingency procedures. However, use of the descriptions given above for Westinghouse contingency procedures provides guidance on the types of events to be considered.

(4) General Electric

- Alternative Level Control
- Emergency Reactor Pressure Vessel (RPV) Depressurization
- Primary Containment Flooding
- Level/Power Control
- RPV Flooding
- Steam Cooling

h. *Simulator Run Time*

A scenario should be designed to run approximately 60 to 90 minutes. However, this does not preclude scenarios taking more or less time. The nominal run time of 60 minutes may not provide sufficient time to conduct a scenario that progresses through several EOPs or requires performance of fairly involved procedural steps. It is possible to conduct very meaningful and involved scenarios in less time, but care should be taken not to place an undue burden on the operators by initiating malfunctions at too rapid a pace. This parameter is one of many that should be considered in assessing the overall quality of a scenario, and as long as the scenario meets the other criteria stated herein, the scenario run time is a secondary concern.

i.   *EOP Run Time*

The time during which the operators are involved in EOPs has a strong relationship to the complexity of the scenario because most critical tasks occur in the EOPs and the actions the operators take have the most potential to affect the health and safety of the public. Therefore, a significant percentage of the time a scenario is progressing should be spent in the EOPs. Usually, more time is required when contingency procedures are in effect, because it generally takes some time for the plant to degrade to a point where critical safety functions are jeopardized. However, operators should be evaluated in EOP activities beyond the point at which an event is diagnosed and initial mitigation actions are taken. Many of the actions taken to stabilize the plant and recover from a transient are safety significant. Therefore, scenarios should be allowed to progress so that these operations can be observed.

Scenarios should not be solely EOP oriented. Valuable assessments can be made within AOPs with the plant at power because of the level of safety significance associated with transients in these conditions.

j.   *Critical Tasks*

Critical tasks range between fairly simplistic but safety-significant tasks (starting the standby liquid control system during an ATWS condition or tripping a reactor coolant pump during a small-break LOCA) and other tasks that require a much higher level of skill involving several crew members (executing a rapid cooldown within predefined limits using steam generator power-operated relief valves or using low-pressure injection systems to maintain the vessel level while cooling the suppression pool). Therefore, the difficulty level must be considered when assessing the appropriateness of the number of CTs in a scenario or scenario set.

Refer to Section D for a detailed explanation of the CT methodology.

D.   Critical Task Methodology

The requalification examination uses CTs to evaluate crew performance on tasks that are safety significant to the plant or the public. As such, the CTs are objective measures for determining whether the performance of an individual or a crew is satisfactory or unsatisfactory. On the initial licensing examinations, CTs provide a basis for the individual operator competency evaluations because they help the examiner to focus on those tasks that have a significant impact on the safety of the plant or the public. Refer to ES-303 and ES-604 for specific instructions on the use of CTs in grading initial and requalification examinations.

1.   **Identification of Critical Tasks**

A CT must include the following elements:

a.   *Safety Significance*

In reviewing each proposed CT, assess the task to ensure that it is essential to safety. A task is essential to safety if its improper performance or omission by an operator will result in direct adverse consequences or significant degradation in the mitigative capability of the plant.

If an automatically actuated plant system would have been required to mitigate the consequences of an individual's incorrect performance, or the performance necessitates the crew taking compensatory action that would complicate the event mitigation strategy, the task is safety significant.

Examples of CTs involving essential safety actions include those for which operation or correct performance prevents the following:

- degradation of any barrier to fission product release

- degraded emergency core cooling system (ECCS) or emergency power capacity

- a violation of a safety limit

- a violation of the facility license condition

- incorrect reactivity control (such as failure to initiate emergency boration or standby liquid control, or manually insert control rods)

- a significant reduction of safety margin beyond that irreparably introduced by the scenario

Examples of CTs involving essential safety actions include those for which a crew demonstrates the following abilities:

- effectively direct or manipulate engineered safety feature (ESF) controls that would prevent any condition described in the previous paragraph

- recognize a failure or an incorrect automatic actuation of an ESF system or component

- take one or more actions that would prevent a challenge to plant safety

- prevent inappropriate actions that create a challenge to plant safety (such as an unintentional reactor protection system (RPS) or ESF actuation)

b.   *Cueing*

For a CT to be valid, an external stimulus must prompt at least one operator to perform the task. A cue prompts the operators to respond by taking certain actions and provides the initial conditions. The cue need not indicate the task as "critical."

Appropriate cues include the following examples:

*   verbal direction by or reports from other crew members

*   procedural steps, such as satisfying entry conditions, flow chart decision points, and "response not obtained" columns

*   indication of a system or a component malfunction (including passive failures) by meters or alarming devices

c.   *Measurable Performance Indicators*

A measurable performance indicator consists of positive actions that an observer can objectively identify taken by at least one member of the crew.

The NRC and facility licensee should review each critical task to ensure that it is objective. For example, "If pressure falls below 1400 psi, start pump xyz," is a performance measure that is not objective. The operator performing this task could conceivably start the pump when pressure reaches zero psi and still not violate the performance measure stated in the procedure, even though the facility licensee expects the operator to start the pump sooner. The NRC and facility licensee should agree in writing that the limits for each CT are acceptable before the requalification examination begins. For the example given above, adding an acceptable pressure tolerance (e.g., within 200 psi) would clarify the standard of performance that is expected.

Measurable performance indicators include the following examples:

*   actions taken as the result of transitioning between EOPs [for example, transitioning to and performing the actions required in FR-S.1 if the reactor does not trip (Westinghouse), or performing an automatic depressurization after confirming indications of high suppression pool temperature (General Electric)]

*   control manipulations (such as a manual reactor trip or the start of an ECCS pump)

*   verbal reports or notifications of abnormal parameters or conditions (such as "all control rods are not inserted" or "containment pressure is greater than 2 psi")

The following are examples of performance indicators that *cannot* be measured objectively during a simulator scenario:

Appendix D, Page 14 of 39

- understanding (such as of the significance of a certain plant response)

- verification that an expected response has occurred

- passive observations (such as monitoring the performance of a system)

d.   *Performance Feedback*

Each CT must provide at least one member of the crew with performance feedback. The feedback provides the crew member with information about the effect of the crew's actions or inaction on the CT. This requirement must be met for all CTs.

2.   Critical Tasks as "Generic" Safety Tasks

Avoid assigning the "CT" designation to generic tasks that have safety significance but do not meet all of the criteria required to identify a critical task.

Although a crew is not performing optimally if it fails to anticipate an automatic action given sufficient time to assess plant behavior, crew members are not required to anticipate an automatic action. A crew member may, at any time, take manual action in advance of an automatic action if, in the crew member's judgement, manual action is needed to place the reactor in a safe condition. If an operator takes an action that the examiners did not expect, the examiners must further evaluate the individual's rationale for taking that action. Such preemptive actions may indicate a misunderstanding of plant conditions or a weakness in integrated plant knowledge that should be clarified with followup questions.

Taking manual control of an automatic safety system qualifies as a CT only if the auto-initiation feature fails to work. It is then safety significant for the crew to take manual actions, as plant conditions clearly indicate that an automatic action should have occurred and did not. Moreover, during scenario development and validation, identification of CTs is based on those actions which, if performed incorrectly or omitted, degrade the mitigation strategy needed in the scenario. If the manual system has also failed and no action will be effective, this should not be identified as a CT. However, if an operator or the crew significantly deviates from or fails to follow procedures that affect the maintenance of basic safety functions, those actions may form the basis of a CT identified in the post-scenario review.

Experience has shown emergency event classification to be an important evaluation area, but generally not a CT. The argument is made that an incorrect classification could adversely affect public health and safety if the appropriate instructions are not given to public service agencies in a timely manner. If a misclassification occurs, the emphasis for corrective action is placed on the facility licensee and an appropriate period allotted for implementation of the corrective actions.

Therefore, although emergency classification is still an area that is to be evaluated, it should not receive the weight of a CT. If a misclassification occurs, the examiners should determine the rationale used to establish the classification in order to determine whether the crew understood the status of the plant and incorporate into the program evaluation those pertinent corrective actions deemed appropriate. If a widespread problem is observed during a program evaluation, the examiner should share this information with other inspection program managers.

E. Competency Descriptions

1. Reactor Operator

a. *Interpret/Diagnose Events and Conditions Based on Alarms, Signals, and Readings*

This competency involves the ability to accurately and promptly *recognize and analyze* off-normal *trends* and *diagnose* plant conditions to guard against and mitigate conditions that are out of specification. It includes the abilities to prioritize one's attention in keeping with the severity and importance of annunciators and alarm signals and to correctly *interpret and verify* that signals are *consistent with plant and system conditions*. It does *not* include knowledge of system operation, such as set points, interlocks, or automatic actions, or the understanding of how one's actions affect the plant and system conditions.

b. *Comply With and Use Procedures, References, and Technical Specifications*

This competency involves the ability to *refer to and comply with* normal, alarm/annunciator, abnormal, emergency, and administrative *procedures* in a timely manner (i.e., in sufficient time to avoid adverse impacts on plant status). It includes the ability to *recognize* emergency operating procedure *entry conditions, carry out immediate actions* without assistance, and *recognize and comply* with required *limiting conditions for operation and action statements*. It also includes the *use* of control room *reference materials*, such as prints, books, and charts, to aid in the diagnosis and classification of events and conditions.

c. *Operate the Control Boards*

This competency involves the ability to *locate and manipulate controls* to attain a desired plant and system response or condition. It includes *knowledge of system operation*, including set points, interlocks, and automatic actions and the ability to *locate* plant and system *instruments* and to understand how one's actions *affect* plant and system conditions. It also includes the ability to take *manual control* of automatic functions, when appropriate.

d. *Communicate and Interact With Other Crew Members*

This competency involves the ability to *provide and receive* pertinent information, both oral and written (e.g., log entries). It includes the ability to *carry out supervisory instructions* and to *interact with other crew members* with respect to conditions affecting safe plant operation, regardless of which applicant's control board is directly affected.

2. <u>Senior Reactor Operator</u>

a.   *Interpret/Diagnose Events and Conditions Based on Alarms, Signals, and Readings*

This competency involves the ability to *diagnose* plant conditions to guard against and mitigate conditions that do not meet specifications. It includes the abilities to prioritize one's attention in keeping with the severity and importance of the annunciators and alarms and to correctly *interpret the significance* of each alarm and *verify* that it is *consistent* with plant and system conditions. It also includes the ability to *recognize and analyze off-normal trends* in an accurate and timely manner. In addition, it includes knowledge of system operation, such as set points, interlocks, or automatic actions, or the understanding of how one's actions affect the plant and system conditions, unless that knowledge is evaluated under Control Board Operations.

b.   *Comply With and Use Procedures and References*

This competency involves the ability to *refer to and comply with* normal, alarm/annunciator, abnormal, emergency, and administrative *procedures* in a timely manner (i.e., in sufficient time to avoid adverse impacts on plant status). It includes the *use* of control room *reference materials*, such as prints, books, and charts, to aid in the diagnosis and classification of events and conditions. It also includes the ability to *use procedures correctly* and ensure correct *implementation by the crew.*

c.   *Operate the Control Boards*

This competency involves the ability to *locate and manipulate* controls to attain a desired plant and system response or condition. It includes *knowledge of system operation*, including set points, interlocks, and automatic actions and the ability to *locate* plant and system *instruments* and to understand how one's actions *affect* plant and system conditions. It also includes the ability to take *manual control* of automatic functions, when appropriate.

d.   *Communicate and Interact With the Crew and Other Personnel*

This competency involves the ability to *provide and receive* pertinent *information* in a clear, easily understood manner. It includes the ability to *keep crew members and personnel outside* the control room *informed* of plant status.

e.   *Direct Shift Operations*

This competency involves the ability to take *timely and decisive actions* in response to problems during both normal and off-normal situations. It includes the ability to provide *timely and well-thought-out directions* that indicate *concern for safety*, to encourage a *team approach* to problem-solving and decision-making by *soliciting and incorporating feedback* from crew members; and to remain in a position of *oversight* to maintain the "big picture." It also includes the ability to ensure that *the crew* carried out *correct and timely activities*.

f.   *Comply With and Use Technical Specifications*

This competency involves the ability to *recognize* when conditions are covered by technical specifications. It includes the ability to *locate* the appropriate TS and *ensure correct compliance* with any limiting conditions for operation and action statements.

F.   Security Considerations for Simulator Operating Tests

Simulators present a unique set of integrity concerns during the development and administration of operating tests. NRC examiners and facility licensees should be aware of the simulator's vulnerabilities and take appropriate measures to ensure that operating test security is maintained in the three areas of (1) the instructor station, (2) the programmers' tools, and (3) the external interconnections. Because facility licensees are more familiar than the NRC examiners with their simulator's unique capabilities, limitations, and vulnerabilities, it is expected that the licensees will take responsibility for determining and implementing whatever measures might be necessary to ensure the integrity of the operating tests.

Most of the instructor station features can be checked through the tableau or graphic interface provided at the instructor's console. The programmers' tools and the external interconnections are not generally apparent to the instructor or the examiner. The simulator staff should be consulted to determine the status of those items.

1.   Instructor's Station Features

- *Snapshots*: All simulators have snapshot capability. Initial conditions (ICs) are recorded for future recall.

- *Backtrack*: Backtrack files are snapshots that are automatically recorded at predetermined intervals (usually up to 1 hour of operation at intervals as frequent as 1 minute). Backtrack files are usually only accessible through the BACKTRACK feature. The files typically can be overwritten by real-time operation, but cannot be erased.

- *Replay/Playback*: The replay/playback feature steps through a series of snapshots and displays the output status (lights, meters, etc.) for each sequentially.

Often, the replay feature uses the backtrack files, although separate replay file storage may be provided.

- *Scripts/Computer-Assisted Exercises*: Many simulators have a feature that allows pre-programmed implementation of malfunctions and remote functions based on time and/or logical conditions. The simulator staff may use scripts to facilitate scenario administration, and can typically store scripts for future use. Stored scripts can also be selected for review and editing from the instructor's station.

- *Initial Conditions Summary*: Snapshots are usually labeled on the IC menu of the instructor's station with date/time recorded, pertinent plant parameter status, and instructor's comments. Even if the comment field has been changed to indicate that a snapshot is available for re-use, the data (scenario initialization) may still be representative of test conditions until the snapshot is overwritten or updated.

- *Malfunction Summary*: Malfunction summary menus display the status of selected malfunctions, both active and inactive. The malfunction summary is usually IC-dependent and, therefore, depicts the malfunctions that were active or staged when an IC (such as a scenario validation) was stored.

- *Monitored Parameters*: Instructors are afforded the capability to define individual or groups of parameters for display or printout. The monitored parameter group assignments can be recalled for review and editing. If used to facilitate scenario validation or examination administration, the monitored parameters can provide insight into the focus of the examination.

- *Trend Recording*: Groups of parameters can be defined and assigned to trend recorders. The recorders may be, but are not necessarily, located at the instructor's station. The recording may also be in file format for presentation on instructor's station screens. Recording sessions are typically activated or deactivated at the instructor's station.

- *Student Performance Monitoring*: Special groups of parameters and simulated plant operating conditions can often be assigned to a tracking and recording function that plots an individual student's performance during training exercises. Recording sessions are typically activated or deactivated at the instructor's station.

- *Video and Audio Recording*: Many simulators are equipped with video and audio recording capability in the control room. Video and audio controls are typically located at the instructor's station.

2. **Programmers' Tools**

- *Software Terminals*: Simulator engineers have access to real-time monitoring and control of simulator and model conditions through software support terminals. These terminals may be located in the computer facility or the engineer's desk.

- *Independent Executives*: The conditions for scenarios can sometimes be replicated off-line using independent executive programs. These programs should not be in communication with the I/O. Independent executives and their associated initialization files may provide an indication of planned exercises if they have been used to resolve problems during scenario validation.

- *Graphical User Interfaces (GUIs)*: Instructor's station graphical user interfaces often display simulated plant conditions and performance in real-time. At remote locations, such as a programmer's desk, the GUI could display the full scenario.

3. **External Interconnections**

- *ESF Feeds*: Many simulators have data links to the ESF and the operations management offices for emergency planning drills. These links can display simulated plant condition to observers outside the simulated control room during scenario validation or examinations.

- *Remote Plant Process Computer and Instructor Station Screens*: Repeater screens in the training area can display scenarios in real time to observers outside the simulated control room.

- *Modems and Remote Simulator Support Systems*: Many simulators are equipped with modems from the instructor's station or simulation computers for outside monitoring and control of simulator status and activities by parties off site.

G. Attachments/Forms

Attachment 1,	"Example Initial Dynamic Simulator Scenarios"
Attachment 2,	"Example Requalification Dynamic Simulator Scenarios"
Form ES-D-1,	"Scenario Outline"
Form ES-D-2,	"Required Operator Actions"

Facility: _____PWR_____   Scenario No.: ____1___   Op-Test No.: ___1___

Examiners: _____   Operators: _____

_____   _____

_____   _____

Initial Conditions: IC-38; 100% power, middle of life; CCP "B" is running; Unit 2 is in Mode 5.

Turnover: The following equipment is out of service: DG "A" (6 hrs); CCW pump "A" (2 days); VCT level transmitter LT-185; the block valve for PORV 456 is inoperable with power removed; MFP "A"; and AFW pump "A" (30 hrs). All required surveillances have been done. A severe thunderstorm warning is in effect.

Event No.	Malf. No.	Event Type*	Event Description
1	XXX, XXX	C(RO) N(BOP) R(RO)	70-gpm tube leak on "A" SG (ramped over 5 min.) with running CCP trip and failure of standby pump to start; requires power reduction
2	XXX	I(RO)	pressurizer level instrument L-459 fails low
3	XXX	C(ALL)	instrument bus 112 inverter failure
4	XXX, XXX	M(ALL) I(BOP)	450-gpm tube rupture on "A" SG (ramped over 3 min.) with an "A" SG pressure transmitter failure causing the PORV to open
5	XXX, XXX, XXX	M(ALL)	concurrent failures of the station auxiliary transformer and the "B" DG result in a loss of all AC power; power remains available through Unit 2
		C(BOP)	TDAFW pump trips on overspeed (can be reset)

*   (N)ormal,   (R)eactivity,   (I)nstrument,   (C)omponent,   (M)ajor

Note: *The scenarios in this attachment are individual examples; they are not intended to represent complete scenario sets/operating tests.*

For each planned event, enter on Form ES-D-2 (or equivalent) a description of the event and detailed actions required by the applicable plant procedures (e.g., normal, abnormal, emergency, and administrative, including the TS and emergency plan) for each operating position (i.e., SRO, RO, BOP) in a manner similar to the first event on the next page.

Op-Test No.: __1__	Scenario No.: __1__	Event No.: __1__	Page _1_ of _5_

Event Description: A 70-gpm tube leak on the "A" SG (ramped over 5 min.), combined with a trip of the running CCP and a failure of the backup CCP to start, forces a reduction in power because RCS leakage exceeds TS limits.

Time	Position	Applicant's Actions or Behavior
	RO/SRO/BOP	Recognize indications of the tube leak on the "A" SG: • air ejector offgas radiation monitor • steam line radiation monitor • charging/letdown mismatch • SG blowdown radiation monitor
	SRO	Direct RO/BOP actions in accordance with AOP-1.2: • monitor and control pzr level and pressure • monitor and control VCT level • verify leakage greater than TS limit • announce possible high radiation in turbine bldg • verify tube leak with SG samples • have health physics verify release calculation • commence unit shutdown • notify NRC • minimize secondary contamination • classify the event in accordance with the EPIPs (unusual event)
	RO/BOP	Execute AOP actions in accordance with SRO directions
	SRO/RO	Recognize running CCP tripped: • no charging flow • pump tripped light • various charging/letdown annunciators
	SRO	May direct RO/BOP per AOP-1.3: • isolate letdown • monitor pressurizer level and pressure • start the standby CCP • reestablish letdown • refer to TS 3.8.1 • initiate repairs
	SRO	Supervise/coordinate power reduction: • review precautions in GOP-3 • ensure delta-I maintained within limits • verify load reduction rate
	RO	Coordinate with BOP to initiate power reduction: • review GOP-3 precautions • calculate/estimate boration required for shutdown • contact load dispatcher • borate and/or insert rods to maintain T-ave within 5F of T-ref and maintain delta-I within limits
	BOP	Coordinate with RO to initiate power reduction: • review GOP-3 precautions • operate turbine controls to maintain unloading rate

Facility: _____PWR_____   Scenario No.: ____2___   Op-Test No.: ___2___

Examiners: _____   Operators: _____

_____   _____

_____   _____

Initial Conditions: IC-20; approximately 100% power, 218 ppm boron (EOL), equilibrium xenon; bank "D" rods are at step 216

Turnover: The operations department is making preparations to shut down the plant as a result of equipment problems. Train "B" CSS logic failed an actuation test last shift; the LCO for TS 3.3.2 was entered 2 hrs ago; I&C is working on the problem. MDAFW pump "B" is out of service to repair an oil leak and should be back in about 45 min. The block valve for PORV 445A is closed and deenergized for leakage control.

Event No.	Malf. No.	Event Type*	Event Description
1	XXX, XXX	I(BOP)	spurious containment spray actuation, phase "B" isolation, and CSS pump "A" failure to auto start (reset malf. to allow equipment restoration and before required stop of RCPs)
2	N/A	N(BOP) R(RO)	begin normal shutdown as a result of CS problems
3	XXX	C(RO)	boric acid filter plugged (100% in 1 min) at start of boration; when asked, filter d/p is 80# (remove when backflushed)
4	XXX	I(RO)	narrow range RCS temperature detector fails high
5	XXX, XXX	C(BOP)	emergency bus 1A-SA normal feeder breaker trips, and DG "A" breaker trips 2 min later
6	XXX, XXX, XXX, XXX	M(ALL) C(BOP) C(RO)	"A" SG line break in containment with auto SI on high containment pressure but failure of reactor and turbine trip; the local manual breaker is operable and the turbine will follow; TDAFW pump overspeed on SI; PORV "B" failure to open in auto or manual

*   (N)ormal,   (R)eactivity,   (I)nstrument,   (C)omponent,   (M)ajor

For each planned event, enter on Form ES-D-2 (or equivalent) a description of the event and detailed actions required by the applicable plant procedures (e.g., normal, abnormal, emergency, and administrative, including the TS and emergency plan) for each operating position (i.e., SRO, RO, BOP) in a manner similar to the first event for the first PWR scenario (page 2 of this Attachment).

Facility: _____BWR_____   Scenario No.: ____1__   Op-Test No.: ___1___

Examiners: _____   Operators: _____

_____   _____

_____   _____

Initial Conditions: IC-11; approximately 90% reactor power at dispatcher request; at power for 28 days, beginning of cycle; core spray pump 2A is out of service to replace a breaker closing coil; APRM F failed downscale last shift and is bypassed

Turnover: Raise power to 100% when contacted by dispatcher; test core spray pump 2A when the clearance is lifted (imminent)

Event No.	Malf. No.	Event Type*	Event Description
1	N/A	R(RO)	raise reactor power to 100% upon load dispatcher's request
2	XXX	N(BOP) C(BOP)	test core spray pump 2A starting at step 7.9.2 of PT-07.2.4a and respond to the motor overload
3	XXX	C(SRO)	individual bus breaker failure (MCC DGD), requiring DG #4 to be declared inoperable and a plant shutdown in accordance with TS 3.0.5
4	XXX	I(RO) C(BOP)	UPS inverter 2A malfunction and loss of UPS (no APRMs, rod positions, or rod control)
5	XXX	C(BOP)	turbine bearing #3 vibration alarm
6	XXX, XXX, XXX, XXX	M(ALL)	turbine trip and reactor scram with very few rods inserted (SLC pump 2A will trip after initiation and the scram discharge volume vents and drains fail to reopen when RPS is reset)
		C(ALL)	bypass valves fail closed after turbine coasts down (no UPS)

*   (N)ormal,   (R)eactivity,   (I)nstrument,   (C)omponent,   (M)ajor

For each planned event, enter on Form ES-D-2 (or equivalent) a description of the event and detailed actions required by the applicable plant procedures (e.g., normal, abnormal, emergency, and administrative, including the TS and emergency plan) for each operating position (i.e., SRO, RO, BOP) in a manner similar to the first event for the first PWR scenario (page 2 of this Attachment).

Facility: _____BWR_____   Scenario No.: ____2___   Op-Test No.: ___2___

Examiners: _____   Operators: _____

                       _____   _____

                       _____   _____

Initial Conditions: IC-17; 100% reactor power; B CRD pump is in service

Turnover: The load dispatcher has asked that power be lowered to 70%, and chemistry requests an SSW surveillance to be run at the beginning of the shift.

Event No.	Malf. No.	Event Type*	Event Description
1	N/A	R(RO)	decrease power to 70%
2	XXX	N(BOP) C(BOP)	perform SSW surveillance in accordance with chemistry request; SSW pump B will trip shortly after start
3	XXX	I(RO)	feedwater master controller fails as is
4	XXX	C(BOP)	loss of power to Division 2 ESF bus
5	XXX, XXX, XXX	M(ALL) C(BOP) M(ALL) C(BOP)	1.5 minutes after event 4, the service transformers lock out, the Division 1 EDG fails to start, and a 5% recirculation loop break develops in the drywell 30 seconds after initiating, the high pressure core spray pump trips

*   (N)ormal,   (R)eactivity,   (I)nstrument,   (C)omponent,   (M)ajor

For each planned event, enter on Form ES-D-2 (or equivalent) a description of the event and detailed actions required by the applicable plant procedures (e.g., normal, abnormal, emergency, and administrative, including the TS and emergency plan) for each operating position (i.e., SRO, RO, BOP) in a manner similar to the first event for the first PWR scenario (page 2 of this Attachment).

The following are two PWR and two BWR simulator scenario outlines that can be used for reference when developing or reviewing requalification examinations.

PWR Scenario One: Loss of Heat Sink

Scenario Objectives

- Evaluate the operators' use of the "Loss of Heat Sink" procedure, FR-H.1.
- Evaluate the crew's performance of a "bleed-and-feed" sequence, using reactor head vents and pressurizer vents.

Scenario Summary

Initial Conditions:

- 75 percent power
- "B" auxiliary feedwater pump inoperable
- One PORV (A) leaking and isolated

Events:

- Feed pump control problem that will eventually trip causing a partial loss of feed
- Total loss of main feedwater
- Loss of all feedwater

Scenario Sequence

- "A" feedwater pump hydraulic control unit problems prompt the crew to reduce power.
- During power reduction, the "A" feedwater pump trips, causing a plant runback.
- Feedline break occurs causing a reactor trip.
- Auxiliary feedwater pumps fail over several minutes, causing a loss of all feedwater, and prompting the crew to initiate a bleed-and-feed procedure.

Event one: Malfunction/loss of feed pump

Crew responds to a problem with the "A" feed pump, which eventually trips causing a runback.

Malfunctions required: 2 (RFP "A" HCU failure and RFP "A" Trip)

Objectives:

- Evaluate the crew's use of normal operating procedures to reduce power when the feed pump starts to fail.
- Evaluate the crew's use of abnormal operating procedures (AOPs) to respond to a partial loss of feed.

Success Path:

- Use the normal operating procedures to reduce power when initial problems occur with the feedwater pump.
- Use the AOPs to respond to the partial loss of feedwater and stabilize the plant to avoid a reactor trip.

Event two: Feedline rupture/reactor trip

Crew responds to a total loss of feed flow with only the remaining motor-driven AFW pump available.

Malfunctions required: 1 (feedline rupture)

Objective:

Evaluate the crew's response to a loss of feed transient requiring a reactor trip by using the reactor trip response and reactor trip recovery EOPs.

Success Path:

- Recognize the impending reactor trip, trip the reactor if time permits, and implement the appropriate immediate actions.
- Make the correct transition to the reactor trip recovery EOP upon completing the immediate and applicable subsequent actions of the reactor trip EOP.

Event three: Loss of all AFW/PORV failure

Crew responds to a total loss of feed flow, eventually implementing a bleed-and-feed procedure with a failed PORV. Evaluators inform the crew that time compression is being used to accelerate the decrease in steam generator level.

Malfunctions required: 2 (failure of all AFW and "B" PORV fails to open)

Objective:

Evaluate the crew's ability to recognize that there is no longer a heat sink, and correctly implement the applicable contingency procedure (loss of heat sink), including performing the bleed-and-feed procedure.

Success Path:

- Implement the EOP for loss of heat sink.
- Attempt to reestablish auxiliary feed flow; when SG levels become too low, initiate the bleed-and-feed procedure.
- Recognize the failure of the available PORV and reenergize, unblock, and open the leaking PORV; open both pressurizer and reactor head vents to ensure adequate bleed flow.

Scenario Recapitulation

Total Malfunctions:	5	
Abnormal Events:	1	
Major Transients:	2	(loss of main feed and total loss of feed)
EOPs Entered:	1	
EOP Contingencies:	1	(loss of heat sink)

PWR Scenario Two: LOCA and Cold Leg Recirculation

Scenario Objectives

- Evaluate the crew's response to unidentified primary leakage.
- Evaluate the crew's response to a circulating water pump trip and a condenser tube leak.
- Evaluate the crew's use of the EOPs during a LOCA with adverse containment conditions.
- Evaluate the crew's sensitivity to key parameters and ability to implement cold leg recirculation.

Scenario Summary

Initial Conditions:

- 100 percent power
- Inoperable "A" diesel generator and "A" instrument air compressor
- Seismic event occurred during last shift

Events:

- Primary leak increases to a point requiring a reactor trip.
- AFW pumps fail to automatically start on reactor trip.
- Leak leads to a safety injection (SI) and high-pressure SI pumps fail to start automatically; LOCA occurs, RWST leak occurs, and crew must initiate cold leg recirculation.

Scenario Sequence

- A small pressurizer steam space leak increases to a point requiring a reactor trip and eventually to the point of SI initiation.
- The high-pressure SI pumps fail to start automatically.
- A LOCA occurs as a result of the seismic event.
- When the SI pumps start, the thermal shock causes a LOCA in the RCS.
- The high pressure of the LOCA causes adverse containment conditions.
- An RWST leak will also occur concurrent with the SI that will eventually prompt the crew to initiate cold leg recirculation.
- RWST level will eventually drop to the point where the crew must initiate cold leg recirculation.

Event one: Unidentified leakage attributable to pressurizer steam space leak

The crew reacts to unidentified primary leakage, eventually requiring a reactor trip.

Malfunctions required: 1 (pressurizer steam space leak)

Objectives:

- Evaluate the crew's use of AOPs and TS to respond to unidentified primary leakage.
- Evaluate the crew's knowledge of parameters in the AOP that require a trip because of primary leakage.

Success Path:

- Use the AOPs, increase reactor make-up and calculate a leak rate.
- Use the NOPs to commence a reactor shutdown in accordance with TS.
- When leakage exceeds the AOP parameters, trip the reactor.

Event two: Reactor trip/AFW pump fails to start automatically

The crew trips the reactor on excessive leakage in accordance with the AOP.
The AFW pumps fail to start automatically, requiring manual initiation.

Malfunctions required: 1 (AFW failure to auto start)

Objective:

Evaluate the crew's use of the EOPs following a reactor trip, with the complication that the AFW pumps fail to start automatically.

Success Path:

- Recognize that the AFW pumps failed to start automatically and manually start the pumps.
- Correctly perform the reactor trip EOP and make the transition to the reactor trip recovery EOP after completing the immediate actions and applicable subsequent actions.

Event three: Increasing pressurizer leak/SI pumps fail to start

The pressurizer leak increases, causing a loss of pressurizer level/pressure requiring an SI. The charging pumps fail to automatically start, requiring manual start.

Malfunctions required: 2 (pressurizer leak increases and charging pumps fail to auto start)

Objectives:

- Evaluate the crew's ability to monitor important parameters in the EOPs and initiate SI when required.
- Evaluate the crew's ability to manually start the charging pumps following an SI signal.

Success Path:

- Initiate SI when pressurizer level and pressure decrease to the values stated in the EOPs.
- Recognize the failure of charging pumps to automatically start, and manually start the required charging pumps to complete the SI initiation sequence.

Event four: LOCA/adverse containment

A LOCA occurs as a result of the seismic event, which leads to adverse containment conditions. RWST level decreases to the point where the crew must enter the EOP for initiating cold leg recirculation. Evaluators inform the crew that time compression is being used to accelerate the decrease in RWST level.

Malfunctions required: 2 (LOCA and RWST leak)

Objectives:

- Evaluate the crew's use of the EOPs with adverse containment.
- Evaluate the crew's ability to recognize the need for and use the cold leg recirculation procedure.

Success Path:

- Correctly enter and use the LOCA EOP and the containment functional recovery EOP using adverse containment criteria.
- When RWST levels reach the low-low alarm and the reactor sump level is high enough, enter and implement the cold leg recirculation EOP.

Scenario Recapitulation

Total Malfunctions:	6	
Abnormal Events:	2	
Major Transients:	2	(leak requiring SI and LOCA with high containment pressure)
EOPs Entered:	4	(enter LOCA EOP twice)
EOP Contingencies:	1	(containment safety)

BWR Scenario One: Loss of Offsite Power with a LOCA

Scenario Objective

Evaluate the operators' use of the "Emergency Depressurization" and "RPV Flooding" EOP contingency procedures.

Scenario Summary

Initial Conditions:

- 98 percent power
- "A" average power range monitor (APRM) failed and bypassed

Events:

- Reactor core isolation cooling (RCIC) becomes isolated during an RCIC flow surveillance.
- Loss of offsite power/Division III diesel generator fails to start, disabling the high-pressure core spray (HPCS).
- Small-break LOCA occurs.
- Adverse containment conditions make the reactor level instrumentation unusable.

Scenario Sequence

- The RCIC system becomes isolated during surveillance testing, rendering the system inoperable.
- Faults in the 345-KV switchyard and reserve auxiliary transformer result in a loss of offsite power and a reactor scram.
- The Division III diesel generator fails to start and will not start manually, disabling the HPCS system.
- The plant transient causes a recirculation line break resulting in a small break LOCA that develops over several minutes.
- Reactor level instrumentation becomes erratic and unusable because of the rapid decrease in pressure and the elevated drywell temperature.

Event one: RCIC isolation

The crew responds to an isolation of the RCIC system during a full-flow test surveillance.

Malfunctions required: 1 (RCIC isolation)

Objective:

Evaluate the crew's use of technical specifications to determine that RCIC is inoperable.

Success Path:

Use technical specifications to recognize that the RCIC system should be declared inoperable until the problem can be investigated and corrected.

Event two: Loss of offsite power with concurrent Division III diesel generator failure (HPCS)

The crew responds to the loss of offsite power, reactor scram, and loss of high-pressure injection sources.

Malfunctions required: 2 (loss of offsite power and HPCS failure)

Objective:

Evaluate the crew's response to a plant transient that causes a reactor scram and a loss of high-pressure injection sources by using the reactor pressure vessel (RPV) and primary containment control EOPs.

Success Path:

- Maintain RPV pressure at less than 1065 psig using the main turbine bypass valves.
- Manually control pressure with safety relief valves (SRVs) upon a loss of electro-hydraulic control (EHC) hydraulic pressure because of the loss of power to the EHC pumps.
- Initiate suppression pool cooling and pump down in accordance with EOPs if the temperature in the suppression pool exceeds 90 °F or the level exceeds 18.5 feet.

Event three: Small-break LOCA

The crew responds to a loss of vessel inventory and an inability to maintain a level greater than the top of the active fuel, eventually implementing emergency depressurization.

Malfunctions required: 1 (LOCA)

Objective:

Evaluate the crew's ability to recognize an inability to maintain reactor water level and correctly implement the applicable contingency procedures, including emergency depressurization.

Success Path:

Execute RPV emergency depressurization so reactor pressure can be decreased to allow injection by the low-pressure ECCS systems.

Event four: Reactor level instrumentation failure

The crew recognizes a loss of reactor level instrumentation and responds in accordance with RPV flooding EOP.

Malfunctions required: 1 (reactor level instrumentation failure)

Objective:

Evaluate the crew's ability to recognize failed reactor level instrumentation and correctly implement the applicable actions of the RPV flooding EOP to ensure adequate core cooling.

Success Path:

Reflood the RPV in accordance with the EOPs and establish adequate core cooling. Adequate core cooling will be ensured when reactor pressure can be maintained greater than 120 psig with at least three SRVs opened by manually controlling the low-pressure ECCS injection flow.

Scenario Recapitulation

Total Malfunctions:	5	
Abnormal Events:	3	
Major Transients:	2	(emergency depressurization and RPV flooding)
EOPs Entered:	2	
EOP Contingencies:	3	(alternate level control, emergency depressurization, and RPV flooding)

BWR SCENARIO TWO: POWER OSCILLATIONS WITH AN ATWS

Scenario Objective

Evaluate the operators' use of the "Level/Power Control" and "Emergency Depressurization" EOP contingency procedures.

Scenario Summary

Initial Conditions:

- 75 percent reactor power
- High-pressure core spray pump out of service
- "B" recirculation pump flow control valve is locked

Events:

- The "A" reactor recirculation pump trips, causing power oscillations, and an SRV fails open during the power oscillations.
- Anticipated transient without scram (ATWS) requiring lowering of level to control power.
- Feed system pumps fail to restart and standby liquid control (SLC) pumps and RCIC pump trip during the transient, complicating recovery from the event.

Scenario Sequence

- The "A" recirculation pump trips, resulting in power oscillations within 5 minutes. The reactor fails to manually scram.
- The safety relief valve (SRV) sticks open during power oscillations.
- Condensate booster and feedwater pumps fail to restart, and the SLC pumps trip after power is reduced less than 3 percent.
- RCIC pump trips after it is restarted by an operator.

Event one: "A" recirc pump trip resulting in power oscillations

The crew responds to a recirculation pump trip and a failure of the reactor scram system.

Malfunctions required: 2 (recirculation pump trip and ATWS)

Objectives:

- Evaluate the crew's use of AOPs and EOPs to respond to an ATWS and to restore the power and flow parameters to acceptable values.
- Evaluate the crew's use of TS that apply to single recirculation loop operation.

Success Path:

- Recognize power to be in Region B or C of the power and flow map, and initiate control rod insertion to reduce thermal power.
- Recognize symptoms of thermal-hydraulic instability, and attempt to manually scram.
- Use the EOP flow charts for RPV level, power, and pressure control.
- Trip the "B" recirculation pump, and initiate actions to achieve control rod insertion and to actuate the standby liquid control system in accordance with the EOPs.

Event two: SRV sticks open during power oscillations

The crew recognizes and responds to the stuck open SRV, eventually implementing the actions of the primary containment control EOP.

Malfunctions required: 1 (SRV sticks open)

Objective:

Evaluate the crew's ability to recognize the failed open SRV and implement the applicable abnormal and emergency procedure actions.

Success Path:

- Initiate actions to close the SRV.
- Use EOPs to initiate suppression pool cooling and reduce the level.
- Terminate all injection into the RPV, except for the control rod drive and SLC systems when suppression pool temperature exceeds 110 °F with reactor power less than 3 percent.

Event three: Failure of injection sources after control rod insertion

The crew responds to a loss of vessel inventory and the inability to maintain a level greater than the top of the active fuel by eventually implementing emergency depressurization.

Malfunctions required: 3 (feedwater system failure, SLC pump trip, and RCIC fails to start)

Objective:

Evaluate the crew's use of EOPs to respond to an inability to maintain the reactor water level and initiate an emergency depressurization.

Success Path:

Execute RPV emergency depressurization to allow for injection by the low-pressure ECCS systems.

Scenario Recapitulation

Total Malfunctions: 6
Abnormal Events: 2
Major Transients: 2   (ATWS and emergency depressurization)
EOPs Entered: 2
EOP Contingencies: 3   (level and power control, alternate level control,
                                     and emergency depressurization)

Facility: _____ Scenario No.: _____ Op-Test No.: _____

Examiners: _____ Operators: _____

_____ _____

_____ _____

Initial Conditions: _____

_____

_____

Turnover: _____

_____

_____

_____

Event No.	Malf. No.	Event Type*	Event Description

* (N)ormal, (R)eactivity, (I)nstrument, (C)omponent, (M)ajor

Op-Test No.: _____   Scenario No.: _____   Event No.: _____   Page ___ of ___

Event Description: _____

_____

_____

Time	Position	Applicant's Actions or Behavior

APPENDIX E
POLICIES AND GUIDELINES FOR TAKING NRC EXAMINATIONS

Each examinee shall be briefed on the policies and guidelines applicable to the examination category (written, operating, walk-through, and/or simulator test) being administered. The examinees may be briefed individually or as a group. Facility licensees are encouraged to distribute a copy of this appendix to every examinee before the examination begins. All items apply to both initial and requalification examinations, except as noted.

Part A: General Guidelines

1. *[Read Verbatim]* Cheating on any part of the examination will result in a denial of your application and/or action against your license.

2. If you have any questions concerning the administration of any part of the examination, do not hesitate to ask them before starting that part of the test.

3. SRO applicants will be tested at the level of responsibility of the senior licensed shift position (i.e., shift supervisor, senior shift supervisor, or whatever the title of the position may be).

4. You must pass every part of the examination to receive a license or to continue performing license duties. Applicants for an SRO-upgrade license may require remedial training in order to continue their RO duties if the examination reveals deficiencies in the required knowledge and abilities.

5. The NRC examiner is not allowed to reveal the results of any part of the examination until they have been reviewed and approved by NRC management. Grades provided by the facility licensee are preliminary until approved by the NRC. You will be informed of the official examination results about 30 days after all the examinations are complete.

Part B: Written Examination Guidelines

1. *[Read Verbatim]* After you complete the examination, sign the statement on the cover sheet indicating that the work is your own and you have not received or given assistance in completing the examination.

2. To pass the examination, you must achieve an overall grade of 80.00 percent or greater, with 70.00 percent or greater on the SRO-only items, if applicable. If you only take the SRO portion of the exam (as a retake or with an upgrade waiver of the RO exam), you must achieve an overall grade of 80.00 percent or better to pass. SRO-upgrade applicants who do take the RO portion of the exam and score below 80.00 percent on that part of the exam can still pass overall, but may require remediation. Grades will not be rounded up to achieve a passing score. Every question is worth one point.

3. For an initial examination, the nominal time limit for completing the examination is 6 hours for the RO exam; 3 hours for the 25-question, SRO-only exam; 8 hours for the combined RO/SRO exam; and 4 hours for the SRO exam limited to fuel handling. Notify the proctor if you need more time.

    For a requalification examination, the time limit for completing both sections of the examination is 3 hours. If both sections are administered in the simulator during a single 3-hour period, you may return to a section of the examination that you already completed or retain both sections of the examination until the allotted time has expired.

4. You may bring pens, pencils, and calculators into the examination room; however, programable memories must be erased. Use black ink to ensure legible copies; dark pencil should be used only if necessary to facilitate machine grading.

5. Print your name in the blank provided on the examination cover sheet **and** the answer sheet. You may be asked to provide the examiner with some form of positive identification.

6. Mark your answers on the answer sheet provided, and do not leave any question blank. Use only the paper provided, and do not write on the back side of the pages. If you are using ink and decide to change your original answer, draw a single line through the error, enter the desired answer, and initial the change. If you are recording your answers on a machine-gradable form that offers more than four answer choices (e.g., "a" through "e"), be careful to mark the correct column.

7. If you have any questions concerning the intent or the initial conditions of a question, do *not* hesitate to ask them before answering the question. Note that questions asked during the examination are taken into consideration during the grading process and when reviewing applicant appeals. Ask questions of the NRC examiner or the designated facility instructor *only*. A dictionary is available if you need it.

    When answering a question, do *not* make assumptions regarding conditions that are not specified in the question unless they occur as a consequence of other conditions that are stated in the question. For example, you should not assume that any alarm has activated unless the question so states or the alarm is expected to activate as a result of the conditions that are stated in the question. Similarly, you should assume that no operator actions have been taken, unless the stem of the question or the answer choices specifically state otherwise. Finally, answer all questions based on actual plant operation, procedures, and references. If you believe that the answer would be different based on simulator operation or training references, you should answer the question based on the *actual plant*.

8. Restroom trips are permitted, but only one applicant at a time will be allowed to leave. Avoid all contact with anyone outside the examination room to eliminate even the appearance or possibility of cheating.

9. When you complete the examination, assemble a package that includes the examination questions, examination aids, answer sheets, and scrap paper, and give it to the NRC examiner or proctor. Remember to sign the statement on the examination cover sheet indicating that the work is your own and that you have neither given nor received assistance in completing the examination. The scrap paper will be disposed of immediately after the examination.

10. After turning in your examination, leave the examination area as defined by the proctor or NRC examiner. If you are found in this area while the examination is still in progress, your license may be denied or revoked.

11. Do you have any questions?

Part C: Generic Operating Test Guidelines

1. If you are asked a question or directed to perform a task that is unclear, you should *not* hesitate to ask for clarification.

2. The examiner will take notes throughout the test to document your performance, and the examiner may sometimes take a short break for this reason. The amount of note-taking does not reflect your level of performance. The examiner is required to document satisfactory as well as less-than-satisfactory performance.

3. The operating test is considered "open reference." The reference materials that are normally available to operators in the facility and control room (including calibration curves, previous log entries, piping and instrumentation diagrams, calculation sheets, and procedures) are also available to you during the operating test. However, you should know from memory certain automatic actions, set points, interlocks, operating characteristics, and the immediate actions of emergency and other procedures, as appropriate to the facility. If you desire to use a reference, you should ask the examiner if it is acceptable to do so for the task or question under consideration.

   You may not solicit technical information from other operators, engineers, or technical advisors.

4. In order to maintain test integrity and fairness, you must not discuss any aspect of your operating test with, or in the presence of, any other examinee who has not completed the applicable portion of the operating test (i.e., the administrative topics, the systems walk-through, or the dynamic simulator test).

Part D: Walk-Through Test Guidelines

1.   The walk-through test covers control room systems, local system operations,
     and administrative requirements. The examiner will evaluate these areas using
     job performance measures (JPMs) and specific followup questions, as necessary.

     The initial walk-through consists of 15 JPMs for RO and SRO(I) applicants
     and 10 for SRO(U) and LSRO applicants. Except for LSROs, most of the JPMs
     will be conducted in the control room or simulator and the remainder will be conducted
     in the plant.

     The requalification walk-through consists of a total of five JPMs, with at least two
     in the control room/simulator and at least two in the plant.

2.   The examiner is a visitor at this facility. When you enter the plant, you may be expected
     to escort the examiner and ensure that he or she complies with safety, security,
     and radiation protection procedures.

3.   You should not operate plant equipment without appropriate permission
     from the operating crew. Nothing the examiner says or asks will be intended
     to violate this principle.

4.   Before beginning each JPM, the examiner will describe the initial conditions,
     explain the task that is to be completed, indicate whether the task is time-critical,
     and explain which steps are to be simulated or discussed. You should perform
     or simulate the required actions as if directed by plant procedures or shift supervision.
     Do not assume that the examiner will accept an oral description of the required action
     unless the examiner indicates otherwise.

5.   Time-critical JPMs have been validated by your facility and must be completed
     within the predetermined time interval in order to obtain a satisfactory grade for that JPM.
     You will be permitted to take whatever time is necessary to complete those JPMs
     that are not time-critical, provided that you are making reasonable progress
     toward achieving the task standard. If the examiner believes that you are not
     making reasonable progress, he will ask you to explain what remains to be done
     and how long it should take before stopping the task. You will be permitted at least
     twice the validated time to complete the JPM, whether you are making progress or not.

6.   When performing JPMs, you are expected to make decisions and take actions
     based on the facility's procedural guidance and the indications available.
     Some of the tasks that the examiner asks you to perform will require implementation
     of an alternative method directed by plant procedures.

7.   If your facility licensee's procedures and practices require the use of procedure readers
     or peer checks, you may ask the NRC examiner to perform those functions.
     However, because the NRC examiner must be able to evaluate your individual
     performance without assistance from others, he or she will simply acknowledge
     your request and proposed actions, regardless of their accuracy or correctness.

8.   As part of the examination, the examiner may ask followup questions to investigate your knowledge of an administrative topic, system, or task. Many of the questions will require you to use plant reference material, while others should be answered without the use of references. If you need to consult a reference to answer a question, ask the examiner if it is acceptable to do so. There is no specific time limit for any question; however, you may be evaluated as unsatisfactory on a question if you are unfamiliar with the subject or reference material and are unable to answer the question in a reasonable period of time. You will not be permitted to conduct unlimited searches of the plant reference material during the examination.

9.   To facilitate the examination and better enable the examiner to assess your level of understanding, please verbalize your actions and observations while performing the JPMs. Also, please inform the examiner when you consider your performance of each JPM and your answer to each question to be complete.

10.   If you need a break during the test, you should ask the examiner.

11.   Do you have any questions before we begin the walk-through test?

Part E: Simulator Test Guidelines

1.   Your primary responsibility is to operate the simulator as if it were the actual plant. If you believe that the simulator is not responding properly, you should make decisions and recommendations on the basis of the indications available, unless directed otherwise by the examiner.

2.   If the examiner asks you a question, you should answer it *only if* doing so will not interfere with simulation facility operations.

3.   Teamwork and communications are evaluated. You can enhance the evaluation process by vocalizing your observations, analyses, and the bases for your actions.

      Requalification examinations evaluate the crew's ability to safely operate the plant and the performance of both the individuals and the crew.

4.   If you recognize, but fail to correct, an erroneous decision, response, answer, analysis, action, or interpretation made by the operating team or crew, the examiner may conclude that you agree with the incorrect item.

      Members of the operating team or crew (whether applicants or surrogates) should perform peer checks in accordance with the facility licensee's procedures and practices; non-crew members and NRC examiners will not perform this function. However, if you begin to make an error that is corrected by a peer checker, you will be held accountable for the consequences of the potential error without regard to mitigation by the crew.

5. You should keep a rough log during each scenario that would be sufficient to complete necessary formal log entries.

6. A designated facility instructor (or an examiner) will act as the auxiliary operators, radiation health and chemistry technicians, maintenance supervisors, plant management, and anyone else needed outside the control room.

7. The facility instructor (or examiner) will provide a shift turnover briefing before the scenario begins. The briefing will cover present plant conditions, power history, equipment that is out of service, abnormal conditions, surveillances that are due, and instructions for the shift.

8. Control board switches may be purposely misaligned to enhance a scenario or transient where appropriate. You will not be required to locate misaligned switches as part of the evaluation. If a switch is misaligned, it will be tagged or otherwise highlighted as appropriate to the facility and will be noted during the shift turnover briefing. The examiners will not misalign switches during the scenario.

9. Time compression may be used to expedite the sequence of events in some scenarios, but it will not preclude you from performing the actions that you would typically be required to perform in response to the events. If time compression is used, you will be so informed during and after the scenario.

10. You will be given sufficient time (normally about 5 minutes) to familiarize yourselves with plant conditions before starting each simulator scenario.

11. The initial test will normally consist of two or three scenarios lasting a total of 3–4 hours. The requalification test will normally consist of two scenarios lasting about 1 hour each. You will be given a short break between scenarios.

12. SRO upgrade applicants who fill the role of an RO or balance-of-plant (BOP) operator during a scenario will be evaluated on their ability to manipulate the controls even though an examiner may not be assigned to directly monitor their performance.

13. Do you have any questions before we begin the simulator test?

APPENDIX F
GLOSSARY

Achievement test: An instrument designed to measure a trainee's skill proficiency or grasp of some body of knowledge.

Annual: In most instances, a period of time equal to 365 days reckoned from any point in a calendar year to the same point in the following calendar year. However, annual requirements in successive years can reach a period of nearly 2 years. "Annual" could encompass a range extending to 729 days depending on when an event occurred in the first calendar year and viewing December 31 of the following calendar year as meeting the annual requirement.

Applicant: Any individual who has submitted an NRC Form 398, "Personal Qualifications Statement — Licensee," in pursuit of an RO or SRO license. For purposes of this and the NRC's other examination standards, "applicant" is synonymous with "candidate."

Applicant license level: The level of operator license (i.e., RO or SRO) for which the applicant has applied.

Aptitude test: An instrument designed to assess an individual's potential for performing some task or skill area.

Average: A score that provides an indication of the typical performance of a group of scores. The mean, median, and mode of a distribution of scores are all commonly used as averages.

Biennial: In most instances, a period of time equal to 730 days and synonymous with "2 years." Biennial requirements can extend beyond 730 days if the requirement is met during the anniversary month of the second year. For example, a biennial medical examination last performed on January 10, 1995, would be due again by January 31, 1997. In this case, January is seen as the anniversary month, and the biennial requirement is satisfied even though the period of time between the two examinations is longer than 730 days.

Bloom's Taxonomy: A classification system that depicts knowledge and information processing in a hierarchy from lowest to highest as fundamental knowledge, comprehension, analysis, synthesis, and evaluation.

Calendar quarter: One of four parts of a calendar year, each consisting of a 3-month segment. In any calendar year, the first quarter is from the first day of January to the last day of March, the second quarter is from the first day of April to the last day of June, the third quarter is from the first day of July to the last day of September, and the fourth quarter is from the first day of October to the last day of December.

Central tendency: A term referring to the most typical performance of a group of individuals; generally the mean, median, or mode

Cognitive: Aspects of a person or test level that refer to knowledge or understanding.

Content validity: The degree to which a test measures the specific objectives or content of a given test.

Correlation coefficient: A numerical value, ranging from -1 to +1, that indicates the relationship between two sets of scores or other measures of each individual in a group. A value of 0 indicates no relationship; +1 or -1 indicates a perfect relationship (either positive or negative).

Criterion: A characteristic or combination of characteristics used as the basis for assessing performance.

Criterion-referenced test: An examination based upon mastery of objectives of content that was or should have been taught and mastered and one that uses an established standard or cutoff score as a measure of acceptable performance.

Cut score: The score at which a trainee is deemed to have met the criteria for an exam.

Designated nuclear control room operator: In accordance with Section C.1.2 of Regulatory Guide 1.8, Revision 3, an individual assigned to a licensed control room operator position identified in either Technical Specification Table 6.2.1 or the table of "Minimum Requirements Per Shift for On-Site Staffing of Nuclear Power Units by Operators and Senior Operators Licensed Under 10 CFR Part 55" in Title 10, Section 50.54(m)(2)(l), of the *Code of Federal Regulations* [10 CFR 50.54(m)(2)(i)].

Diagnostic test: An instrument that is designed to identify an individual's strengths and weaknesses in a given content area.

Difficulty index: A numerical index, ranging from 0.00 to 1.00, that indicates the percentage of trainees who correctly answer a test item. An index of 0.00 indicates that no one correctly answered the test item, while an index of 1.00 indicates that all individuals correctly answered the item.

Discrimination index: A measure of a test item's ability to differentiate between good and poor trainees. A high discrimination index indicates that more high performers than low performers correctly answered the item. (High and low are typically determined by overall test scores, but may also be established by external criteria.)

Discrimination validity: Setting the item difficulty at an estimated level around the cut score.

Distractor: An incorrect alternative among the possible answers for a test item.

Error of measurement: Any difference between an obtained score and a true score on a test. The actual error of measurement can only be estimated, since it is impossible to know the true score.

Equivalent forms: Two or more exams that test the same objectives using different test items or the same test items in a different sequence.

Frequency distribution: A graphic display listing scores or score intervals on one axis of a graph, and the number of trainees at that score or in that interval on the other.

Item analysis: A set of procedures performed on examination items to determine their difficulty and discriminating power.

Item bank: A group of test items covering a defined area. Items for a test can be chosen from this source.

Item stem: The part of a test item that presents the problem or situation to be solved. The item stem may be a question requiring a response, or a statement that is followed by the alternatives from which the trainee must choose the best answer.

Job performance measure (JPM): An evaluation tool that is based on tasks contained in the facility's job task analysis (JTA) or the applicable NRC Knowledge and Abilities Catalog (NUREG-1122 or 1123) and requires the applicant to perform (or simulate) a task that is applicable to the license level of the examination.

Job task analysis (JTA): A systematic analysis of the knowledge, skills, and abilities required to perform a particular occupation.

Learning objective: A statement of the behavior a trainee is expected to exhibit following instruction.

Low-power: In accordance with NUREG-1449, "NRC Staff Evaluation of Shutdown and Low-Power Operation," the range of reactor power from criticality to 5 percent.

Mastery test: A term synonymous with "criterion-referenced test" (i.e., one that evaluates the expected behavior following instruction).

Mean: An indication of "central tendency." Mean usually refers to the arithmetic mean, which is computed by summing all the scores of a group, and dividing that sum by the number of scores in the group.

Median: A measure of "central tendency"; the point on a scale of scores that splits the scores in half, with 50 percent of the scores below this point, and 50 percent of the scores above this point.

Mode: The least reliable of the common measure of "central tendency"; the "mode" is the most frequently occurring score in a distribution of scores.

Multiple-choice item: A test item that is composed of an item stem and several alternatives from which the trainee must select the best answer.

Normal distribution: A theoretical frequency distribution represented by a symmetrical bell-shaped curve; sometimes referred to as the bell curve.

Norm-referenced: A score interpretation based on the comparison of an individual's score with a comparable reference group.

Nuclear power plant experience: As defined in Section 2 of ANSI/ANS-3.1-1993, "American National Standard for Selection, Qualification, and Training of Personnel for Nuclear Power Plants," applicable work performed in a nuclear-fueled electric power production plant during pre-operational, startup testing, or operational activities. Observation of others performing work does not constitute experience.

Objective test: A test that can be scored without subjective judgment in the scoring.

On-the-job training: Participation in nuclear power plant startup, operation, maintenance, or technical services as a trainee under the direction of experienced personnel.

Operating test: That portion of the operator licensing examination that is based on direct interaction between an examiner and an applicant. The operating test assesses the applicants' knowledge of the design and operation of the reactor and its associated plant systems, both inside and outside the control room. It is administered in a plant walk-through and a simulation facility.

Operational validity: A test item that (1) relates to the operations of the job and appears reasonable to ask and (2) is expressed in an operational context that requires the candidate to mentally or physically perform through understanding or analysis.

Performance test: Any test that requires the trainee to demonstrate either mental performance through knowledge testing or skill by actual operation or manipulation of tools and equipment. Typically, performance tests connote the meaning of skill testing.

Plant-referenced simulator: As defined in 10 CFR 55.4, means a simulator modeling the systems of the reference plant with which the operator interfaces in the control room, including operating consoles, and which permits use of the reference plant's procedures. A plant-referenced simulator used to administer operating tests (under 10 CFR 55.45(b)) or to meet experience requirements (under 10 CFR 55.31(a)(5)) must be designed and implemented in accordance with 10 CFR 55.46.

Power plant experience: As defined in Section 2 of ANSI/ANS-3.1-1993, applicable work performed in a fossil-fueled or nuclear-fueled electric power production plant during pre-operational, startup testing, or operational activities. Observation of others performing work does not constitute experience.

Predictive validity evidence: The ability of a test to forecast future performance on a subsequent measure.

Psychomotor: The domain of human performance that relates to physical performance based on mental activity.

Range: The smallest interval on a scale of scores that will include all scores; mathematically defined as the largest score minus the smallest score plus one.

Raw score: The numerical score first assigned when scoring a test before conversion to a derived score.

Reactor operator applicant: An unlicensed individual who is applying for an RO license.

Reference plant: As defined in 10 CFR 55.4, the specific nuclear power plant from which a simulation facility's control room configuration, system control arrangements, and design data are derived.

Related experience: In accordance with Section C.1.1 of Regulatory Guide 1.8, Revision 3, experience in performing job duties in the discipline for which the individual seeks qualification; such experience may or may not be at a nuclear power plant.

Related technical training: Formal training beyond the high school level in technical subjects associated with the position in question, such as acquired in training schools or programs conducted by the Military, industry, utilities, universities, vocational schools, or others. Such training programs shall be of a scheduled and planned length and include textual material and lectures.

Reliability: The consistency or repeatability of any measure as an indicator of confidence in that measure.

Responsible nuclear power plant experience (RNPPE): As defined in Section C.1.3 of Regulatory Guide 1.8, Revision 3, a senior operator applicant has actively performed as a designated nuclear control room operator or as a power plant staff engineer involved in the day-to-day activities of the facility. Time spent in academic or related technical training may fulfill the requirement for RNPPE, on a one-for-one basis, up to a maximum of 1 year.

Scenario: An integrated group of events that simulates a set of plant malfunctions and evolutions at a simulation facility.

Scenario set: A group of scenarios that constitutes a complete simulator test (i.e., "Integrated Plant Operations," of the operating test).

Score: A numerical indication of the performance an individual displays on a test.

Senior reactor operator upgrade (SRO-U) applicant: A licensed RO who is applying for an SRO license on the same unit(s).

Senior reactor operator instant (SRO-I) applicant: An unlicensed individual who is applying for an SRO license.

Simulation facility: As defined in 10 CFR 55.4, one or more of the following components, alone or in combination, used for the partial conduct of operating tests for operators, senior operators, and applicants [under 10 CFR 55.45(b)] or to establish on-the-job training and experience prerequisites for operator license eligibility [under 10 CFR 55.31(a)(5)]:
(1)   a plant-referenced simulator
(2)   a Commission-approved simulator under 10 CFR 55.46(b)
(3)   another simulation device, including part-task and limited-scope simulation devices, approved under 10 CFR 55.46(b)

Staff engineer: In accordance with Section C.1.4 of Regulatory Guide 1.8, Revision 3, an individual in a technical support position (i.e., personnel covered in Sections 4.4.10 and 4.6 of ANSI/ANS3.1-1993) who is responsible for the coordination and implementation of plant equipment control; integrated operation procedures; operations, maintenance, and radiological support; and/or review of modification and maintenance plans for plant systems.

Standard deviation: A measure of variability of a set of scores around the group mean. The standard deviation is mathematically defined as the square root of the mean of the squared deviations of the scores from the mean of the distribution.

Standard error of measurement: An estimate of the standard deviation of the errors of measurement associated with the scores in a given test.

Standardized test: A test that has the directions, time limits, and conditions of administration made consistent for all offerings of the test; this test is usually norm-referenced.

Statistic: A numerical value computed on a sample of data.

Technical Specifications: A document that identifies the plant-specific safety limits, system operability and surveillance testing requirements, and administrative controls. Whether stated or not, references to the technical specifications in this NUREG include those administrative controls that have been moved to other technical requirements documents.

Test: A measurement instrument; examination.

True score: The ideal or correct score for an individual. Its value cannot be known, but it can be estimated when assumptions regarding error of measurement are made.

Validity: The degree to which a test measures what it purports to measure.

NRC FORM 335
(9-2004)
NRCMD 3.7

U.S. NUCLEAR REGULATORY COMMISSION

BIBLIOGRAPHIC DATA SHEET

(See instructions on the reverse)

1. REPORT NUMBER
(Assigned by NRC, Add Vol., Supp., Rev.,
and Addendum Numbers, if any.)

NUREG-1021, Revision 9,
Supplement 1

2. TITLE AND SUBTITLE	3. DATE REPORT PUBLISHED	
Operator Licensing Examination Standards for Power Reactors	MONTH	YEAR
	October	2007
	4. FIN OR GRANT NUMBER	

5. AUTHOR(S)	6. TYPE OF REPORT
D. Muller	Technical
	7. PERIOD COVERED *(Inclusive Dates)*

8. PERFORMING ORGANIZATION - NAME AND ADDRESS *(If NRC, provide Division, Office or Region, U.S. Nuclear Regulatory Commission, and mailing address; if contractor, provide name and mailing address.)*

Division of Inpsection and Regional Support
Office of Nuclear Reactor Regulation
U.S. Nuclear Regulatory Commission
Washington, DC 20555-0001

9. SPONSORING ORGANIZATION - NAME AND ADDRESS *(If NRC, type "Same as above"; if contractor, provide NRC Division, Office or Region, U.S. Nuclear Regulatory Commission, and mailing address.)*

Same as above

10. SUPPLEMENTARY NOTES

11. ABSTRACT *(200 words or less)*

NUREG-1021, "Operator Licensing Examination Standards for Power Reactors," provides policy and guidance for the development, administration, and grading of examinations used for licensing operators at nuclear power plants pursuant to the Commission's regulations in 10 CFR 55, "Operators' Licenses." NUREG-1021 also provides guidance for maintaining operators' licenses, and for the NRC to conduct requalification examinations when necessary.

Supplement 1 to Revision 9 of NUREG-1021 includes a number of minor changes that are intended to: (1) clarify licensed operator medical requirements, including the use of prescription medications; (2) clarify the use of surrogate operators during dynamic simulator scenarios; (3) clarify the selection process for generic knowledge and ability (K/A) statements; (4) qualify the NRC review of post-examination comments; (5) provide additional guidance for maintaining an active license (watchstander proficiency) and license reactivation; and (6) conform with proposed updates to NUREGs-1122 and -1123.

12. KEY WORDS/DESCRIPTORS *(List words or phrases that will assist researchers in locating the report.)*	13. AVA LAB LITY STATEMENT
Operator Licensing Examinations	unlimited
	14. SECURITY CLASSIFICATION
	(This Page) unclassified
	(This Report) unclassified
	15. NUMBER OF PAGES
	16. PRICE

NRC FORM 335 (9-2004)
PRINTED ON RECYCLED PAPER

www.ingramcontent.com/pod-product-compliance
Lightning Source LLC
Chambersburg PA
CBHW081428170526
45166CB00008B/2129